THE ULTIMATE ASVAB STUDY GUIDE

Crush the Test on Your First Try in 7 Days With the Innovative Ultra-Fast Learning Method and Exclusive Practice Tests

OPTIMAL PREP PUBLISHING

TABLE OF CONTENTS

INTRODUCTION

The Army Services Vocational Aptitude Battery, better known as the ASVAB, is a test that everyone needs to take if they want to join any branch of the US military:

- Air Force
- Army
- Coast Guard
- Marines
- Navy
- National Guard

This is a guide that will help you to study so that you can to get the best grade, as well as help you determine which branch is the right one for you.

JOINING THE MILITARY

Joining the military is a big decision, so you wan to make sure that you find the branch that is right for you. Regardless of which branch you choose, we need bright, capable adults who can assume the responsibility that comes with protecting the US. The ASVAB helps the military to determine who is likely to do well in the military, and who may be able to thrive in the much more regimental environment.

Each of the different branches has something unique to offer. Some give you the potential to travel the world, while others target solutions that help people within a smaller community.

HOW THIS GUIDE WILL HELP YOU

This book has been planned around the knowledge you need to pass the test, and to stand out from the other people working to get into the military. The book is broken down into chapters that give you the time to understand and study the different areas you need to know when you take the test. It also includes some resources in the Appendices that will help you to study, even when you are on the go.

Whether you just want to refresh your current knowledge, improve your chances the next time around, or are just starting to consider the military, this guide will help you to focus on whatever you need.

How to Use This Guide

This guide outlines what the ASVAB is and a bit of history. It then has you take an example test so you can get an idea of how you will perform on the test. This will help you to see where you need to focus your studies. This does not mean that you should entirely skip the other sections because you need to make sure that you don't slip in other areas.

Don't feel discouraged by your initial test results from this guide. The whole purpose of the early test is to show you what it will be like, the kinds of questions you can anticipate, and to give you motivation to prepare. Think of it as establishing the baseline for where you are so you can better map where you need to go.

Once you finish grading and reviewing your initial test

- Create a list of the areas where you need significant improvement, then create a scale for the areas where you should improve, then the areas where you just need to maintain your current knowledge.
- Spend time reviewing your weaker areas.
- Regularly take practice tests to see where you are improving and where you still need to make improvements.

Over time, you'll start to see how much you've improved. You can also start to work out some of the nerve you may have when you take the test because you have already taken similar tests. It will start to feel more familiar.

Preparing to Test

Use the weeks leading up to the test wisely. You don't want to end up cramming or getting nervous. Those last couple of days should be spent letting yourself relax so that you can mentally prepare – you aren't likely to remember much from crunch sessions or fevered studying last minute.

- You can talk to a recruiter to learn more about the test and what to expect. You can ask for tips for how to prepare and what you can expect when you actually sit for the test.
- Regularly set aside time to review and take practice tests. The more you get acclimated to the tests, the easier it will be to keep your calm when you take the actual test.
- Set aside the day or two before the test to do something you enjoy. Don't think about the test or worry about it until the night before.
- Eat a healthy meal, get some good exercise, and make sure that you get plenty of sleep (the exercise can help with that).
- When you sit to take the test, focus on answering, not on scoring. Someone else is going to score you, so what matters now is focusing on is answering what's in front of you.
- Take yourself out for a treat when it is over. There's nothing like planning something you enjoy to make you feel that there's light at the end of the tunnel.

CHAPTER 1

UNDERSTANDING THE ASVAB

The test isn't quite as intimidating or nerve racking as the SAT because the military is looking for people who are well-rounded, not just good at testing and book smarts. While those things are important, studying isn't for everyone. The military understands that, and there are many roles where having a more mechanical or practical aptitude is definitely preferred. For example, mechanics need to be adept at working with a wide range of vehicles, and be able to work quickly, and sometimes without having all of the tools you need. For example, if you end up on a submarine and there is a maintenance issue, you may not have everything onboard to make the fixes.

That's why the Armed Services Vocational Aptitude Battery, better known as the ASVAB, looks at more than just what you learn for school – it looks for people who have aptitude in other areas too.

A BIT ABOUT THE ASVAB

The ASVB is meant to assess how well test takers perform across multiple areas that are critical to the military. According to the ASVAB Career Exploration Program, the subtests can be broken down into the four categories:

- Science and Technology
- Math
- Verbal
- Mechanical

Those categories are further broken down into subcategories, called subtests.

Subtest	What It Tests	# of Questions (P&P)	# of Questions (CEP)
General Scient (GS)	Life science, earth and space science, and physical science	25	16
Arithmetic Reasoning (AR)	Solve basic arithmetic word problems	30	16
Word Knowledge (WK)	Ability to understand the meaning of words through synonyms	35	16
Paragraph Comprehension (PC)	Ability to obtain information from written material	15	11
Mathematics Knowledge (MK)	Mathematical concepts and applications	25	16
Electronics Information (EI)	Electrical current, circuits, devices and electronic systems	20	16
Mechanical Comprehension (MC)	Principles of mechanical devices, structural support and properties of materials	25	16
Auto & Shop Information (AS)	Automotive maintenance and repair and wood and metal shop practices	25	11
Assembling Objects (AO)	How an object will look when its parts are put together		11

Science and Technology

Math

Verbal

Mechanical

Note: The Auto and Shop Information areas are considered one subject on the P&P ASVAB, but two on the CEP iCAT.

Note: The table shows which subtests are included under the different categories.

By measuring your performance for each of these areas, the different military branches can determine your suitability to join the forces and understand your current abilities to be trained in the different positions.

There are two versions of the ASVAB:

- Written, which requires a pencil, identified on the table as the P&P ASVAB
- Digital, better known as the Computer Adaptive Test ASVAB, identified on the table as the CEP iCAT

Roughly 70% of people who take the test are given the CAT-ASVAB, so this is very likely to be the test you are given. If you would like, you can ask the recruiter which version you will take so you can mentally prepare for it.

The Armed Forces Qualification Test - AFQT

Taking the ASVAB is not like other tests. They aren't interested in your overall score because the primary focus is in seeing where you are most skilled. As stated on the Air Force websites:

The Armed Forces Qualification Test (AFQT) is not a separate test. If you meet the requirements, you may be able to join the U.S. Air Force based on a different score that is obtained from four ASVAB subtests in addition to other factors.

- US Airforce

In other words, the military focuses on the total score you get for the four subtests under the Math and Verbal categories:

- Paragraph Comprehension (PC) and Word Knowledge (WK) are added together to determine your Verbal Expression score.
- Mathematics Knowledge (MK) score
- Arithmetic Knowledge (AK) score

These four scores determine your eligibility for each of the different military branches.

The following is the formula they use to determine your AFQT:

2 (PC + WK) + AR + MK = Raw AF!Q Score

Once your raw score is determined, they calculate your percentile performance against roughly 6,000 other candidates to see how your scores compared. So if you scored a 75% then it means your score was higher than 75% of the group.

The next step is to see how your percentile fits with the requirements for each of the military branches.

Scoring for the Different Military Branches

If you know what branch you want to pursue, this section will help you to focus on what score you need to achieve. These scores do periodically change, so you if you have your heart set on a particular branch or profession within the military, you should check out the ASVAB website for your preferred military branch to see what score you need at the time you take the test.

Your AFQT is also important to determine what occupation you are best suited to fil within the military. Most occupations that are highly desired require a higher score, and those who score hire are more likely to receive enlistment bonuses as a result of the sores. While not covered by this guide, some occupations do require high school diplomas or GED to qualify for them. If you have a GED or other equivalency test (TASC or HiSET, they apply more stringent requirements than those who have a diploma. Failing to have a high school diploma or having passed one of the equivalency test may result in disqualification, and people who do not have one of these are usually accepted under special circumstances.

There are also a couple of considerations that could result in your lower skills being waived. If you have special skills or a more experience in a particular area, this could result in the assessment waiving your minimum scores. For example, if you are fluent in more than one language, especially if you are fluent in a language that is particularly useful to the miliary, then your lower scores could be waived. Similarly, if you know computer programming, cyber security, car mechanics (or any type of vehicle mechanics, such as boat or plane), or other less common skill/experience, it will make you a more highly desirable choice for the military. It's also possible that your lower AFQT scores could be waived if you score particularly high on some of the subtests that are not included in the AFQT but are valuable knowledge for occupations that are considered recruiting targets.

It is best not to count on your scores being waived. You want to perform to your best so that you don't require a waiver.

The following are the requirements for the different branches at this time in 2024. However, there are also links to the different military branch requirements so you can verify that the requirements have not changed at the time you are preparing to take the test. Keep in mind , these are the AFQT *minimum* scores for people who have a high school diploma. If you have a GED or other equivalency test score, the minimum requirement is higher (it is usually around 50).

Military Branch	Minimum AFQT Score
Air Force	31
Army	31
Coast Guard	36
Marine Corps	31
Navy	35

Notice that the Coast Guard and Navy do have higher requirements than the other three branches. These are also the scores required for regular enlistment. If you are interested in becoming an officer, you can check out the military branch website to see what the requirements are.

REGISTERING AND PREPARING TO TAKE THE ASVAB

If you are still in high school, you may be able to take the test at your school. The military does send personnel out through the Department of Defense Career Exploration Program to conduct the test at schools. You can ask your school administrators if the ASVAB is scheduled for your school and when. Usually, the test is conducted during the school day, so you will miss some class to attend and take the test.

You can also call your local recruiter for the branch you are most interested in pursuing to ask about the test. They can get the recruitment process started. When you've met the enlistment qualifications, you will get to schedule your testing time, and you will learn which type of test you will take.

If you are given the PiCAT, it can be done at your own pace using an access code provided by your recruiter. This test can be quickly scored, so. Your recruiter can likely tell you how you performed not long after you finish. You will need to do a followed-up test to verify your scores, but this one will take less than 30 minutes.

The military is also trying a new method of assessing potential recruits through the tailored Adaptive Personality Assessment System (TAPAS). It is a computer-administered personality test, making it inclusive of other factors, such as your motivations and character. This test is meant to try to predict how successful you will be in the military. The test has 120 questions, and it is a companion test to the ASVAB (it is not currently a replacement for it). The main purpose is to see if people who didn't perform well on the ASVAB could still be successful in the military.

Once you are done, the results have a two year life-span. If you did well, you will not need to take it again if you want to take some time to consider if you want to join the military or which branch you want to join. If you don't enlist by the end of the two years, you will need to retake it before trying again.

You don't have to accept your first test if you were not happy with the results.

Times You Take the Test	Duration to Retake
1	No time frame
2nd time	1 calendar month
3rd time	1 calendar month
4th time on	6 months

This means that you can take the test three time over a three-month period. After the third test, you will need to wait 6 months before you can take the ASVAB again.

UNDERSTANDING YOUR SCORE

Once your test is scored, you will be able to talk about it with either a school counselor or a recruiter. All of these scores matter in terms of what role you will serve in the military, but some will be less important to you as an individual if the focus is in an area where you aren't interested. In other words, some of the scores will have a greater effect ton your military future.

The following details what the different sections of your ASVAB score means:

Section	Meaning
AFQT Score	The overall score for the four subtests
Standard Score	The result for each of the subtests. These are compare your raw score to the national sample used for the raw score. The Department of Defense has indicated that about half of the test takers get a 50 for this, while roughly 16% get higher than 59.

Service Composite Scores	Also known as the Line Scores, these are combined to determine if you have the aptitude for training into different roles within the different military branches. The Navy Engineering Aid (EA) adds twice the MK score with the AR and GS scores. You must have a minimum score for the role to qualify for consideration to be trained. The Army's. Electronics require you to have a score that combines the GS, AR, MK, and Electronics Score to qualify. This is something you should discuss with a recruiter to understand what scores you need if you have a particular occupation in mind.
Career Exploration	You will have this on your assessment if you took the ASVAB at school. These are a composite result of the Science, Math, Technical, and Verbal scores. They are scored both as standard scores and as percentages relative to your current year in school and your gender.

UNDERSTANDING THE MULTIPLE VERSIONS OF THE ASVAB

There are two considerations when it comes to the different version. First, you can take a different version based on your current age and situation. Second, the version may vary based on if you use a computer or pencil and paper. This section explains the differences in the versions of the test currently available.

The ASVAB Based on Your Current Situation

Based on your motivation and the time in your life when you take the test, you may get a different version of the ASVAB. The following table details the variety of tests and when they are issued to potential enlisters.

ASVAB Version	When It Is Offered
Enlistment Testing Programs / Production Version	This is given to people for enlistment purposes and is given at the different military branches. If you take this test, it will help the recruiters determine if you have the required aptitude for that particular branch. If you do have the aptitude, it also determines which occupations you are suited for in that branch.
Career Exploration Program	This is the test given in high schools and secondary schools, in addition to the interest inventory. The test is administered by the Department of Defense Career Exploration Program representative. The test is mostly the same as the Enlistment test, but it is done with a pencil and paper – you don't take a digital version. There isn't an Assembling Objects subset on this version either.
Armed Forces Classification Test (AFCT) / In-Service	This test is administered to people who are current military personnel who want to change their military role. It is like the low tech version.

Comparing the Digital Test (CAT) against the Traditional Test (Pencil and Paper)

The CAT and traditional tests are different because the traditional paper test is going to be the same no matter how you've performed up to that point. The CAT adapts the questions to better fit your performance.

Some people think that they can improve their score by testing slower, verifying their answers as they go, which then improves their final score. This assumes that they will perform better, even if they don't complete the test. Unfortunately, this doesn't help improve their score as unanswered questions earn you a penalty for each question you don't complete. This is similar to how your tests are scored in class. To ensure that you are not penalized for not finishing, it is best to complete the section within the designated time. After you finish, if you have time remaining, you can return to the questions you weren't sure you answered correctly. That doesn't mean rush through the test, but you should establish a pace that means you will finish in the time given. If you aren't sure, you can guess, then make sure to return to that question later.

The method of taking the test may be different, but the current studies indicate that people tend to have comparable results, regardless of if they take the traditional or the CAT version. As more people because used to digital testing, they do tend to have a greater advantage with the CAT. This doesn't mean that you have a choice, so it is important to find out which version you will be taking so you can plan ahead.

The following are generally considered the advantages to taking the CAT (if it is an option)

- It has fewer question to answer, so you have more time to complete each question.
- You don't have to wait long for the score because it is nearly immediate after you finish.
- The people giving these test are more likely to be flexible with when you can take the test.
- You won't be distracted by other questions since you only see one at a time.
- You can't inadvertently answer the wrong question since they are presented one at a time (not all at once where you can misalign your answer with the question), meaning you won't lose points for this mistake.

The following are generally considered the advantages to taking the traditional test:

- You can skip an answer and return to it later after you've finished the rest of the question. That improves the chance that you will finish the questions you are more likely to get right, getting them out of the way quickly.
- You can go back and check your work. If you find something wrong, you can change it until time is called for the test.
- It is more effective when you want to eliminate wrong answers. You have a better chance of guessing right if you eliminate two of four answers first. This is much harder to track on a CAT since you have to scribble on a piece of scrap paper, which can get messy and confusing.

PLANNING YOUR TESTING STRATEGY

The following table will help you plan how to take the test so that you can increase the likelihood of scoring better on each section.

Strategy	Why It's Important
Take the time to become familiar with the different test categories.	There are no surprises or trick questions with this test. They want to know what you know, not see if you can be outsmarted by a test. The best way to perform better is to become familiar with the test. The format and sections are known – it doesn't change much from year to year. Also, if you are familiar with the test when you start, it will feel more comfortable as you go. Consider how much easier it is two months into school compared to your first day. That familiarity helps you to focus on the questions instead of on being anxious about what you are doing.
Practice as often as possible, being careful not to burnout on testing.	Time is on your side, and you have a wealth of practice tests you can take. There are many free online tests you can take to ensure you know your stuff. You can also go and study the subjects on your own to be more knowledgeable in the different sections. You can see improvements and gain confidence as you see your practice test score improve. You can also recognize when the test has patterns or when they are offering trap answers that are close to the real answer, but just a little off for those who don't properly read each answer. The ASVAB doesn't have trick questions, but it definitely has trick answers. Learn what kinds of tricks they are likely to use so you can avoid them.
Make the most of the fact the test is multiple choice.	The ASVAB is a multiple choice test, and that is something that you should learn to use to your advantage. Even if you have to guess, you have a 1 in 4 chance of guessing the right answer. If you can entirely eliminate one or two of the answers, you improve your chance of guessing currently if you don't know the right answer. It's also possible that eliminating one or two answers will make it easier to see the answer. Keep this one in mind when you feel like you want to skip a questions – guess now and return later to see if you can further eliminate answers.
Answer all of the questions in each section.	This means guessing if you don't know. Remember, you aren't penalized for guessing, but you are penalized for not finishing questions. You can always return at the end and try to determine the correct answer if you have time.

The previous table is good for both types of tests. The following list provides a few strategies that can help you if you are taking the CAT.

- Pace yourself. This can be trickier on a computer as you don't see all of the questions, making it easier to get hung up on one question for longer than you should. Questions at the beginning of the test tend to be more impactful on your total score, but if you spend too much time on these, you are more likely to have harder questions later. You will have less time to finish those harder questions. This will mean that all of that hard work at the beginning will be nullified when you are penalized for not completing all of the questions.
- Since you can't skip any question, you will have to guess to progress on the test. Try to eliminate as many as possible before you guess. Be strategic in how you answer, even if you feel incredibly uncertain in your answer. Once you know you can't eliminate any other answers, make a choice quickly and move on so that you don't spend a lot of time on what is literal guesswork.
- As you reach the midway point of the section, try not to get too many wrong at a time since having too many wrong in a row can really hurt your score. In other words, if you had no idea how to answer one question, try to put a little more time on the next one to try to get it right.
- Don't let a difficult question rattle you. The test is meant to get harder the better you are doing and as you progress through the test. If questions get noticeably harder, that means you are doing well, so use that to feel more confident, not shaken by the harder questions. The test will continue to adapt to your performance until about halfway through you answering correctly. If you feel like you are starting to get as many answers wrong as right, it means that the test is working as intended.

Make sure to keep these in mind as you go through the test. It can be harder as you go, but if you can practice thinking about these when you do practice tests, it can make it almost second nature during the actual test.

PREPARING TO TAKE THE ASVAB

Now that you have the basics of the test, it's time to start diving into the specifics. And it means diving into the deep end because you need to know where you are starting to plan your path forward. The next chapter is going to delve into that first test you should take long before you take the actual ASVAB. From there, you will be able to learn more about the different sections so that you will better understand what they are asking for each question.

Make sure you complete the diagnostic test and score it to see how you perform before you move into the rest of the guide.

Each of the chapters in this guide provides strategies for answering the different questions. The four subtests that make up the AFQT do have a greater emphasis and coverage as they are the ones that you need to do well on to be enlisted. That doesn't mean that you should skimp on the other sections, but you should get comfortable with those first. Then you can delve into the other sections, especially if they will improve your chance of getting into the branch and occupation you want.

Don't forget there are a lot of free practice tests online that can help you to continue seeing how well you are doing.

Finally, it's important not to burn out on testing, especially the night before a test. There is such thing as testing too much so that you are burnt out and have trouble focusing on the test itself.

CHAPTER 2

TAKING THE DIAGNOSTIC TEST

Before diving into the details each of the subsections on the test, you need to get a baseline of where you are in your skills for the test. This is the purpose of the diagnostic test – to act as a map from where you are to where you want to be.

LEARNING ABOUT YOURSELF

The diagnostic test actually teaches you a lot about yourself. In other words, the diagnostic test is a learning experience all on its own, with you being the primary focus.

When you take the diagnostic test, it is essentially a simulation test. You will see how the test is written, the order of the sections, the kinds of questions they ask, and how long it takes you to complete each section. At the same time, it provides you with a sense of familiarity that will make it easier to actually take the test.

Once you finish the test, you will need to score yourself and start to understand how you performed on each section. Start noting what kind of questions you have trouble with so you can focus on how to deal with those as you go through the res t of the guide.

Over time, you can take other tests to see how and where you are improving, and where you still need to work on improving.

Obviously, the diagnostic test isn't going to be like taking the CAT, at least not in the method of testing – this is a paper and pencil test. That means that you will have more questions to answer, but you will have the advantages of being able to skip around so that you can find questions you have a better chance of answering correctly. This can train you how to guess more effectively over time, so if you end up taking the CAT, you will better know how to guess efficiently, increasing the odds that you will be able to finish the test. Also, having more questions to answer means having more practice – and a better idea of what kind of questions you need to prepare to answer.

Finally, you can start seeing how long it takes before you start feeling test fatigue. This is definitely a thing, and the longer you test, the less focused you are likely to be. Also, the less care you will have as you answer. You can actually improve your focus and attention span through practice.

GETTING YOURSELF SETUP TO TEST

Here is what you need to do to setup to take the diagnostic test.

- Set aside time where you won't be interrupted to take this diagnostic test. Have a pencil ready, along with some scratch paper for working on your answer.
- Have a timer to simulate the amount of time you have for each section.
- Take the test in the order it is presented. Do NOT skip around the different sections because you won't be able to do this during the actual test.
- Give yourself short breaks between sections. The breaks shouldn't be longer than 5 minutes. Use this time to stand up, go to the bathroom, or just stretch.
- If you don't finish before the timer goes off for a section, take note because that means you need to improve how quickly you answer questions in that section. Make sure to set a better pace for subsequent sections.

If you do this for your first diagnostic test, it will be as close to a simulation as you can get using this guide. It will give you a more accurate assessment to your current knowledge and testing ability.

When you start noticing that you are experience testing fatigue before you reach the end, establish a weekly testing schedule to build up your endurance to taking the test. That is the most effective way to deal with the problem.

COMPLETING THE TEST

Go to Appendix A and complete the test.

Once you are done, score your answers using the Answer Sheet at the end of the test.

Do NOT move on to the next chapter until you have at least finished and scored the diagnostic text (Appendix A).

All of the following chapters will make more sense once you've been through the test and seen what kinds of questions you will be asked.

FURTHER TESTING

Every chapter has two practice tests at the end. Since you may feel that you need additional practice, you can go to Appendix B to do two more additional practice tests for the four primary sections that make up the raw test score. That gives you four tests for chapters 3 to 6.

Once you've exhausted all of the practice tests in this book, you can go online to find some more free tests. Several of the military branches provide short versions of the test, giving you another way to simulate the

CHAPTER 3

THE WORD KNOWLEDGE SECTION

The ASVAB begins with the Word Knowledge (WK) section as the first part of the language portion. The purpose of this section is to assess your ability to determine the meaning of a word. If don't know a word, you should use clues, such as pre-fixes and suffixes, to narrow down the meaning of the word. There may be words that you know without having to use any clues, but it's almost certain that you'll encounter words you know that are new to you.

The thing about English is that as a native speaker, you often have most of the clues you need to determine the meaning without a dictionary. Even if you aren't 100% sure of the meaning, you will be able to narrow down the potential meanings form the multiple choice questions.

Go back to your diagnostic test and see how you performed. Were there questions you didn't know? Were you able to narrow down the choices?

This chapter will help you build the knowledge and better eliminate the answers that don't fit what you've determined based on parts of the word.

A BIT MORE ABOUT THIS SECTION OF THE ASVAB

You've experienced what it will be like to take the traditional ASVAB. If you are taking the CAT, here's what you need to know.

- With 15 questions and 9 minutes to go through them, you will need to finish nearly 2 questions a minute. That seems fast, but it's not as bad as it sounds. The traditional version is actually much worse (you have 35 questions with 11 minutes, so you need to finish more than 3 questions a minute).
- Like the diagnostic test, a lite over half of your answers will be determined based on the word alone – you won't have context to help you. Here's an example.

  Egregious most nearly means:
  o friendly
  o nefarious
  o conscious
  o gorgeous

  If you know that word, the choice is probably fairly easy to make as the four words are all very different. You can finish this one in a couple of second and move to the next question. If you don't know the answer off the top of your head, this chapter is going to provide a decoder for you to narrow the choices so that you at least have a good chance of guessing correctly.

- The rest of the test include context clues because the word is used in a sentence. Here's an example.
  His <u>frugal</u> approach to finances meant he would have a good retirement.
    - careless
    - private
    - thrifty
    - opulent

    Again, if you know the word, it's a quick answer. However, if you don't know the meaning, it's a lot easier to narrow down the choices because you can use the rest of the sentence to eliminate at least two of the options without losing much time. This chapter will give you a bit more information for using the rest of the sentence to rule out that third option, or at least make you feel more confident in your final choice.

This chapter provides a consistent way of approaching each question, no matter of it is a standalone word or one with a bit more context.

THREE STEPS TO REMEMBER

Use the following three steps to complete the WK section:

1. Determine if it is a word you know. If not, take a few seconds to look for decoders to the meaning.
2. Formulate an initial definition for the word and write it down.
3. Look for the word or a similar word in the choices. Strike out words that you feel are wrong based on your initial assessment and definition.

The way you use these will depend on the type of question you are reading. The following sections demonstrate how to use the steps based on if you have context or just a single word.

DECODING WORDS WITHOUT CONTEXT

As you've seen words that have no context are much harder to figure out. You can use the three steps to decode the words, but you will need to spend more time actually dissecting the word. Here's how the steps look for standalone words.

1. Dissect the word by identifying the following:
    a. The word's root
    b. Prefixes
    c. Suffixes
2. Use what you've found to formulate the definition of the word.
3. Find the word that matches what you've determined.

Yes, this process takes a while to master, especially since there are a lot of prefixes and suffixes that you need to learn. Once you know those, you can start to decode words a little faster. The really tricky part is figuring out the meeting of the root word. Since English is heavily influenced by Latin and Greek, it does mean memorizing a good bit of root words that are frequently used as the base of many English words.

Dissecting the Word

This will take a bit of time to learn, but once you start to memorize the repetitive prefixes and suffixes of English, this process will become much easier to do. For example, you can probably list five words off the top of your head that start with *dis-*

- Dissatisfied
- Disassociated
- Disfigured
- Dismantled
- Disgusted

This is one of the most popular prefixes, and as you probably noticed when reading the list, it has negative connotations. It means "opposite of" or "not." Now look back at the words, and you will quickly see the roots for four out of the five words (disgusted is different, but *dis*-still serves the same purpose, we just don't use gusted as a root word on its own).

You can do the same things with suffixes. One of the most popular is *-tion*.

- Revolution

- Evolution
- Constitution
- Precipitation
- Tarnation

Again, you can tell there's a pattern to it, but suffixes are a little trickier. Instead of indicating something like a positive or negative, suffixes generally determine what type of word it is. In this case, the addition of -*tion* to a word means that the word is a noun. So if one or two of your choices for a word ending in -*tion* is an adjective or a verb, you can quickly eliminate it.

Nearly all 10 words have a root that you can identify, and a number of them have both a prefix and a suffix.

Let's test your ability to start dissecting words. You don't need to try to determine the meaning of the word, just see if you can identify the different parts of each word. Try to write out what part of the word you think fits each category. The words may only have two of the three parts.

Word to Dissect	Prefix	Root	Suffix
Antibiotic			
Dejection			
Homogeneously			
Submarine			
Ulterior			

Have you finished? Let's take a look to see how you did.

Word to Dissect	Prefix	Root	Suffix
Antibiotic	anti	bio	tic
Dejection	de	ject	ion
Homogeneously	homo	genos	ly
Submarine	sub	marine	
Ulterior		ulter	ior

You've probably noticed that it's not always obvious what is a root and what is a prefix or a suffix. Appendix C gives provides tables to help you memorize

Focus on Prefixes

Let's take a look at 10 of the most common prefixes in English.

Prefix	Meaning	Examples
anti-	against	antibiotic, antifreeze, antisocial
de-	opposite	defrost, demote, demonetize
dis-	not, opposite of, apart	disagreement, disease, disinterest
en-/em-	cause to	encase, encourage, embrace, empower

in-/im-/il-/ir-	not	inadequate, injustice impossible, illiterate, irregular
mis-	wrong, wrongly	misfire, misread, misunderstand
pre-	before	prefix, pretend, prevent
re-	again	repeat, return, rewind
semi-	half	semiannual, semicircle, semiautomatic
un-	not	uncertain, unfriendly, unavailable

Now let's practice determining the meaning of a word based on these prefixes. Some of these you almost certainly know what they mean – this is when it can really help you to dissect the word and see how the prefix helps to determine what the word means.

Word to Dissect	What You Predict It Means
Antipathy	Strong feeling of dislike
Depict	to show
Disabuse	persuade, idea, belief
Endure	undergo go through
Emanate	flow, pour
Inspiration	creativity
Imitate	take or follow as a model
Misconstrue	misunderstand
Predict	to say what will happen
Respect	regard
Semibreve	Same value of 2 half notes
Unafraid	not afraid

Now take a few minutes to look up the words and their etymology. This will help you to see how the prefix works with the root to change the meaning of that root.

With this knowledge, let's see how knowing the meaning of a root helps you understand a word when the prefix is added.

The root *scrib* and *script* means to write. Let's add prefixes to the word to see how it changes what that room means.

Word to Dissect	What You Predict It Means
Describe	
Prescribe	
Unsubscribe	

Most of these words are probably familiar. Don't worry, you are going to get a chance to work through words you are less likely to know in a bit. For now, focusing on words you know will illustrate how much a word changes just because it has a few extra letters at the beginning.

Suffixes have a similar effect, but in a very different way.

Focus on Suffixes

A suffix does a lot of hard work because it not only changes the meaning of a word, it can change what kind of word it is (and usually does). Before getting into suffixes, let's do a quick review of the four parts of speech most often used with suffixes so you can more easily identify how a suffix changes this basic function of a word.

Part of Speech	What It Does
Noun	a person, place, thing, or idea
Verb	shows action or state of being
Adjective	a descriptive word that modifies nouns
Adverb	a descriptive word that modifies verbs

There are other parts of speech, but they rarely have suffixes added to them. Pronouns, prepositions, and conjunctions are standalone words that link sentences together. You don't change they, of, or but because those words are what they are – there are no prefixes or suffixes that attach to them.

Like with prefixes, there are many suffixes. Appendix C provides a long list of them, but for now, let's look at some of the most common suffixes.

Suffix	Meaning	Examples
-able/-ible	possible, can be done	comfortable, invisible
-ed	past tense for of the verb	happened, shadowed
-ful	full	graceful, mindful, spiteful
-ic	has the characteristics of	civic, despotic, diabolic, prophetic
-ing	present participle of the verb	airing, harping, walking
-ion/-tion/-ation/-ition	act or process	deliberation, inculpation, nonparticipation
-ity/-ty	state of	absurdity, ingenuity, reality
-ive/-ative/-itive	changes a noun to an adjective	detective, narrative, supportive
-ology	branch of scientific field	biology, philosophy
-ly	characteristic of	attentively, exclusively, naively
-s/-es	plural form of a noun	apples, oranges, sharks
-y	characterized by	classy, happy, misty, rocky

Now let's practice determining the meaning of a word based on these suffixes. We aren't going to add the suffixes that change verb tenses or pluralize nouns – you probably already know those without having to think about it.

Word to Dissect	What You Predict It Means
Convertible	
Awful	
Synthetic	
Participation	
Activity	
Accusative	
Bewilderingly	
Grumpy	

Now take a few minutes to look up the words and their etymology. Also, see which words also have prefixes.

With this knowledge, let's see how knowing the meaning of a root helps you understand a word when the suffix is added.

The root and *cycle* means to circle. Let's add suffixes to the word to see how it changes what that room means.

Word to Dissect	What You Predict It Means
Cyclable	
Cyclic	
Cyclicity	

Most of these words are probably familiar. Don't worry, you are going to get a chance to work through words you are less likely to know in a bit. For now, focusing on words you know will illustrate how much a word changes just because it has a few extra letters at the beginning.

You've had a chance to see how root words are changed because both prefixes and suffixes are predictable and often easy to spot. Roots are a bit more of a challenge.

Focus on Roots

Many of the roots in English come from Greek and Latin, so to better guess what a word means, you are going to have to start learning a bit of Greek and Latin. Don't worry, it isn't as hard as you think because many of these roots are used in words you know. Once you remove the prefixes and suffixes from them, you can better guess the meaning of the root word. Appendix C details a lot of the roots that you should memorize to help you perform better on the KW section.

Let's take a look at 10 commonly used roots.

Prefix	Meaning	Examples
cent	one hundred	century, centimeter, percent
form	shape	conform, uniform, reform
multi	many	multiage, multiply, multifid
sent	to feel, to send	assent, insentient, resent
voc	to call, voice	advocate, vocation, vocalize

graph	writing	graphic, paragraph, telegraph
hypo	below, beneath	hypothyroidism, hypothetical
micro	small	microscopic, microbial, microcosm
phobia	Fear	arachnophobia, claustrophobia, phobic
scope	viewing tool	borescope, stethoscope, telescope

Now let's practice determining the meaning of a word based on these prefixes. Some of these you almost certainly know what they mean – this is when it can really help you to dissect the word and see how the prefix helps to determine what the word means.

Word to Dissect	What You Predict It Means
Accentuate	
Information	
Multifaceted	
Sensible	
Advocate	
Graphite	
Hypocritical	
Microphone	
Bibliophobia	
Photophobia	
Periscope	

Now take a few minutes to look up the words and their etymology. This will help you to see how you can more easily understand a word just by finding the root.

With this knowledge, let's see how knowing the meaning of a root helps you understand a word when the prefix is added.

A Few Tips and Tricks

As you've probably noticed, you don't have to have a definitive meaning to be able to get the gist of what a word means. Knowing the common prefixes, suffices, and roots can help you narrow down the meaning of a word. With many of these word parts having positive, neutral, and negative connotations, you can quickly find the right meaning if the other three options don't have that same sense of meaning.

See if you can determine if the following words have positive, neutral, or negative connotations.

Word	Is it Positive, Neutral, or Negative?
burdensome	
currently	
decision	
destitute	
diabolic	
fanfare	
harmony	

swampy	
zestful	

What was it about each word that made you decide it was positive, neutral, or negative? Did you already know the word, or did you get a feeling about the word that gave you a certain vibe?

Now let's check to see if your reaction to those words was accurate.

Word	Type	Definition
burdensome	negative	heavy or tiresome
currently	neutral	happening now
decision	neutral	reach a conclusion, form a judgement
destitute	negative	impoverished, penniless, poor
diabolic	negative	nefarious, suggestive of the devil, evil
fanfare	positive	ceremonial or happy tune played to celebrate or as a marketing tool
harmony	positive	having a sense of peace or agreement; musical chords that create a pleasing effect
swampy	negative	uncomfortably wet or humid; having the characteristics of a swamp
zestful	positive	full of energy, excitement, and enthusiasm

If you don't immediately know the meaning of a word, you can also think back to when you have heard it used.

Consider the word _broach_. It is a word that does _not_ have a Latin or Greek root, so memorizing those isn't going to help you get the meaning of the word. However, you've probably heard someone in real life, on a stream, or on TV say that they need to "broach a topic/subject carefully." That provides a bit of context for the word. If your choices are work, introduce, play, and find, the only one that really makes sense in that context is introduce.

See if you can recall hearing these words to see if you can think of a possible context for the word.

Word	How Was It Used? (Write the sentence if you can remember it.)	Your Rough Guess What It Means
guzzle		
hybrid		
jiffy		
vicarious		

These words probably tickled you mind, if you didn't just know them. While we don't know how you heard it, we can tell you what they actually mean.

Word	Meaning
guzzle	to eat or drink quickly or greedily
hybrid	the result of combining two different things
jiffy	a brief period of time, happening momentarily

vicarious	experiencing through someone else; imagining going through something someone is experiencing

USING CONTEXT TO DETERMINE A WORD'S MEANING

Now that you have a way of dissecting words to determine their meanings, you have a lot of useful tools. However, if you have a sentence round the word, you can use that to help you better narrow down the meaning. All you have to do is plug the word in and see if the sentence has the same meaning. Ok, it's not quite that easy, but you can use context to get a rough meaning of the word before you break it down. Here's an example.

> The birds <u>thronged</u> around the woman on the bench like children at an ice cream stand.

The word like indicates a comparison between the birds and children. And if you know anything about children, you can imagine they are flocking or swarming to that ice cream stand. Both flocked and swarmed are words that are similar to thronged.

See if you can determine the meaning of the underlined word based on the clues.

The Word in Context	Words Suggesting Similarities	Predict the Meaning
The fans greeted the popstar with an enthusiasm that was almost a <u>fervor</u>.		
Discussing the budget was difficult, I would say it was very nearly a <u>quagmire</u>.		
The prodigy of today is tomorrow's <u>virtuoso</u>.		

How do you feel about what you found? Could you identify words that hinted at the meaning?

Let's see how you did.

The Word in Context	Words Suggesting Similarities	Definition
The fans greeted the popstar with an enthusiasm that was almost a <u>fervor</u>.	that was almost a	Intense, passionate emotions
Discussing the budget was difficult, I would say it was very nearly a <u>quagmire</u>.	it was very nearly	a bog or area that is swampy; a hazardous or complicated situation
The prodigy of today is tomorrow's <u>virtuoso</u>.	the two are equated to each other	person who is exceptionally creative, particularly musicians

Not all context is nearly so obvious though. Sometimes you need to look at the full sentence to get a feel for the word. Consider the following sentence.

> The best teachers provide support, but they are also <u>exacting</u>.

The familiar word "but" lets you know that the word "exacting" is very different from supportive. It isn't the opposite of supportive, since a teacher should be both. This suggests that the words seem dissimilar, but they help a student to do better.

"Exacting" actually means "demanding" or "requiring precision and attention."

See if you can determine the meaning of the underlined word based on the clues.

The Word in Context	Words Suggesting Contrasts	Predict the Meaning
You can learn some actions, while others are <u>instinctive</u>.		
My mom prefers action movies, but was willing to sit through a <u>cloying</u> chick-flick with me last night.		
Unlike the rest of us, Tina found weeding to be enjoyable instead of <u>onerous</u>.		

How do you feel about what you found? Could you identify words that hinted at the meaning?

Let's see how you did.

The Word in Context	Words Suggesting Contrast	Definition
You can learn some actions, while others are <u>instinctive</u>.	while	automatic, without prompting or thinking about it
My mom prefers action movies, but was willing to sit through a <u>cloying</u> chick-flick with me last night.	but	Sickening or excessively sweet or sentimental
Unlike the rest of us, Tina found weeding to be enjoyable instead of <u>onerous</u>.	unlike … instead	burdensome or difficult

Sometimes you can get a feel for the word just based on how it is used. Like prefixes and suffixes, you can find hints that are positive, neutral, or negative. Here's a short example.

> Elise was excited for her well-deserved <u>respite</u>.

There are two hints:

1. She is excited for the respite.
2. The respite is deserved.

Using those clues, it's safe to assume that respite has a positive meaning. And that assumption is correct since the word means "a rest or break." When you sit down to play video games, you are taking a respite from work and chores. You are doing something you enjoy after doing a lot of things you didn't enjoy as much.

See if you can determine the meaning of the underlined word based on the clues.

The Word in Context	Positive or Negative?	Predict the Meaning
I realized throwing more money at the problem would only <u>exacerbate</u> our financial situation.		
My eyes kept trying to shut as the teacher <u>droned</u> on about stocks.		
The <u>injudicious</u> plan would create more issues than it solved.		

How do you feel about what you found? Could you identify words that hinted at the meaning?

Let's see how you did.

The Word in Context	Words Suggesting Contrast	Definition
I realized throwing more money at the problem would only <u>exacerbate</u> our financial situation.	throwing, situation	to worsen a situation; to make something worse
I fought sleep as the teacher <u>droned</u> on about stocks.	fought	a humming noise; speaking in a monotonous tone
The <u>injudicious</u> plan would create more issues than it solved.	Issues (it's creating those issues)	using poor judgement; unwise

Remember how we started with the word egregious? Let's use the 3 steps to use a sentence to understand the meaning.

The most egregious problem with the setup was with the leaking pipes near the computer.

1. The word is an active for problem, and given that there is water dripping around a computer, it seems like a very serious problem.
2. Look for a word that indicates something particularly bad.
3. The word nefarious seems a bit extreme, but it is definitely the word that most closely indicates something particularly bad.

See if you can determine the meaning of the underlined word based on the clues.

The Word in Context	Predict the Meaning
The new students found the Calculous material <u>opaque</u>.	
The bandit carefully hid her tracks while going to the <u>clandestine</u> meeting.	
My initial <u>bewilderment</u> was clear the professor.	

How do you feel about what you found? Could you identify words that hinted at the meaning?

Let's see how you did.

The Word in Context	Definition
The new students found the Calculous material <u>opaque</u>.	unclear, not transparent; does not allow light through
The bandit carefully hid her tracks while going to the <u>clandestine</u> meeting.	secret, concealed
My initial <u>bewilderment</u> was clear the professor.	perplexed, confused

The best way to deal with these is using the three steps, but you should also plug the choices into the sentence in place of the underlined word to see if it has the same meaning. Here's an example.

Meridith <u>feigned</u> joy at the couple's engagement announcement.

- o   feinted
- o   created
- o   hid
- o   pretended

The first one is actually a trick answer because it looks the same. However, based on the rest of the sentence, it isn't likely that Meridith actually feel unconscious, so that one is quickly eliminated.

The second one doesn't make sense since she isn't making anything.

The third one also doesn't make sense either because why would she hide the fact that she's happy?

That just leave pretended. "Meridith pretended joy" would probably be Meridith pretended she felt joy, but it definitely has the same sense when it replaces feigned. And that is the correct answer.

USING LOGIC TO ELIMINATE WORDS

The previous section gave a lot of ways to determine the meaning of a word based on different components of the word and the sentence. What if there isn't anything that obvious though?

Consider the following:

Exult most closely means

- o jubilant
- o pessimistic
- o blasé
- o energetic

Using steps 1 and 2 doesn't help at all because there is no obvious breakdown and not context.

With that in mind, look at the options. Pessimistic and blasé are fairly similar, so you can rule those out. That leaves jubilant and energetic, which seem fairly similar too.

Consider if you have heard the word "exult" used, or if you have heard it used with prefixes or suffixes. Have you heard the word? If you know the Christmas song "O Come, All Y Faithful," then, yes you have heard it.

"Oh sing, choirs of angels, sing in exultation."

The song is a celebration of the religious event that is considered to be joyful. Or jubilant.

If there are two choices that appear to be opposites, there is a good chance that one of them is the correct one. This is just something that test makers do, adding a word with the opposite meaning. If you see that, look at those two words more closely. If you don't have time, you can guess, knowing that you had almost a 50/50 chance of getting it right. Unless you've heard the word used before, you probably aren't going to be able to narrow it down farther, but at least you have a much better chance of choosing the right answer.

See if you can use that approach with the following two examples.

Demur most closely means

- study
- argue
- believe
- trust

Believe and trust have nearly the same meaning, so you can eliminate those. At this point you ca guess if you haven't heard the word used in a sentence. In this case, argue is the closest word as "demur" means to disagree or object to something.

Logy most closely means

- circular
- wearily
- actively
- chubby

The two words that have opposite meanings are actively and wearily. If you've never heard the word "logy" before, then you can guess with good odds that you have a 50/50 chance. The word actually means sluggish, which makes sense because it's hard to manage an active log.

BUILDING YOUR VOCABULARY

Take the time to get these processes down because they will make it much easier to take the test with a better chance of getting the right answer. It does mean a lot of memorization, but it's not wasted. You'll be able to use what you learn during your memorization in everyday life. You can even practice what you learn with people around you.

Here are a few ways to expand your vocabulary for those words that don't have an easy method of dissecting or determining their meaning.

- Read books and news articles to see how words are used. This can help reinforce words you know or to introduce you to new words and give them context. You are more likely to remember words when you've heard them used and have context behind them.
- If you hear or read a word you don't know, take the time to look it up. You should see if you can determine the meaning, but always take the time to verify what you thought. That way you don't have the wrong association with the word.
- Try to use new vocabulary when talking.
- Learn between 5 and 10 words every day. make sure you never learn less than 5, but try to learn more on days when you have time. The more you learn, the more likely you will recognize words on the test.
- Create flashcards every day with the new words (or if you can use an app for vocabulary, that is even better as you will almost always have your phone on you). Make sure to practice the words every day. You don't have to go through all of them, but you should regularly review words so you don't forget them.
- Write a log of the words you learn and use those words. You can create a work of fiction and add the words, or you can write about your life using the words. As long as you associate the words with something personal, your brain is much more likely to connect to that word.

Don't forget about Appendix C too. Knowing prefixes, suffixes, and roots are going to do a lot to help you understand words without having heard them used.

PRACTICING WHAT YOU'VE LEARNED

Now that you've got the basics, you can see test your newly acquired information and put it to the test. Here are two practice sets with 15 words each, which is how many you will answer if you take the CAT. Set your timer to 9 minutes before you start to see how well you do. The last section of this chapter details the right answer, and how you could have reached that answer.

Practice 1

Abundant most closely means

- A. plentiful
- B. many
- C. risky
- D. empty

Edifice most closely means

- A. rubble
- B. unimpressive
- C. plumbing
- D. building

My child's excitement about the theme park abated quickly after standing in line for 10 minutes.

- A. waned
- B. prioritize
- C. sleepy
- D. sunken

Protagonist most closely means

- A. villain
- B. climax
- C. main character
- D. plot point

Introspection most closely means

- A. self-reflection
- B. introduction
- C. revelation
- D. internal conflict

He was despondent after being passed over for the promotion

- A. despairing
- B. sad
- C. upset
- D. unresponsive

Unanimous most closely means

- A. union
- B. ambiguous
- C. anonymous
- D. everyone in agreement

Dialogue most closely means

- A. writing
- B. dial tone
- C. conversation
- D. derogatory

The entire group believed in the fallacy that the earth is flat.

- A. a false belief
- B. false idol
- C. face reveal
- D. facet

Laconic most closely means

- A. few words
- B. lactose
- C. iconic
- D. lore driven

Symbiosis most closely means

- A. rival
- B. symbol
- C. oasis
- D. close relationship

There was an uproar when she handed in her resignation without warning.

- A. handout
- B. relief
- C. departure from a position
- D. escape

I tried to alleviate the pain after my wife went into labor.

- A. allocate
- B. relieve

Practice 2

Mosaic most closely means

- A. type of architecture
- B. decorative tile or glass
- C. sewer system
- D. complication

Undulate most closely means

- A. up-down motion
- B. positioning
- C. pistons
- D. movement

Phloem most closely means

- A. biology
- B. herbology
- C. plant sugar system
- D. how flowers pollinate

Embargo most closely means

- A. a video game move
- B. embarrassing yourself
- C. a trade ban
- D. a type of cargo load

Analogy most closely means

- A. comparing two things
- B. putting together pieces
- C. puzzling
- D. a frustrating job

Repose most closely means

- A. chilling a room
- B. taking a stroll
- C. state of rest
- D. scheduling the day

- C. cast upon
- D. irritant

My dog was able to quickly discern that I had once again not thrown the ball.

- A. prescribe
- B. fooled
- C. perceive
- D. preformed

After moving into my first house, I cleaned with a new vigor.

- A. intensity
- B. fascination
- C. low key
- D. annoyance

Tumult most closely means

- A. a tea party
- B. a marathon
- C. a type of sport
- D. an uproar

Tranquility most closely means

- A. a state of calm
- B. a koi pond
- C. a forest grove
- D. a library room

The outrageous situation was like something out of a story, not real life.

- A. shockingly bad
- B. break out
- C. draconic
- D. blithering

The nocturnal nature of my cat means I don't always feel well rested in the morning.

- A. nightly patrols
- B. a type of octave
- C. active at night
- D. a moment after dusk

As a voracious reader, my mother often had a hard time finding a new book to read.

- A. disinterested
- B. eager
- C. laconic
- D. weary

I cannot divulge my sources.

- A. keep secret
- B. have an idea
- C. bring out
- D. make known

My dog is more of an <u>omnivore</u> than I am because I avoid vegetables at all costs.

    A.   eats everything
    B.   likes trash
    C.   consumes plants and animals
    D.   has an allergy

I used to <u>aspire</u> to become a doctor.

    A.   to attain a goal
    B.   become beloved
    C.   needs attention
    D.   strong desire

The <u>jovial</u> atmosphere quickly lifted me out of my bad mood.

    A.   monotony
    B.   cheerful
    C.   languish
    D.   bouncy

CHECKING YOUR ANSWERS

Now it's time to see how well you did.

Practice 1

Abundant means plentiful (A).

An edifice is a building (D), usually an impressive one.

If something abates, it is waning (A).

The protagonist is the main character (C) of a story.

Introspection is self-reflection (A).

If someone is despondent, they are sad (B) or depressed.

If something is unanimous, it means everyone agreed (D).

Dialogue is another word for conversation. (C)

A fallacy is a false (or wrong) belief (A).

Laconic means few words (A).

Symbiosis means a close relationship (D).

A resignation is given to depart from a position (C).

To alleviate something means to relieve (B) it.

To discern something means to perceive (C) it.

If you do something with vigor, you are doing it with energy and intensity (A).

Practice 2

A mosaic is decorative tile or glass (B).

If something is undulating it is moving smoothly moving up and down (A).

Phloem is a plant sugar system (C).

An embargo is a trade ban (C).

An analogy compares two things (A) against each other.

If something is in repose, it is a state of rest (C).

If something is in tumult, there is an uproar (D).

Tranquility means that something is a in a state of calm (A).

If something is outrageous, it is shockingly bad (A).

Nocturnal means active at night (C).

Voracious usually means eating with gusto, but it can mean to do eagerly (B) do something.

To divulge something means to make something known (D).

An omnivore is a create that consumes plants and animals (C).

If you aspire to something, you've set a goal you want to attain (A).

Jovial means cheerful (B).

CHAPTER 4

PARAGRAPH COMPREHENSION

The ASVAB begins with the Paraph Comprehension (PC) section. The purpose of this section is to assess your ability to understand what you've read. They provide a short passage, then ask questions about the passage. You don't need to memorize anything for this, and there is nothing more you need than the passage in front of you. It is a very self-contained section since you probably won't even need scratch paper.

This chapter will help you to better answer questions about what you just read. In theory, this is easy. When you are taking a test and have your nerves going, and are waiting for trick questions, you are more likely to miss something you need to answer correctly. By the end of this chapter, you will have a consistent approach that you can apply as you answer questions about passages.

A BIT MORE ABOUT THIS SECTION OF THE ASVAB

If you take the traditional test, you will have 13 minutes to go through 13 or 14 passages. There will be 15 questions total about the passages. If

If you take the CAT, you will have 27 minutes for 10 questions. You get more time because it can be more difficult to read on a monitor, and the questions get harder to answer the better you do. This is not a problem if you are using pencil and paper.

Regardless of which version you take, you can use the same method to be successful. And that starts with focusing on what is in front of you. You don't need to learn anything outside of what's on the test – you don't need to make guesses or know formulas, prefixes, or a plant life cycle to find the answer. After the other sections, this can be rather refreshing.

THE FOUR STEPS OF READING COMPREHENSION

There are four steps you can apply to this portion of the test to feel more confident that you got the answers correct.

1. Read the question or questions first. That way when you read the passage, you will know what you are supposed to get out of reading the passage. Don't read the passage first because you will probably have to read it again after you read the question as you may not remember where the information it wants is located. You really don't have time to read the passage several times. Don't worry about the answers, the point is to know when the passage answers the question. That way you can mark it or note it.

2. Read the passage, keeping the question or questions in mind. When you find the place where you think the question is answered, mark it or note it, then keep reading to the end of the passage.

3. Read the question again, and try to recall what the answer was in the passage. You can check back to where you marked to verify what you think the answer should be. It's possible that you already know the answer because you were focused on it when you read the passage (and that's why you should know the questions first). Sometimes, you may have a general idea or aren't entirely sure of what the answer should be. That's ok. Predict what you think the answer should be.

4. Finally, look at the answers. Select the one that best matches your prediction and understanding of the passage. If you aren't sure which is the best answer, start eliminating the ones that you know are wrong. Like with the KW answers, if you have to guess, you have a better chance of guessing right by having fewer choices. Some of the answers will be extreme or will change details, so it will be obvious that they are wrong. You have a 50/50 chance of being right if you eliminate two obviously wrong answers.

This is the best way to make sure that you don't have to read the passage several times. This helps you to get your mind on just what you are reading, and drones out the world around you. If you are a nervous test-taker, this is can be incredibly beneficial as your mind will start to focus on the test and not on the situation.

UNDERSTANDING THE THEME OF A PASSAGE

When it comes to reading comprehension, the types of question you will answer are fairly predictable. Perhaps the most common type of question is to determine what the main theme of the passage is. For this type of question, you do need to fully read the passage to make sure that you understand the point of the passage. Here are the ways these questions often read.

-   What is the main idea/purpose of the passage?
-   The tone of the passage can be characterized as…
-   The primary purpose of the passage is …
-   The passage is mostly concerned with …

To determine the main purpose, you have to understand the tone and the value the author gives to the written statements. The passage always contains supporting details for the passage's purpose. This can include simply information, expressed opinions, and background details. All of these support the point that the writer is trying to make, and that is usually included as strong opinions and statements, a prediction, or a recommendation or a rebuttal to other ideas.

Consider this claim: "Cryptocurrency is a great way to have a high yield investment." If you've paid attention to any of the news, you know that simply isn't true. It can be a high yield investment, but it is also one of the riskiest investment strategies. At best, it's a recommendation, and not even a particularly good one. The author could supply reasons for why the statement is true. For example, the passage could provide details about how much people have gotten by selling Bitcoin, compared to what they paid for it back in 2008. Pointing out that new cryptocurrency is often incredibly inexpensive could also help to bolster the argument.

When you read a passage look for the following details.

A sentence is …	Question It Answers
The Main Idea	What is the author expressing? What belief is provided?
A Supporting Detail	What does the author believe?

An author may present a strong opinion, in which case the main idea is often easy to determine. The following are some ways that this can be expressed in a very obvious way:

-   I think/believe
-   I recommend/suggest
-   Thus/therefore

All of these are clear indicators of an opinion or belief.

Sometimes authors are more subtle, making the main idea less obvious. You need to be more discerning for these passages, and it will require a bit more focus. Fortunately, if you focus on the structure of the passage, you are more likely to be able to find the contrasts to determine the main focus. The following are contrast indicators:

-   Although
-   But
-   However

- Though
- Yet

The information that follows these indicators often provides the author's thoughts and opinions, especially if the piece is a rebuttal or disagrees with what someone else has expressed.

You can also use the process of elimination to reduce how many options you have to consider. Even if the author is being subtle, the answers are sometimes very blatantly wrong. Look for signs that one of the choices is wrong.

- The answer provides to much detail, making the answer more of a supporting detail, not the main point of the passage.
- The answer is too generic and general, bringing up points that are outside of the point of the passage.
- The answer may contradict the point of the passage, which is obvious if it disagrees with what the passage says.
- The passage may overstate what the author was expression, resulting in an answer that is too extreme based on what you read.

Here's a passage you can use to start applying the steps and elimination process. Make sure you start by reading the question before you read the passage.

> With the introduction of alternative light sources that are more energy efficient, there has been a move around the world to stop using the once popular incandescent light bulb. Numerous nations have even mandated the change. The problem is that replacing the older light bulb with fluorescent libs is that fluorescents use mercury, which is very toxic. Another action that countries have taken is to require the use of ethanol in gasoline because it creates less pollution. However, ethanol is a biproduct by crops like corn. This means having to grow and harvest these crops, then convert them into the more energy efficient fuel, but is creates just as much pollution and uses just as much energy over the entire process. It has the added problem of increasing chemicals often found in fertilizers, that then contaminate the surrounding water.
>
> What's the main theme of this passage?
>
> A. Environmental regulations should be a higher priority than economic issues.
> B. Countries that regulate light bulbs and ethanol are misguided.
> C. Environmental regulations may have negative results.
> D. Fluorescent lights have mercury.

Now determine what you think the right answer is and why.

____

Ok, now that you've had time to think about it, let's go through the elimination process.

- The last one is far too specific, so that one is eliminated.
- The tope one covers things not discussed in the passage – economic factors are not even mentioned.
- B has some truth, but at there is nothing to say that the countries are misguided.

That leaves C. If you look at the passage again, that does appear the point of the passage – intention doesn't guarantee the results you want.

Let's try another one.

> Alchemy is the science of trying to chemically alter common metals into more valuable metals, with the end goal usually being gold or silver. Scientist today call it a pseudoscience as it relies on belief system that doesn't have any real scientific basis. It has close ties to both astrology and occult teaching. As a result, people who call themselves alchemists are seen as scammers and charlatans who are looking for a quick way of getting rich. However, this is the modern impression of alchemy. For centuries, people who practiced it were well-respected. Ironically, the people in pursuit of turning common metals into rarer metals actually created a lot of the tools used in labs now. The methods they used to research and test their ideas are the foundation of chemistry.
>
> What's the main theme of this passage?
>
> A. Alchemists are scammers.
> B. Alchemists will never succeed in turning metals into gold.
> C. Overtime, alchemy evolved into what we know as chemistry.
> D. One of the dreams of those who practiced alchemy was becoming rich.

Now determine what you think the right answer is and why.

____

The first sentence tells you what the focus is – alchemy. The next few sentences focus on the history of alchemy and how it is perceived today. Notice the introduction of the word "Ironically" toward the end, that is a contrast indicator. It then talks about how we benefited from this now obsolete type of science.

Now let's look at each answer.

A.   That was a detail, not the main point of the passage.
B.   It's probably true, but it isn't the point of the passage because it ignores nearly everything after that one sentence.
C.   This looks really promising as it takes the full passage into account.
D.   This was true, but again, it wasn't the point of the passage.

The answer is C.

ANSWERING DETAILED QUESTIONS

When you are trying to determine the theme of a passage, you are looking for the big picture. However, you may also get questions that want you to pay attention to the details. The following are examples of these types of questions.

-   According to the passage, what kind of curtain is best for helping sleep during the day?
-   When completing the third step, make sure to …
-   According to the author, what is one of the benefits of having a dog?

Unlike theme based questions, these questions focus on a small part of the passage. Answering these questions is usually pretty straightforward as you don' t need analyze anything to find the answer. You pretty much just need to find the sentence with the detail it wants. Sometimes the answer paraphrases the portion of the passage, other times it uses the verbiage verbatim.

What you have to look out for is the answer to the specific question. The trick answers are usually other details that look correct if you only glance over the passage – that's why you have to read the full article. Before you make you look at the choices, make sure you predict what the answer should be. That way you don't fall into the trap of choose a similar, but wrong answer.

Here's a passage you can use to start applying the steps and elimination process. Make sure you start by reading the question before you read the passage.

> The musician Wolfgang Amadeus Mozart is regarded as histories most outstanding prodigies who went on to be one of our most popular classical composers. He completed his first musical composition at just five years old. He wrote his first symphony at the tender age of 8 years old. By the time of his death at just 35 years old, he had written over 600 pieces. Over 200 years later, he is still one of the most well-known musicians in human history.
>
> How old was Mozart when he wrote his first composition?
>
> A.   35 years old
> B.   5 years old
> C.   8 years old
> D.   Not stated

Now determine what you think the right answer is and why.

____

Ok, now that you've had time to think about it, let's go through the elimination process.

Three ages are provided. If you simply skimmed over the passage, you may have chosen the wrong answer. He wrote his first *symphony* at 8 years old, but he wrote his first *composition* at 5 years old. It's an easy question to answer, but also it is very easy to pick the wrong answer if you don't focus.

Let's try another one.

> Martha Graham is a celebrated pioneering figure in dance. She revolutionized the art form with her innovative approach, breaking away from tradition. Over five decades, she crafted over 170 pieces, spanning from intimate solos to grand productions, and she was willing to perform them herself. Deciding to break away from classical ballet, Graham began forging a new movement during the early

part of the 1920s. Her new movements and form were a better reflection of how the world was changing following the end of World War I.

Martha Graham's decision to establish newer techniques was based on ….

    A. A desire to leave the past behind.
    B. A need to draw more attention to her work.
    C. A desire to capture the emotions at that time through movement
    D. A need to demonstrate that the conventional moves were too rigid.

Now determine what you think the right answer is and why.

____

Only one answer really makes sense because the details are pretty clear. She wanted to capture what people were feeling at the time, and that wasn't possible with the traditional techniques.

DETERMINING WHAT IS BEING INFERRED

Probably the most difficult of all of the questions, this type wants to draw a conclusion based on what is presented. The following are some common types of questions that require you to infer what is being said in the passage.

- What does the passage imply?
- It seems that the author thinks that …
- Based on the passage, you can infer that …

Inference questions seldom direct you to a single part of the passage, so you can quickly find the answer. Nor do they focus solely on the passage's primary theme, so you aren't looking for a single sentence summary of what you should take away from reading it. Instead, you are asked to interpret what different sections say to assess what the author believes. Since you can infer a lot from a piece of writing, and what you infer may be different from what someone else infer, it isn't nearly as straightforward as the other two types of questions. That means you won't need to stop and try to predict the final answer. Instead, you will keep in mind what the question is, then paraphrase what you read. Once you have an idea of what you feel the piece is saying, you can review the questions and see which one best aligns with your paraphrase of the text. Remember, you are meant to choose the answer that is supported by the passage.

This type of question is much more guess work than the other kinds as the answer isn't stated in an obvious way. The correct answer is the one that has the most support from what you just read. This means that you can remove choices that include one of the following problems:

- The answer has contradictory details from the test.
- The answer includes details not provided in the text itself. This is similar to questions asking about the theme of the passage.
- The answer distorts what the passage says, making it wrong, not an inference
- The answer has a claim that is extreme based on the passage, meaning it is close, but goes too far. The claim is not fully supported by the text.

Let's go through a passage that has an inference question.

Mary and Lou are a married couple who have three adult children – Nick, Wendy, and Hank – and they have three grandchildren. Wendy often watches Nick's twin sons, and Hank often watches Wendy's kid.

If this passage is true, what also much be true?

    A. All of the grandkids are girls.
    B. Hank does not have any children.
    C. Hank also watches Nick's kids.
    D. Lou has at least one granddaughter.

Now determine what you think the right answer is and why.

____

Yeah, this kind of problem is more of a process of elimination. Let's apply the steps to figure this one out.

1. It's an inference question, so there are a lot of unknown details that aren't covered and cannot be assumed.
2. You'll probably read it a couple of times because it throws a lot of names at you.

3. Write down the people and their relationships. Make a family tree and add the names or details. For example, the gender of one of the children is not provided, so the only detail you have is that the child is Nick's kid.
4. Refer to the family tree as you check each answer.

Here's how you determine the right answer.

A. Since one of the genders isn't revealed, so you can't make this determination. It isn't supported.
B. Looking at the chart, this is true.
C. There is no mention if this is true or not, so you have to assume, which means it's probably not the right answer.
D. Again, the gender isn't provided, so you don't know if one of the kids is a granddaughter.

It's tiring, but necessary. Use the steps to make sure you are looking at it from the facts, not what you think *should* be true.

> Every year for five years, my city's courts awarded settlements over $10 million. This has created a significant burden on the system. To help reduce the strain on the city's budget, _________________.

Choose the answer that best completes the blank.

A. People should start fewer lawsuits.
B. Lawsuits should include multiple defendants.
C. Settlements should only compensate based on the plaintiff's needs.
D. The courts need to stop awarding such massive settlements.

Now determine what you think the right answer is and why.

____

This one is actually a bit easier because you aren't trying to untangle relationships. Here's how to solve it.

1. You are supposed to complete the sentence with a logical conclusion based on the information given during the rest of the passage.
2. The passage starts by stating that every year has seen an excessive amount of money awarded because of lawsuits. To the point that it is creating a "significant burden" on the system.
3. The passage makes it clear that the awards are the problem as they are excessive every year, so it would be reasonable for the courts to award more reasonable amounts of money. This would reduce the burden.
4. The answer is D because the courts are going overboard in their awards. The first three answers don't have any information that support them in the passage. And they don't address the problem that the amount of money being awarded is the problem.

USING CONTEXT TO DETERMINE WHAT A WORD MEANS

This one is familiar because you just finished Chapter 3 where you used this for single sentences. Now you have an entire passage to help you figure out the meaning. It's going to be much easier to do after the inference questions. You should be able to plug in the right word and get the same meaning.

The best way to work with these questions is to keep the word in mind that you need to define. Make sure to note it in the passage and how it is used. Once you finish reading, come up with words that mean the same thing or words you think would work in its palce.

Once you've made your prediction, look for the answer that most closely matches the words you though could replace the word. When you've found the word you think will work, plug it into the passage and read through that part of the passage again. Determine if it has the same meaning.

Yes, this is pretty much exactly the same as what you did in the previous chapter.

Let's look at another one.

> You are probably familiar with voting, and you know that during each election, you cast a single vote for each candidate. However, there are different types of votes that you can cast, depending on the situation. You may be asked to participate in a rank voting. This voting scheme has the voter rank their choices from first to last. If there are five candidates, you rank them between 1 and 5, with one person taking each individual rank. It takes longer as it requires calculations to determine the final answer. The Associated Press uses this method of voting to determine college sport teams rankings. It is also used in Australia for their elections.

Which of the following words could be substituted for *scheme* in the passage?

A. Collaboration
B. Belief
C. Method
D. Plot

Now determine what you think the right answer is and why.

____

You'll quickly notice it's a vocabulary question, and the word that you need to replace is *scheme*. Follow the same steps you used in the last chapter. In this case, the only answer that makes sense is (C) method.

Let's look at another one.

When Edith sat down to study for her final, she knew that she needed to review all of her notes from the last year. They had started the year by looking at Ancient Greek and they recently finished with the Fall of Rome. There were many historically significant dates, figures, myths, and catastrophes. As she sat down to pour through her notes, quizzes, and tests, she knew that she was in for a long couple of days.

What is the meaning of "pour through" in the passage?

A. Searching through
B. Taking notes
C. Filtering through
D. Reading carefully

Now determine what you think the right answer is and why.

____

This one is actually really easy because there are so many clues. There's only one answer that actually makes sense when you plug it into the passage – (D) Reading carefully. When you try plugging in the other three answers, they simply don't work.

PRACTICING WHAT YOU'VE LEARNED

Now that you've got the basics, you can see test your newly acquired information and put it to the test.

Here are two practice sets with 10 passages, each with their own question. This is similar to how the CAT is setup, although obviously your questions aren't going to change the better you do.

The last section of this chapter details the right answer, and how you could have reached that answer.

Practice 1

Read the question, then reach the paragraph. Once you finish select the answer that best fits.

1. The Amazon Rainforest is the largest tropical rainforest in the world, covering an area of over 6 million square kilometers across nine countries in South America. It is home to an estimated 10% of the world's known species and plays a crucial role in regulating the Earth's climate. Deforestation, primarily driven by agriculture and logging, poses a significant threat to the Amazon Rainforest and its biodiversity.

What is the main theme of the passage?

(A) The geography of South America
(B) The importance of biodiversity
(C) The history of deforestation
(D) The impact of climate change

2. Emily had always been fascinated by astronomy. She spent countless nights gazing at the stars through her telescope, marveling at the vastness of the universe. Her favorite celestial body was Saturn, with its majestic rings and mysterious moons. One summer evening, while observing Saturn, Emily noticed a faint glow in the sky. As she focused her telescope on the source of the light, she realized that she was witnessing a rare meteor shower.

What is Emily's favorite celestial body?

(A) Jupiter
(B) Mars
(C) Saturn
(D) Venus

3. Sarah loved to read and was frequently immersed in a book's pages. Her favorite genre was fantasy, and she loved immersing herself in magical worlds filled with dragons, wizards, and mythical creatures. One day, while browsing the shelves of her local bookstore, Sarah stumbled upon a dusty old tome hidden behind a row of books. Curious, she picked up the book and began to read, unaware that its pages held the key to unlocking a fantastical adventure of her own.

What is Sarah's favorite genre of book?

(A) Mystery
(B) Romance
(C) Fantasy
(D) Science fiction

4. By superimposing digital data over the actual world, augmented reality (AR) improves the user's perception of their surroundings. Real-time augmented reality applications, like Pokemon Go and Snapchat filters, combine virtual and real-world aspects. What distinguishes virtual reality from augmented reality? What distinguishes virtual reality from augmented reality?

How does augmented reality differ from virtual reality?

(A) AR requires specialized headsets, while VR does not
(B) AR overlays digital information onto the real world
(C) AR isolates users from their physical environment
(D) AR focuses on creating entirely virtual environments

5. There are five cars—Audi, BMW, Chevrolet, Ford, and Honda—each parked in a different spot. The Audi is parked next to the Chevrolet. The BMW is parked next to the Ford. The Ford is not parked in the first or last spot. Which car is parked in the second spot?

Which car is parked in the second spot?

(A) Audi
(B) BMW
(C) Chevrolet
(D) Honda

6. DNA, or deoxyribonucleic acid, is the molecule that contains the genetic instructions for life. It consists of four chemical bases: adenine (A), cytosine (C), guanine (G), and thymine (T). The sequence of these bases determines an organism's traits and characteristics. Understanding DNA is fundamental to fields such as genetics, medicine, and biotechnology.

What is the main theme of the passage?

(A) The functions of genetic instructions
(B) The structure of DNA
(C) The process of cell division
(D) The history of genetics

7. Through pay-as-you-go internet connectivity, cloud computing enables customers to access computational resources including servers, storage, and apps. This technology is perfect for organizations of all sizes since it is affordable, flexible, and scalable. In what ways may cloud computing help businesses?

(A) By limiting access to computing resources
(B) By increasing capital investment in hardware
(C) By providing scalability and cost-effectiveness
(D) By restricting data storage to physical servers

8. Six friends—Ally, Rob, Karol, Dave, Emile, and Frank—are seated around a circular table. Alice is sitting between Bob and Carol. Emily is sitting opposite David. Who is sitting opposite to Alice?

Who is sitting opposite to Ally?

(A) Rob
(B) Karol
(C) Dave
(D) Frank

9. The goal of sustainable development is to satisfy current needs without endangering the capacity of future generations to satisfy their own. In order to build a more just and resilient society, it entails striking a balance between social advancement, economic expansion, and environmental preservation. In order to address global concerns like poverty, inequality, and climate change, sustainable development necessitates collaboration across industries and countries.

What is the main theme of the passage?

(A) The history of economic development
(B) The importance of environmental conservation
(C) The challenges of global cooperation
(D) The concept of sustainable development

10. The human immune system is a complex network of cells, tissues, and organs that defends the body against harmful pathogens like bacteria, viruses, and fungus. It has two primary branches: the innate immune system, which offers immediate defense against infections, and the adaptive immune system, which responds to specific diseases. Vaccines are critical instruments for boosting the immune system and avoiding infectious diseases.

What is the main theme of the passage?

(A) The structure of pathogens
(B) The function of vaccines
(C) The components of the immune system
(D) The history of infectious diseases

Practice 2

Read the question, then reach the paragraph. Once you finish select the answer that best fits.

1. Alex had a passion for photography and loved capturing the beauty of nature through his camera lens. One day, while hiking in the mountains, he stumbled upon a breathtaking vista overlooking a tranquil valley. Inspired by the scene before him, Alex snapped a series of photographs, each one more stunning than the last.

What is Alex passionate about?

(A) Writing
(B) Cooking
(C) Gardening
(D) Photography

2. The Industrial Revolution, which began in the 18th century, caused a dramatic upheaval in human history. It was defined by the transition from rural to industrialized society, which was fueled by technological and manufacturing developments. The Industrial Revolution brought about significant social, economic, and environmental changes, including urbanization, the emergence of capitalism, and increased pollution. What is the passage's major theme?

What is the main theme of the passage?

(A) The history of technology
(B) The consequences of the Industrial Revolution
(C) The impact of urbanization
(D) The development of capitalism

3. The human brain is a complex organ responsible for controlling bodily functions, processing information, and regulating emotions. It consists of billions of neurons that communicate through electrical impulses and chemical signals. Understanding the structure and function of the brain is essential for advancing neuroscience and developing treatments for neurological disorders.

What is the main theme of the passage?

(A) The structure of the human brain
(B) The functions of the nervous system
(C) The importance of mental health
(D) The history of neuroscience

4. Anna, Brian, and Claire are three friends who have different pets. Anna has a cat, Brian has a dog, and Claire has a fish. If the statements are true, which pet does Brian have?

What pet does Brian have?

(A) Cat
(B) Dog
(C) Fish
(D) Cannot be determined

5. Oprah Winfrey is an American media mogul, actress, and philanthropist known for her influential talk show, "The Oprah Winfrey Show." Through her media empire and philanthropic initiatives, Winfrey has become one of the most influential figures in the world. What is Oprah Winfrey primarily known for?

What is Oprah Winfrey primarily known for?

(A) Her achievements in professional sports
(B) Her advocacy for environmental conservation
(C) Her contributions to culinary arts
(D) Her influential talk show and philanthropy

6. The compass, invented by ancient Chinese civilization, revolutionized navigation by enabling sailors to determine direction accurately. This magnetic instrument played a crucial role in maritime exploration and trade routes, shaping the course of human history. What invention revolutionized navigation by enabling sailors to determine direction accurately?

What invention revolutionized navigation by enabling sailors to determine direction accurately?

(A) The telephone
(B) The printing press
(C) The compass
(D) The internet

7. The Great Barrier Reef is the world's biggest coral reef system, stretching over 2,300 kilometers off the coast of Queensland, Australia. It supports a wide variety of marine life, including around 1,500 fish species and 400 coral species. The reef is a UNESCO World Heritage Site that draws millions of people annually for snorkeling, scuba diving, and eco-tours.

What is the main theme of the passage?

(A) The geography of Australia
(B) The importance of marine biodiversity
(C) The history of coral reefs
(D) The tourism industry in Queensland

8. The process of evolution is how genetic variety, natural selection, and adaption cause species to change throughout time. It is the central concept in biology and explains how life on Earth has diversified and evolved over billions of years. The theory of evolution, first proposed by Charles Darwin in the 19th century, has since been supported by a vast body of scientific evidence.

What is the main theme of the passage?

(A) The history of biology
(B) The theory of natural selection
(C) The process of evolution
(D) The contributions of Charles Darwin

9. In a group of friends—James, Katie, Liam, Megan, and Noah—each person has a different favorite sport: basketball, football, soccer, tennis, and volleyball. Katie's favorite sport is soccer. Megan's favorite sport is basketball. Liam's favorite sport is not volleyball or soccer. Whose favorite sport is volleyball?

Whose favorite sport is volleyball?

(A) James
(B) Katie
(C) Liam

(D) Noah

10. Vaccines are critical tools for preventing infectious diseases and protecting public health. They work by stimulating the immune system to recognize and fight off harmful pathogens, such as bacteria and viruses. Vaccines have played an important role in eliminating illnesses like smallpox and lowering the prevalence of others like polio and measles.

What is the main theme of the passage?

(A) The history of infectious diseases
(B) The development of vaccines
(C) The importance of public health
(D) The role of the immune system

CHECKING YOUR ANSWERS

Now it's time to see how well you did.

Practice 1

1. (B) The importance of biodiversity

2. (C) Saturn

3. (C) Fantasy

4. (B) AR overlays digital information onto the real world

5. (C) Chevrolet

6. (A) The functions of genetic instructions

7. (C) By providing scalability and cost-effectiveness

8. (D) Frank

9. (D) The concept of sustainable development

10. (C) The components of the immune system

Practice 2

1. (D) Photography

2. (B) The consequences of the Industrial Revolution

3. (A) The structure of the human brain

4. (B) Dog

5. (D) Her influential talk show and philanthropy

6. (C) The compass

7. (B) The importance of marine biodiversity

8. (C) The process of evolution

9. (D) Noah

10. (B) The development of vaccines

CHAPTER 5

MATH STRATEGIES FOR THE ASVAB

As illogical as it seems, answering a math problem correctly requires more than just doing the math problem – there's a bit more strategy to it. This is definitely true with the ASVAB because you have to do everything yourself – there is no calculator to help you. You will not receive partial credit for presenting your work. You either pick the correct answer, or you don't. You get full credit or no credit for each answer. On the other hand, if you guess right, you get full credit. This doesn't mean that you should just go through guessing on the test, but it does ease the stress and anxiety that may feel because you don't have to spend a lot of time slaving over each question trying to make sure you write every step. If you have a quick way to solve math problems, you won't be punished for it.

A BIT MORE ABOUT THIS SECTION OF THE ASVAB

There are two sections of the ASVAB that require you to prove your math skills:

- Arithmetic Reasoning (AR)
- Mathematics Knowledge (MK)

These are the other important components of the AFQT, so you need to make sure to study and practice these two sections regularly. There is an expectation that you will have decent math skills, and doing poorly on these two sections can result in your disqualification.

The AR section focuses on your basic ability to do math word problems.

- The AR section of the traditional version of the test has 30 questions and gives you 36 minutes to complete them.
- The AR section of the CAT version of the test has 15 questions and gives you 55 minutes to complete them.

Obviously, the CAT provides more time, but that's because they do get harder if you do well. In both versions, the AR section is designed to see how well you can reason based on your ability to solve problems that require you to use basic math. The following are the kinds of problems you will solve:

- Word problems
- Ratios
- Percentages
- Rates
- Number properties
- Averages

- Conversions
- Proportions

The MR section focuses on more complex mathematical concepts, like algebra and geometry.

- The MR section of the traditional version of the test has 25 questions and gives you 24 minutes to complete them. That is more than one problem per minutes, so make sure you don't spend too much time on any one problem.
- The MR section of the CAT version of the test has 15 questions and gives you 23 minutes to complete them.

Again, the CAT section provides more time, but the problem can definitely get difficult if you do well for the first half. You may have some word problems, but most of the focus is on getting more direct answers and math problems. Because of this, you get less time per problems for both types of test.

For both the AR and the MR, being quick is essential to giving yourself the best chance of performing well.

THE STEPS FOR ANSWERING MATH QUESTIONS

Since the process of solving problems is more likely to result in the best results, you can use the following four steps for all math problems to work faster.

1. Analyze the question and the problem you are expected to solve.
   The means fully reading the question without trying to solve the problem. This first step is important because you need to take the time to focus on the problem, not on what you need to do. You need to comprehend what you are reading. Once you do, the rest tends to be a bit easier and faster.
2. Understand what it is the problem is asking you to do.
   You are verifying what you read to make sure you know what kind of answer you need to provide. You should be determining what kind of answer set you will have before you actually look at them. This helps prevent you trying to just guess and move on. If the question is an equation with two variable , take the time to understand if you are solving for one or both variables, or if the question is asking you to solve something else.
   This step is the one that helps you to avoid the trick answers. They do this to see if you have taking the time to comprehend what they are asking, and if not, you make that clear by selecting the trick answer. Usually, this means they give you what looks like the right answer, but you are answering the wrong question. In other words, you are answering a question that they aren't asking. For example, they may ask for the answer to what y represents, but the more obvious and instinctive answer is to answer x. It's an easy trap to avoid by taking the time to verify you understand what they want, and not just running on automatic based on what you've experienced when taking tests in a class.
3. Solve the problem.
   When you feel certain that you know what the focus is, start strategizing how to reach the answer the question wants. Based on what you determined for steps 1 and 2, look for the most effective and efficient way to answer the question. This may mean that you need to do math (such as adding and dividing), or you may need to make a strategy to reach the answer (such as using an equation or studying an figure). You want to use the strategy that reaches the answer as quickly as possible without just guessing your way through it.
4. Confirm the answer with the multiple choice.
   After you have what you think is the answer, go back and read the question to make sure that you have answered the question, not what you guessed it was asking. It's a good chance to see if you missed something or if you lost focus during the process. If you find you did get the question wrong, repeat step 3 to answer the question before you loo kat the answers.

Here's a chance to try it for yourself.

Say you have a table with three baskets on it. One basket has 7 apples, one has 3 apples, and the third has 14 apples. What is the average of the number of apples in the baskets?

- (A) 24
- (B) 7
- (C) 8
- (D) 3

What is are you meant to answer? ____

What pieces of information do you have? ____

What is your strategy to answer the question? ____

This is a question where there are a couple of trick answers. If you do a quick read and jump into answer it, you are probably going to add the number together, and you'll get an answer, but it isn't the right one. In this case, you need to add the three numbers together, then divide it by the number of baskets. The answer is C; the average is 8 apples.

Here's how you reach that answer.

7 + 3 + 14 = 24 total apples

24 / 3 baskets = 8 apples on a

PREPARING YOUR STRATEGIES FOR THE MATH SECTIONS

Solving problems is a source of anxiety for a lot of people. There are four strategies that you can apply as you go through the ASVAB to help you speed up the process so that you can finish before you run out of time. We recommend these four strategies.

- Working backwards
- Number Picking
- Guessing strategical through logic
- Combining

A couple of these solutions are good when you don't know where to begin or if you know that you are running out of time. Sometimes you aren't given enough time to go through all of the steps to reach the answer because some of the math processes are pretty time intensive. That's what a couple of these solutions helps you to make your best guess so that you don't have to leave a question unanswered.

You probably haven't used working backwards or picking numbers because most math tests aren't multiple choice. They want you to find the answer, not give you a chance to guess them. However, these two methods are great when you have answers and don't think you have time to go through all of the steps. You can start to see things with more clarity, so that you can more easily reach the answer, or at least narrow down your choices.

The other two methods, Guessing and Combining different approaches give you shortcuts to calculate the answer faster.

Remember, the focus is on answering quickly and moving on to the next problem so that you can answer all of the problems in the amount of time given. Being able to answer questions with high accuracy is important, so any shortcut that helps reach the right answer or quickly narrow your choices is going to help you at the time of the test.

Now let's look what those four methods of solving problems.

Working backwards

When you have a difficult math problem and a short period of time to solve the problem, it can be faster working backwards using the answers that they provide. After all, they give you four answers and you know that one of them has to be correct. From the beginning, you have a 1 in 4 chance of guessing correctly, which is good odds if you aren't particularly good at math. If you are able to quickly reach one of those answers by using it at the start to work backwards, then you can save yourself a lot of time.

When you are working on a time, that can be incredibly beneficial.

Here's how you use this method for the test. The answers are often listed from smallest to largest or from largest to smallest. This means that B or C will is likely to fall in the middle. If you plug these into the problem, and it works, then you are done. If not, you will know if you need a larger or smaller answer, eliminating a lot of the other answers. This could mean that there's only one option that works because it is the only one that is either smaller or larger than the one you tried. That means by working the problem backwards with one answer, you could find the right answer any way through the process of elimination. Select the right answer, then go on to the next question.

This also means that you should only have to work backwards twice – at the most – unless you want to verify your final answer.

Let's take a look at an example.

A garden shop put an apple tree on sale for 20% off, but they raised the price by 10% after putting it on sale. If the tree cost $70.40, what did it originally cost?

- ○ $20
- ○ $50
- ○ $80

   o $110

Now let's solve the problem.

1. You know that the price was reduced by 20% before being raised 10%, resulting in a final cost of $70.40.
2. You are trying to find what the original cost was.
3. The way to solve this is with algebra, but that can take a while. When you don't have much time, you don't want to spend a lot of time muddling through it, especially if you know that one of the four answers is correct. Instead of doing the math trying to find the answer, use one of the answers to find the answer – it's much faster.
   Let's try B, $50.
   Calculate 20% off $50.
   > 50 * 0.2 = $10 off
   > $50 - $10 = $40
   Calculate 10% more than $40.
   > $40 * 0.10 = $4
   > $40 + $4 = $44
   That's less than the final cost, so you can eliminate both A and B.
   Let's try C, $80.
   Calculate 20% off $80.
   > $80 * 0.2 = $16 off
   > $80 - $16 = $64
   Calculate 10% more than $64.
   > $64 * 0.10 = $6.40
   > $64 + $6.40 = $70.40

That's the final cost! That means the answer is C. If you had used D instead of C, you would still know that the answer is C because it would have been the only answer you had not eliminated.

If you work backward, always start with one of the two answers that falls in the middle between the highest and lowest numbers so that you can eliminate two answers at the same time (if you don't pick the right answer the first time). If you find that the answer is too large and only one answer is smaller, then you know that is the right answer.

Let's try another one, but let's use a different type of problem. Before you start, know that right triangles have a hypotenuse (the longest side of the triable) is always equal to the square root of the sum of the shorter sides of the triable. The way this is represented is Side 1^2 + Side 2^2 = Hypotenuse^2. This is usually represented by $a^2 + b^2 = c^2$.

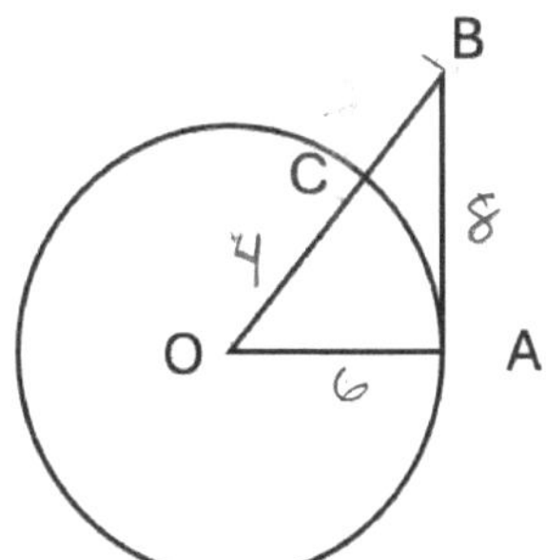

The circle in the image has a center labeled O and has a radius of 4. This can be written as AO = 6. If AB = 8 and ∠OAB = 90° (because it is a right angle triangle). Determine the length of BC.

   o   2(√17 -3)
   o   4
   o   2(√34 -3)
   o   6

What is your answer?

Now let's see what the right answer is and how you can work backward to read it.

1. You know that one side of the triangle is 4 because it is the radius of the circle , as well as a side of the triangle. Since the hypotenuse, OB, is longer than both sides, you know that it has to be larger than 4. Consider that OC is 4, and OB is longer than that.
2. You need to determine what BC is.

3. Use the Pythagorean Theorem to calculate OB. This can be time consuming considering you have to figure out a lot before you get the final answer. Instead, work backwards.

   Let's start with B since it's a whole number, making it easier to use. Since you are trying to figure out BC, you have OC + BC = OB, or 6 + 4 = 10. The problem said that AB is 10, and the hypotenuse has to be the longest line. This means you can eliminate both answer A and B.

   Now let's try C.

   6 + 6 = 12.

   Now let's put that into the Pythagorean Theorem.

   $6^2 + 10^2 = 12^2$ 

   $36 + 100 = 136$

   $12^2 = 144$

   136 does not equal 144.

This eliminates D, so the answer must be C. Thankfully you don't have to try that one out because it would be a little more difficult. You were able to use the easy numbers to determine that is the right one.

Number Picking

Working backwards isn't always the easiest way to solve a problem. Sometimes it's easier to use this method because trying to work backwards doesn't make sense because you don't have enough details to do that. Where there is too much missing information, you can use this method. You can simply chose a number to plug in to see if you can reach the solution. Make sure that you follow any rules, using a number that makes sense in context. You can also do this if all of the answers are also variables.

The recommendation is to avoid using 1 and 0 since those numbers have special properties that can negatively affect the answer.

Let's take a look at the method in use.

When x is divided by 12, the remainder is 9. What is the remainder when x is divided by 6?

- ○ 2
- ○ 3
- ◉ 4
- ○ 5

Now let's use the number picking method to find the answer.

- You know that x is unknown, and that it is 9 larger than a multiple of 12.
- The answer is what's left when you divide x by 9.
- Pick a number of x that is easy to work with. Let's use 16 (you get that by adding 12 and 9, the remainder and the number used to divide x).
- Do the math.
- 16 / 6 = 2 with a remainder of 4.
- The answer is C

Let's do another one.

If a cyclist speeds up by 20%, then speeds up by another 10%, what percent of the original speed is the total increase in speed?

- (A) 12%
- (B) 20%
- (C) 30%
- (D) 32%

What answer did you get by just picking a number and plugging it in?

Now let's work on it together to check your results.

1. The problem tells you that the cyclists speed up twice.
2. The answer is the total increase in speed from the original speed.
3. You aren't told what the end speed is, so you can pick any random number to do the problem. When working with percentages, 100 is always easiest. So, let's say the cyclist managed to start at 100 mph (although that is not all a realistic number, it is easy for you to use to get an answer quickly).

   20% more of 100 is 30. So, the new speed is 100 + 20, or 120 mph.

   10% of 120 is 12. So ,the speed after the second increase is 120 + 12, or 132 mph.

   132 – 100 = 32. In other words, the total increase is 32%.

4. The answer is D.

Notice that answers B and C seem to be too easy. 20% was how much the person sped up the first time, but doesn't take into account the second increase. Nor is it as easy as adding the first increase percentage with the second increase. This is because the person was going faster when they put on the second burst. The easy answer was C, but the correct answer is 32. When you pick an easy number to work with like 100, it makes finding the right answer pretty easy because percentages are based on 100.

If there are variables, you have the option to substitute numbers for those variables. Start by assessing the expression in the question stem using the numbers you chose. Then, apply the same numbers to evaluate each answer choice. Your objective is to identify the answer that produces an identical numerical outcome to the one obtained from your initial calculation. When employing this approach, evaluate all answer choices. In cases where multiple choices yield the same answer, opt for a new set of numbers to assess only the remaining choices that produced matching solutions with the first set.

Always start with permissible numbers that are easy to use when you need to calculate answers. This will make reaching the answer easier, as you've seen with the two examples.

Here is another example.

Linda spent x dollars on groceries every week for y weeks, and z percentage of that spending was on fruit. How much money did she spend on items other than fruits?

- $xy\left(1 - \frac{z}{100}\right)$
- $\frac{xyz}{100}$
- $\frac{100}{xyz}$
- $100z(1 - yz)$

There are no numbers in the example, so you are working with just variables.

- x = total money spent every week on food
- y = the number of weeks Linda when grocery shopping
- z = the percent of money Linda spent on fruit

You are trying to find how much money went to food that wasn't fruit.

Let's assign easy numbers to those variables.

- x = $20
- y = 5 weeks
- z = $10

Now let's do the math. $20 were spent every week for 5 weeks. That's $100. Of that $100, $10 went to fruit. $100 - $10 = $90. Linda spent $90 on food other than fruit.

Since there is really no elimination based on numbers, we'll start by plugging numbers into the equations. First up is A.

$$(20 * 5)\left(1 - \frac{10}{100}\right)$$

100 * 0.90 = 90

That's the right answer! You lucked out since it is the first one. However, you can plug the numbers into the other answers to verify it.

(B) (20*5*10) / 100 = 10 – No, that is not correct. Clearly Linda didn't spend just $10 on groceries that weren't fruit over that period.

(C) 100 / (20 *5 * 10) = 0.10 – That's even worse. Clearly, Linda didn't spend just 10 cents on nonfruit groceries.

(D) 100 * 20 (1 – 5*10) = This results in a negative amount because 1 minus anything larger than 1 is a negative number, and then you are multiplying, which means you are just getting a larger negative number. It is safe to say that she wasn't getting money to buy groceries.

Let's do another one.

For all a, b, c, and d, what does a(c + d) − b(c + d) equal?

- (a + b)(c + d)
- (a - b)(c - d)
- (a + b)(c - d)
- (a - b)(c + d)

Take some time to solve it by picking easy to use numbers. What was your answer?

Now let's go through it one step at a time.

1. You aren't given any numbers, just an algebraic expression and asked to find an expression that means the same thing.
2. You are looking to simplify the expression they gave you.
3. Let's pick some easy numbers.

   a = 4

   b = 3

   c = 2

   d = 1

   Now plug them into the equation.

   4(2+1) − 3(2+1) = 12 − 9 =3

   Now plug those numbers into the equation.

   o   (4 + 3)(2 + 1) = 7 * 3 = 21 − Not the right answer.

   o   (4 - 3) (2 - 1) = 1 * 1 = 1 − Not the right answer.

   o   (4 + 3) (2 - 1) = 7 * 1 = 7 − Not the right answer.

   o   (4 - 3) (2 + 1) = 1 * 3 = 3 − Yes! This is the right answer.

There was only one answer that was correct. Sometimes, you can get two answers that are right. You'll need to pick a new number set to see which of those answers is actually correct. However, you can eliminate the other two as you've already seen they don't yield the right results with the first number set.

This method works to make sense out of a lot of variables because math is easier for most of us when numbers are used instead of letters. Since you've plugged in actual numbers, you also get a chance to see that the answer you chose works with real numbers.

Guessing strategical through logic

This method is fantastic when you can use it because it means using logic instead of math to pick the right answer. Here's an example.

After using 25 percent of the gingerbread, John had 42 panels left for the gingerbread house. How many gingerbread panels did John have when he started?

(A) 51
(B) 54
(C) 56
(D) 59

1. The problem wants you to find how many gingerbread panels there were in the beginning.
2. The answer needs to be larger because it was what John had at the beginning, and he's now used 25% of the gingerbread panels.
3. Since 25% is a quarter, it has to be a whole number that is divisible by 4. That means that the starting number has to be divisible by 4 without a remainder. The only one that fits is (C) 56.
4. Do the math. 56 / 4 = 14. 56 − 14 = 42.

All you had to do was find the one that is divisible by 4 because the problem said that a quarter was used and still left a whole number. That isn't true with any of the answers beside C. You can then quickly plug in the numbers they gave you to verify the answer.

Let's try another one.

Emily can build a dollhouse in 3 hours, and Hank can build an identical dollhouse in 4 hours. How many house would it take for Emily and Hank to pain one dollhouse if they work together at their set paces?

(A) 3 / 2
(B) 12 / 7
(C) 3
(D) 4

You know that there are two people who work at different paces.

The answer should show how long it takes both people to paint a birdhouse if they work together without changing the pace for either of them.

Since Emily paints faster, she finishes in 3 hours. If someone is helping her, the project will take less than 3 hours. That means that C and D are not correct.

You can then eliminate A because that would require that they both work at the same speed, finishing the task in half the time. A is what you get if you have two people working at the same speed, or 3 hours divided by 2 people.

That means the answer has to be B.

You don't need to use any actual math. Just logic, and that can save you so much time.

Combining Methods

There will probably be a number of times when you end up having to use two or more methods to solve math. That's kind of the way it tends to work with numbers. Let's take a look at a problem where you can combine methods.

Mary can either walk or bike home. Her walking pace is 1 block a minute, and her cycling pace is 1 block per 20 seconds. If it takes her 10 minutes longer to walk than to cycle to work, how many blocks is she from her workplace from where she lives?

- o   7
- o   10
- o   15
- o   20

The problem has two different paces using two different units of measurement for how long it takes. To solve this, you need to change it so there is only one unit, so let's convert her cycling rate to minutes. There are 60 seconds in a minute, so it takes Mary 20/60 or 1/3 minute per block. Also, time to walk minus time to bike is 10 minutes.

The answer is the number of blocks Mary lives from where she works.

It's possible to answer this with algebra, but you can use logic, then work backwards to get the answer faster.

Since Mary walks 1 block for every minute, and it takes her 10 minutes longer to walk than ride to work, the answer has to be more than 10 blocks. If not, cycling to work would take zero minutes to get there.

That means you've eliminated A and B.

Now let's work backwards starting with C. If Mary lives 15 blocks from work, walking would take her 15 minutes. Riding would take a third of that time, or 5 minutes.

15 − 5 = 10

And that is the answer you need because it takes 10 minutes more to walk than to ride. If you had tried 20 first, you would have eliminated it, leaving C.

Let's do another one.

Heidi got a new tablet, stylist, and memory drive. She spent $149.99 on the tablet, $19.99 on the stylist, and $29.99 on the drive. If the sales tax on the items was 9.5%, which answer is closest to how much she spent?

- o $175
- o $200
- o $205
- o $220

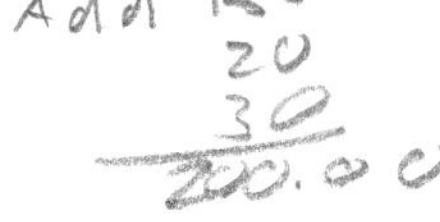

You know the price of all of the items and the tax, so you could actually do the math. However, that's probably going to take a good bit of time. Let's see how we can combine methods to solve it faster. Notice it doesn't say an exact amount, just the closest amount. That's pretty much an invitation to take shortcuts, including estimating.

Let's start by rounding up:

- - $150 for the tablet
- - $20 for the stylist
- - $30 for the drive

By summing these up, you get $200, and that's before you add the tax. That eliminates A and B.

Let's round the tax up to 8%, and then we see that 8% of $200 is $216. The closest number to $216 is $220, so D is the answer.

Take a few moments to verify the estimated math. You want to verify that you got the right answer – you don't want to oversimplify the problem and get it wrong.

Similarly, you should never be entirely guessing you way through problems. If you find yourself guessing too much, step back. The first step should always be using logic to eliminate answers that don't work. You want to have to worry about the fewest number of answers as is possible.

PRACTICING WHAT YOU'VE LEARNED

The ASVAB actually makes math a littler easier because either the question or the choices indicate what strategy you should use.

Let's do a few more to see what methods you use, then we'll go over how we recommend answering the problems.

1. A train uses 79% of its fuel source to travel 1,496 miles. If a train's tank hold 996 gallons of diesel, how many gallons would it use to travel 3,016 miles?

- o   800
- o   1000
- o   1200
- o   1600

Which method would you use answer this problem and why?

2. If $|2x + 4| = 18$, what is x?

- (A) 2
- (B) 4
- (C) 7
- (D) 9

Which method would you use answer this problem and why?

3. Sara spent 30% of her earnings on rent, and 40% on her car. What percentage of her earnings were left after she paid for rent and cars?

- (A) 60%
- (B) 40%
- (C) 30%
- (D) 20%
- (E) Which method would you use answer this problem and why?

CHECKING YOUR ANSWERS

Now it's time to see how well you did.

1. This one is easiest to solve combining a few methods. Start by rounding because it's much easier to work with rounded numbers. D is the closest answer.

2. You can use logic to eliminate D because that is half of 18, and clearly you need a smaller number. From there, you can plug in the answers. It won't take long to eliminate A and B. 2 * 7 = 14 + 4 = 18. C the right answer.

3. This one is straight addition. You don't always have to overthink. She spends 30% on rent and 40% on her car, or 70% total. Apart from the fact that she's spending too much on a car, she has 30% for her other bills and enjoyment, making C the right answer.

CHAPTER 6

ARITHMETIC REASONING

The AR section is the first math portion of the ASVAB. This chapter goes into more detail about how best to answer the types of questions that you have to answer in this section. This section and the MK section are important and you need to make sure that you do your best and answer every question. The better your performance on this section and the next one, the more opportunities you will have in the military. Many specialized roles require higher math skills, so if you don't perform well, you won't be in consideration for some occupations.

A BIT MORE ABOUT THIS SECTION OF THE ASVAB

The following is what to expect based on which type of test you take.

- For the traditional test, you will have 30 problems and 36 minutes to finish, so that's just a little over a minute for each problem. That means you won't have a lot of time to do the math. That's why the shortcuts are necessary to finish the full section.
- For the CAT version, you have 15 problems and 42 minutes to complete it. That means you have nearly 3 minutes a problem, but you don't want to take too long in the beginning. Remember, the problems get harder the more questions you answer correctly.

This chapter covers the following math concepts that are covered as a part of the AR portion of the ASVAB.

- Absolute values
- Decimals, fractions, and scientific notation
- Exponents and radicals
- Factorials
- Factors, multiples, primes, and remainders
- Number properties
- Math definitions
- Rules of dividing

Don't worry if these terms don't immediately seem clear to you – this chapter will help you to understand the concepts.

You will learn more about applied math:

- Averages and percentages
- Combined work
- Mean, median, and mode
- Order of operations
- Sequences and probability

You will need to do some word problems, but they will focus mostly on the following:

- Formulas

- Translation

Finally, you will have a chance to practice what you learn. If you feel that you need more practice, there are additional AR practice problems in Appendix B.

UNDERSTANDING THE AR DEFINITIONS.

You need to make sure that your understanding of the terms in this section align with how they are used in the ASVAB. Use the following table to review the terms, their definition, and some examples.

Term	Definition	Examples
Positive and Negative	Positives are any number larger than 0. Negatives are any number less than 0. 0 is neutral (not positive or negative)	Positive: $\frac{1}{3}$, 1, 11, 50, 958, 1,1111 Negative: $\frac{-1}{3}$, -1, -11, -50, -958, -1,1111
Odd and Even Numbers	Even numbers are any number that is a multiple of 2. Odd numbers are any number that are not multiples of 2. Fractions and mixed numbers are not even or odd.	Even: -10, -4, 0, 4, 8, 158 Odd: -101, -15, -1, 1, 13, 51, 2,001
Integer	A whole number (no decimals or fractions); it can be positive or negative	-1,000, -7, 0, 1, 2, 15, 999
Fraction	A numerical quantity representing a part of a whole, expressed as one integer (the numerator) divided by another integer (the denominator)	$\frac{-5}{8}, \frac{-3}{23}, \frac{1}{3}, \frac{101}{110}$
Improper Fraction	Fraction with a larger numerator than denominator	$\frac{-9}{7}, \frac{-53}{23}, \frac{10}{3}, \frac{111}{110}$
Mixed Number	A whole number and a fraction (no decimals); an improper fraction can be converted to a mixed number, and a mixed number can be converted to an improper fraction	$-3\frac{3}{25}, 1\frac{101}{111}$ $\frac{-9}{7} = -1\frac{2}{7}$ $\frac{10}{3} = 3\frac{1}{3}$
Prime Number	A whole number greater than 1 that is only divisible evenly by 1 and itself Note: No positive even number besides 2 is a prime number Note: Not all positive negative numbers are prime	2, 3, 5, 7, 11, 43 Not prime: 4 – divisible by 1, 2, 4 15 – divisible by 1, 3, 5, and 15 21 – divisible by 1, 3, 7, and 21 27 – divisible by 1, 3, 9, and 27
Consecutive Numbers	Counting (a sequence of numbers in order without skipping any numbers after the first one)	Consecutive integers: 1, 2, 3, 4, 5 Consecutive even: 2, 4, 6, 8 Consecutive multiples of -3: -3, -6, -9, -12, -15
Factor	A number that is positive and evenly divisible to a given number (when divided, it does not have any remainders)	Factors of 4: 1, 2, 4 Factors of 12: 1, 2, 3, 4, 6, 12 Factors of 15: 1, 3, 5, 15 Factors of 20: 1, 2, 4, 5, 20, 20

Term	Definition	Examples
Multiple	The product of the same number and any integer	Multiples of 2: 0, 2, 4, 6, 10 Multiples of 3: 0, 3, 6, 9, 12 Multiples of 5, 0, 5, 10, 15, 20 Multiples of 10: 0, 10, 20, 30, 40

Number Properties

The table showed you most of the number properties. Some number properties have firm rules that don't vary. These rules can help you quickly solve problems as you go through the ASVAB.

For positive and negative numbers, remember the following to ensure you get the right answer:

- Adding a negative is the same as saying you are subtracting from a positive number.
  - o   3 + (-2) = 3 − 2 the answer is 1
  - o   2 + (-3) = 2 − 3 the answer is -1
  - o   -2 + (-3) = -2 − 3 the answer is -5
- Subtracting a negative is the same you are adding that number.
  - o   3 - (-2) = 3 + 2 the answer is 5
  - o   2 - (-3) = 2 + 3 the answer is 5
  - o   -3 - (-3) = -3 − 3 the answer is 0
  - o   -5 − (-3) = -5 + 3 the answer I -2

For multiplying and dividing with positives and negatives have the following hard rules to determine if the answer is positive or negative.

- Two positives yield a positive answer. $2 * 4 = 8$
- Two negatives yield a positive answer. $-2 * -4 = 8$
- A negative and a positive yield a negative answer. $-2 * 4 = -8$ $2 * -4 = -8$

The same rule applies for division.

$8 \div 2 = 4$
$-8 \div -2 = 4$
$-8 \div 2 = -4$ $8 \div -2 = -4$

The rules for working with integers are fairly straightforward:

- Adding, subtracting, and multiplying two integers always results in an integer.
- Dividing with two integers may result in an integer or a fraction.
  - o   If the numerator is a multiple of the denominator, you will get an integer.
    4 is a multiple of 2, so $4 \div 2 = 2$
    2 is an integer.
  - o   If the numerator is not a multiple of the denominator, you will get a fraction.
    3 is not a multiple of 2, so $3 \div 2 = \frac{3}{2}$
    $\frac{3}{2}$ is not an integer.

There are several rules for working with even and odd numbers. The following are for adding and subtracting with even and odd numbers.

- When you add and subtract with two even numbers, you get an even number.
  $2 + 4 = 6$ $8 − 4 = 6$
- When you add and subtract with two odd numbers, you get an even number.
  $3 + 9 = 12$ $9 − 5 = 4$
- When you add and subtract with one even and one odd number, you get an odd number.
  $3 + 8 = 11$ $13 − 6 = 7$

The following applies for multiplying and dividing with even and odd numbers.

- When you multiply with even numbers, you get an even number.
  $2 * 4 = 8$

- When you multiply with an even and an odd number, you get an even number.
  4 * 6 = 48
- When you multiple two odd numbers, you get an odd number.
  5 * 9 = 45

There are no set rules for dividing even and odd numbers because you may not get an integer. If you have a fraction or decimals, the number is not even or odd.

Let's take a look at a problem and apply the rules.

If x is an integer, what is 2(x) + 5?

- (A) An odd integer
- (B) An even integer
- (C) Possibly even, possibly odd
- (D) An even non-integer

Let's work through the problem with

1. You know that x is an integer.
2. The value of the provided formula must be ascertained.
3. Go through the rules to determine the value:
    a. A number multiplied by an even number results in an even number, so 2 * x = an even number.
    b. A number plus an odd number results in an odd number, so 2x + 5 = an odd number.
4. The answer is A.

Absolute Values

The use of the following two vertical lines mean that you need to find the absolute value:

| |

An absolute value sign means that you need to use the positive of the number in the brackets.

- If it is a positive number, you don't need to change anything.
- If it is a negative number, you just need to use the positive version of that same number.

In other words, | 3 | is the same as | -3 |. They both equal 3.

Usually, you'll need to figure out the number in brackets in an equation. Given that it can be either positive or negative, you will need to answer with both positive and negative versions of a number.

Let's take a look at an example.

If | a | = 7, then what does a = ?

You know that a has to be either 7 or -7, since the only thing that changes with an absolute value is possibly turning a negative into a positive. However, you have no way of determining if the a is 7 or -7.

The answer is a = 7 and -7.

Let's take a look at another one.

If | b -7 | = 7, then b = ?

This time you are going to have to setup two different equations because b could be either positive or negative. Use the following two equations to determine the correct answers.

    b − 7 = 7   b = 14
    b − 7 = -7   b = 0

Both of these answers is correct. Let's verify that:

    | 14 − 7 | = | 7 | = 7
    | 0 − 7 | = | -7 | = 7

Yes, both 0 and 14 are correct answers for the problem.

The ASVAB will have problems with absolute values, and you will always need to solve for both positive and negative answers. Let's have you try a few.

If | x – 4 | + 2 = 10, which of the following is a potential answer for x?

(A) -8
(B) -4
(C) 2
(D) 4

The question is looking for the possible value of x. Notice it is not asking for both answers. That means you will not need to know both values.

First, let's get rid of the 2 but subtracting it from 10.

| x – 4 | + 2 = 10 | x – 4 | =8

Now let's look at the answer for the positive and negative values.

X – 4 = 8   x = 12

X – 4 = -8   x = -4

12 isn't one of the answers, so the correct answer is -4. Double check it before moving to the next problem.

| -4 – 4 | + 2 = 10

| -8 | + 2 = 10

8 + 2 does equal 10, so that is one of the two correct possible answers, and the only correct answer from the choices.

Factors, Multiples, and Prime Numbers

To find the factor of any given number, you need to find all pairs of positive integers that can be multiplied to equal that numbers.

- Always start with 1 and the number itself. Every number is divisible by itself and 1.
- If the number is even, you know that it is also divisible by 2. You will then need to determine what other number times 2 equals the number.
- Keep moving up in numbers, determining if the number is divisible by 3, 4, 5, and up until you reach the number you are checking.

Let's find the factors of 24.

- 1 and 24
- It's even, so it is divisible by 2. 2 * 12
- 3 * 8
- 4 * 6

The next number that goes into 24 is 6, and you've already found that one, so you now have all of the factors of 24.

Multiples are almost the complete opposite. Instead of finding the numbers that make up a given number, you are finding out what numbers that number can go into without leaving a remainder.

- 1, 2, 3, 4, 6, 8, 12, and 24 are all factors of 24.
- 24 is a multiple of 1, 2, 3, 4, 6, 8, 12, and 24. In other words, 2 and 12 can be multiplied together to make 24.

The two concepts are related, and they can be confusing. Here's one way to remember the difference:

- Multiples are always equal to or larger than the number you are starting with because you are multiplying two numbers together to reach that number.
- Factors are smaller because you are dividing your starting number to see what numbers go into it.

All numbers have an infinite number of multiples. They have a limited number of factors.

Prime numbers are fairly easy because they are numbers that only have two factors, 1 and the number. However, you have an infinite number of possible multiples of a prime number. Consider this:

2 is a prime number divisible only by 1 and 2. All even numbers (and there are an infinite number of even numbers because 2 goes into half of all positive numbers (it doesn't go into odd numbers).

In other words, prime numbers have only two factors, but infinite number of multiples.

3 and 7 are prime numbers. 3 * 7 = 21, so even though it is odd, 21 is not a prime number. It has four factors (1, 3, 7, and 21).

To determine the prime factorization for any number, you have to determine the number expressed with only multiplication with only primes, even if the primes repeat. To calculate the prime factorization, divide by all of the prime numbers until you have can't find any more non-prime factors.

It's probably easier to see an example of this:

What is the prime factorization of 124?

122 is an even number, so you know it is divisible by 2. 2 * 62 = 124.

62 is also an even number, so you can divide it by 2 again. 2 * 31 = 64.

31 is not even, so it is not divisible by 2. In fact, it isn't divisible by anything other than 1 and 31, so it is a prime number.

To help you see all of the prime factors, you can create a tree:

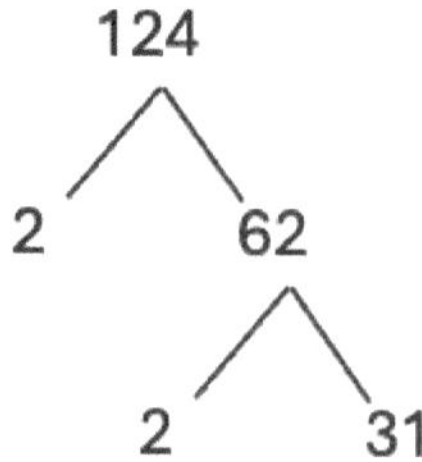

You can use the tree to write the prime factorization for 122 as follows:

2 * 2 * 31

The prime factors do include repeat numbers, but they have to be listed because you can only reach 124 by multiplying those prime numbers.

2 * 2 * 31 = 4 * 31 = 124

This can actually be very helpful on the ASVAB, such as when you need to determine the greatest common factor, or GCF, for two integers. If you find the prime factors for both integers, then multiply all of the common prime factors together, you have your answer. Let's take a look.

Find the greatest common factor of 30 and 130.

1. You have your starting numbers, 30 and 130, and have to find the greatest common factor for them.
   a. 30 prime numbers = 2, 3, 5
   b. 130 prime numbers = 2, 5, 13
2. Both of these have 2 and 5 as prime numbers.
3. 2 * 5 = 10. 10 is the GCF.
4. Check your math.
   a. 30 = 3 * 10.
   b. 130 = 13 * 10

If you are asked to reduce a fraction, the GCF can help you to then divide the top and bottom of that fraction by the number you calculated.

To find the common multiple of two integers, it's a lot easier – just multiply them together. However, the least common multiple, or LCM, may be a lower number. You can calculate the prime factors to help you find the LCMs.

Let's look at an example.

Find the least common multiple of 25 and 18.

1. Your starting numbers are 25 and 18, and you need to find their LCM.
2. Determine the prime factors:
   a. 25 = 5 * 5
   b. 18 = 2 *3 * 3
3. If an integer that is a multiple of 25, it must have two prime factors of 5. If an integer that is a multiple of 18 must have a prime factor of 2 and two 3s.

In other words, if any number that is a multiple for 18 and 25 has to have a prime factor of 2, two prime factors of 3, and two prime factors of 5. This is because there is no repetition between the LCM for the two numbers. You can then combine those to find the LCM.

2 * 3 * 3 * 5 *5 = 450

In this case, multiplying the two numbers together does get you the same number as calculating the prime factors. However, if you are given the two numbers 16 and 20, the LCM is not 320 (the number that you get when you multiply them); it's 80. Work the problem yourself to see how you can get the right, much lower number.

Let's try a few problems.

Frank order parts for his shop. The supplier sometimes meets his orders by sending boxes with 12 of the ordered parts, and sometimes he sends boxes with 16, depending on what is currently stocked. The supplier will only send full boxes, and will only send one size box for an order.

What is the minimum number of parts Frank can order to make sure he gets the exact number of parts he needs, no matter what box size is sent?

- (A) 24
- (B) 36
- (C) 48
- (D) 192

This is a question where you need to determine the lowest number of parts that you can have filled with boxes of 12 parts and 16 part. That means you need to find the LCM for those numbers.

- 12 LCM = 2 * 2 * 3
- 16 LCM = 2 * 2 * 2 * 2

Now multiply the four 2s with the 3.

2 * 2 * 2 * 2 * 3 = 48

Double check your work.

- 12 * 4 = 48
- 16 * 3 = 48

C is the correct answer.

Division

Division is the quickest way to verify factors and multiples. The following rules can help to use division effectively on the test.

- 2 goes into all even numbers.
- Divide the last two digits of the number by 2.
    - o   If you get an even number, the starting number is a multiple of 4.
    - o   If you get an odd number, the starting number is not a multiple of 4.
- If a number ends in 5 or 0, it is divisible by 5.
- If an even number is divisible by 3, then it is also divisible by 6. If it is an odd number, it is not divisible by 6.
- If you can divide a number by 2 twice, then that number is divisible by 8 as well.
- Add the number's digits together. The initial number is divisible by three if the sum is divisible by three. For example, for 21, 1 + 2 = 3, and 21 is divisible by 3 and 7. Or 97, 9 + 7 = 16, which isn't divisible by 3, so 97 is not a multiple of 3. Or 111, 1 + 1 + 1 = 3, so it is a multipole of 3 (3 * 37 = 111).
- Add the digits to the number. If the total is divisible by nine, then so is the original number. For example, if you have the number 108, 1 + 0 + 8 = 9. 9 is divisible by 9, so 108 is also divisible by 9.

Let's take a look at a problem and try these rules.

What are the prime numbers between 50 and 65?

- (A) 50, 51, 62, 63
- (B) 51, 55, 57, 59, 63
- (C) 53, 59, 61
- (D) 47, 53, 61
- 1.   You have a range of numbers and you need to find all of the prime numbers between 50 and 65.
- 2.   Apply the divisibility rules to eliminate the numbers that you know aren't prime.

    a.   You can immediately eliminate 50, 55, and 62. That means the answer is not A or B.
    b.   While all of the numbers in D are prime, 47 is not within the range.
3.   The answer is C, 53, 59, and 61.

Decimals and Fractions

When working with fractions, it is strongly recommended to reduce them as much as possible. The quickest way to do that is to use factors to reduce both the numerator and denominator.

Here's are a couple of examples:

$$\frac{24}{6} = \frac{4*6}{1*6} = \frac{4}{1} = 4$$

$$\frac{35}{10} = \frac{7*5}{2*5} = \frac{7}{2}$$

When you reduce a fraction, make sure to always use the same number to reduce the numerator and the denominator. In the first example, both numbers were divisible by 6. You may also notice that 6 goes into 24. When you divide the denominator evenly into the denominator, you get a whole number.

For the second example, both numbers are divisible by 5, so you need to reduce both the numerator and denominator by dividing them by 5.

When you need to add or subtract with fractions, you have to convert the fractions to have the same denominator. That means multiplying them so that the denominator is the same for both, essentially making it the opposite process of reducing a fraction. You want to find the LCM. Sometimes that means you need to multiply the numbers together, but it could mean finding a number that is lower. After you find a denominator, you need to multiple the numerator by the same number.

Let's take a look at a couple of examples of finding the LCM to add and subtract fractions.

$$\frac{1}{5} + \frac{2}{4} = \frac{1*4}{5*4} + \frac{2*5}{4*5} = \frac{4}{20} + \frac{10}{20} = \frac{14}{20}$$

The LCM for 5 and 4 is 20. However notice that the end result can be further reduced because both numbers are positive. The final answer should be $\frac{7}{10}$.

Let's look at an example with subtracting those fractions.

$$\frac{1}{5} - \frac{2}{4} = \frac{1*4}{5*4} - \frac{2*5}{4*5} = \frac{4}{20} - \frac{10}{20} = \frac{-6}{20}$$

Again, you should reduce the fraction to $\frac{-3}{10}$.

If you need to multiple fractions, you can just multiple the two numerators and then the denominators. When you are done, make sure to reduce it using the rules of division.

Fractions must be divided by multiplying them after flipping the second fraction's numerator and denominator. Let's look at example:

$$\frac{2}{3} \div \frac{2}{5} = \frac{\frac{2}{3}*5}{2} = \frac{2*5}{3*2} = \frac{10}{6} = \frac{5}{3}$$

You may be asked to compare fractions to determine which is bigger or which has a greater value. The quickest way to do that is to convert the fractions so that they have a common denominator. The wone with the larger numerator has the greater value. Let's take a look at an example.

Compare $\frac{1}{3}$ and $\frac{5}{8}$

The lowest common denominator (LCD) for 3 and 8 is 24. Now, you convert them to fractions that have a denominator with 24.

$$\frac{\frac{1}{3}*8}{8} = \frac{8}{24}$$

$$\frac{\frac{5}{8}*3}{3} = \frac{15}{24}$$

When you look at the numbers now, it is clear that $\frac{15}{24}$ is larger than $\frac{8}{24}$. That means that $\frac{5}{8}$ is the larger number.

Here are a couple of practice problems.

$$\frac{3}{4} + \frac{5}{6}$$

(A) $\frac{19}{12}$

(B) $\frac{23}{24}$

(C) $\frac{19}{24}$

(D) $\frac{1}{2}$

First, start with finding the LCD. Both 4 and 6 go into 12, so you need to convert them into fractions with 12 in the denominator.

$$\frac{\frac{3}{4} * 3}{3} = \frac{9}{12}$$

$$\frac{\frac{5}{6} * 2}{2} = \frac{10}{12}$$

Now add them together. $\frac{9}{12} + \frac{10}{12} = \frac{19}{12}$.

The correct answer is A.

Let's try another one.

What does the following equal?

$$\frac{\frac{1}{4} + \frac{3}{8}}{\frac{2}{5}}$$

(A) $\frac{2}{15}$

(B) $1\frac{11}{40}$

(C) $\frac{2}{15}$

(D) $1\frac{9}{16}$

You have to use the order of operations to solve this problem. Let's start by getting rid of the level of division.

$$\frac{\left(\frac{1}{4} + \frac{3}{8}\right) * 5}{2}$$

Now, you need to find the LCD for the two fractions in parentheses. Both 4 and 8 go int 24.

$$\frac{\left(\frac{6}{24} + \frac{9}{24}\right) * 5}{2}$$

Now you can add the two fractions in parentheses.

$$\frac{\frac{15}{24} * 5}{2}$$

Finally, you can multiply the remaining fractions:

$$\frac{75}{48}$$

That isn't one of the answers, so you know that you need to reduce the fraction. Both 75 and 48 are divisible by 3, resulting in $\frac{25}{16}$, or $1.\frac{9}{16}$. The answer is D.

Let's do one more.

0.005 = ?

(A) $\frac{5}{1,000}$

(B) $\frac{5}{100}$

(C) $\frac{5}{20}$

(D) $\frac{1}{4}$

If you've worked with decimals, you know that dividing a numerator by 1,000 moves the decimal 3 places. 5 becomes 0.005. A is the answer. You can use your knowledge of decimals and logic to answer this one without having to go into too much math.

Dividing with Decimals

The ASVAB is going to include decimals, just like that last question. Dividing with decimals can be challenging, so it is best to remove the decimals before you do the math. Multiply by 10, 100, 1,000, or other numbers to get rid of the decimal.

0.034 – multiply by 1,000 to remove the decimal: 0.034 * 1,000 = 34

1.14 – multiply by 100 to remove the decimal: 1.14 * 100 = 114

When you are asked to divide with a decimal, turn the number into a fraction:

$6.5 \div 0.02 = \frac{6.5}{0.02}$

Now you can move the decimal for the top and the bottom, making sure to move by the same number of space. In this case, move the decimal by two places.

$\frac{6.5}{0.02} = \frac{650}{2}$

Now, you can divide 650 by 2 to get the answer, 325.

Now let's do a couple of problems.

Shela is designing her dream home. The scale she used has the guest bathroom floor being 0.16 inches on paper. She needs to use 20 tiles over that width. What is the width of one tile.

    (A) 8
    (B) 0.8
    (C) 0.08
    (D) 0.008

You know that if the design is to scale, the floor is 0.16 on paper, but will take 20 tiles for the actual flooring. For a floor that is to scale, you can do the math to determine how wide each tile is.

To start, divide the design size by the number of tiles.

$$\frac{0.16}{20}$$

Now remove the decimals by moving the decimal point for both the numerator and the denominator.

$$\frac{16}{2000}$$

Since the answers have decimals, you will need to convert it again later. For now, determine if there are common factors between 16 and 2000. In this case, 16 goes into itself and 2,000 evenly.

$$\frac{1}{125}$$

Now you need divide 1 by 125. The answer is 0.008, or D.

You can use the scientific method to simplify problems. The use of scientific notation lets you write small and large numbers in a way that makes them easier to work with. The first part of this method means writing the main number to the left of the decimal. The rest of the number is on the right side of the decimal. Finally, you add * 10 to a certain power to indicate how many time you moved the decimal to have the first number appear to the left of the decimal.

10^4 is 10,000.

Here are a couple of examples:

1.15 * 10^4 =11,500 – You move the decimal to the right because the exponent is positive. The number should be larger.

1.15 * 10^{-4} = 0.000115 – You move the decimal to the left because the exponent is negative. The number should be smaller.

For scientific notation, only one number should be to the left of the decimal.

Here's an example problem.

Scott is working on a scientific article, so he wants to write 15,748,000 more easily. How can he write this number using scientific notation?

    (A) $15.748 * 10^6$
    (B) $1.5748 * 10^7$
    (C) $15.748 * 10^{-6}$
    (D) $1.5748 * 10^{-7}$

First, you are moving the decimal to the left, that means you will have a positive exponent. That eliminates C and D.

Next, count how many times you need to move the decimal until there is only one number to the left of the decimal. In this case, only 1 should be to the left of the decimal. B is the correct answer.

Exponents and Radicals

As the last problem showed, exponents are shown through raising a number to it is superscript to a number. It is always to the right of the number, with the larger number called the base number. The exponent shows how many times the decimal has been moved for scientific notation. For other instances, it indicates that the number is multiplied by itself that many times. There are two rules to keep in mind:

- If the exponent is 1, the answer is the base number. $7^1 = 7$
- If the exponent is 0, the answer is 1 . $7^0 = 1$

If you have 7^3, it means $7 * 7 * 7$, so $7^3 = 343$

You can add these when the base number is the same. If you also have variable, the variable and base must be the same.

$2x^3 + x^3$

$5x^8 - 3x^8$

You cannot combine the following:

$2x^3 + x^5$ cannot be further worked because the exponents are not the same.

$5x^8 - 3y^8$ cannot be further worked because the variables are not the same

You can multiply with exponents as long as the base is the same. Then all you have to do is add the exponents.

$2^3 * 2^4$ means $(2 * 2 * 2) * (2 * 2 * 2 * 2)$, or 2^{3+4}, or 2^7

Division is similar — all you have to do is subtract the two numbers.

$2^4 * 2^3 = 2^{4-3}$, or 2^1

When you have exponents with parentheses, you multiply the two exponents:

$(x^2)^3 = x^{2*3} = x^6$

A coefficient is a number or letter representing a letter that is multiplied by a variable. For example, $3x^3$, the coefficient is 3.

To multiply terms with coefficients and exponents with the same base variable, multiply the coefficients, then add the exponents.

$$5x^3 \times 2x^4 = (5 \times 2)(x^{3+4}) = 10x^7$$

To divide terms with coefficients and exponents with the same base variable, divide the coefficients, then subtract the exponents.

$$4x^5 \div 2x^3 = (4 \div 2)(x^{5-3}) = 2x^2$$

The ASVAB will include some negative exponents, but don't worry, they aren't nearly so difficult as you may initially think. The positive and negative exponents don't really differ from one another.

$$2^{-4} = \frac{1}{2^4} = \frac{1}{16}$$

Here are a couple of problems to try:

$$\left(\frac{(x^3)*(x^2)}{x^4}\right)^2 + \frac{2}{x^{-5}} = ?$$

(A) $3x^2$
(B) $x^2 + 2x^5$
(C) $2x^2 + x^3$
(D) $x^2 + 5x^2$

You have a complicated equation, and it asks you to simplify it. There is only one variable, so that should make it easier. You can pick a number to use with this problem, then solve with that number. Once you have an answer, plug that same number into each of the answers to see which one is correct.

Or you can use the exponent rules to answer this question.

$$\left(\frac{(x^{3+2})}{x^4}\right)\square^2 + \frac{2}{x^{-5}} = \left(\frac{(x^5)}{x^4}\right)\square^2 + \frac{2}{x^{-5}} = (x|5-4)^2 + \frac{2}{x^{-5}} = x^2 + \frac{2}{x^{-5}} = x^2 + 2x^5$$

The answer is B.

You can try both methods to see which one you prefer, picking a number and plugging it in or using the exponential rules.

When you have perfect squares, it is must easier because they are integers that are the result of multiplying a number by itself. For example, 225 is the perfect square of 15 because 15 * 15 = 225.

Which is not a perfect square?

(A) 1
(B) 25
(C) 36
(D) 88

You need to find the answer that is not the result of multiplying a number by itself.

1 * 1 = 1

5 * 5 = 25

6 * 6 = 36

The answer is D.

If you take the square root of something, it means you find the integer that results in that number. For example, the square root of 16 is 4, and of 36 is 6. The square root of a number is always a positive.

You can add and subtract with radicals, which is when the equation asks for a square root but there isn't one. For example, $\sqrt{2}$ is a radical because there is no square root (that is what the $\sqrt{}$ represents). You can add these together if the radical is the same.

$2\sqrt{2} + 4\sqrt{2} = 6\sqrt{2}$

However, you can't add radicals if they aren't the same. You cannot add the following.

$2\sqrt{2} + 4\sqrt{3}$

Radicals can be simplified by finding square roots and removing them from under the radical sign.

$\sqrt{50} = \sqrt{25*2} = 5\sqrt{2}$

Notice that the 25 is removed because the square root of 25 is 5. That then moves outside of the radical, leaving the 2 under it. The ASVAB is not going to ask for you to further reduce the number.

You can also multiply and divide with radicals and variables.

$\sqrt{x} * \sqrt{y} = \sqrt{xy}$

$2\sqrt{3} * 4\sqrt{7} = 8\sqrt{21}$

$$\frac{\sqrt{x}}{\sqrt{y}} = \sqrt{\frac{x}{y}}$$

$$\frac{8\sqrt{21}}{2\sqrt{3}} = 4\sqrt{7}$$

To find the square root of a fraction, break it down to see if the numerator and denominator have square roots.

$$\sqrt{\frac{9}{25}} = \frac{\sqrt{9}}{\sqrt{25}} = \frac{3}{5}$$

Here are a couple of problems to try.

Simplify $\sqrt{\frac{(2*50)}{(8*18)}}$

    (A) $\frac{12}{10}$

    (B) $\frac{5}{6}$

    (C) 5

    (D) $\frac{5}{12}$

First, determine if you can divide anything from both the top and bottom. Unfortunately, the radical sign makes that difficult.

Next, look to see if any of the numbers are perfect squares. Unfortunately, not in this case.

Next, do the multiplication:

$$\sqrt{\frac{100}{144}}$$

Do either of these have square roots? Yes, both of them do:

$$\sqrt{\frac{100}{144}} = \sqrt{\frac{10*10}{12*12}} = \frac{10}{12}$$

That is not one of the answers, so you need to simplify it to find the answer. They are both divisible by 2, so B is the answer.

Factorials

You will probably have to answer a few factorial questions, and you'll quickly recognize them because factorials are marked with an exclamation point (!).

Factorials are product of the number and all numbers before it until the number one. \

5! = 5 * 4 *3 *2 *1 = 120

6! = 6 * 5 * 4 *3 *2 *1 =720

9! = 9 * 8 * 7 * 6 * 5 * 4 *3 *2 *1 = 362,880

Obviously, the higher the starting number, the larger the result, and it doesn't take long to get to a very high number.

You can multiply and divide factorials. Here's an example.

Solve for $\frac{8!}{5!}$

Let's write it out to show all of the numbers that are included.

$$\frac{8*7*6*5*4*3*2*1}{5*4*3*2*1}$$

You can cancel all of the repeated numbers. In this case, everything from 5 to 1 is the same, so you can work with the remainder: 8 * 7 * 6 = 336. Since you eliminated all of the numbers in the denominator, it leave a 1, so you get a whole number, 336, for the answer.

Always start with simplifying before you move farther with these problems.

APPLIED MATH

That comes most of the arithmetic problems you are likely to find, but you will probably have word problems that will require you to apply math to solve them. The ASVAB requires you to use math principles to solve for the kinds of problems you could hear in real life. Applied arithmetic includes the following concepts:

-    Averages
-    Percentages
-    Probability

- Rates
- Ratios

Order of Operations

Math does have very strict rules for the order in which you should do the different math functions:

- Parentheses first
- Exponents second
- Multiplication and division third and fourth, working left to right
- Addition and subtraction fifth and sixth, working left to right

The common acronym used for this is PEMDAS, or Please Excuse My Dear Aunt Sally.

You've seen a few of these in previous problems, but here's another example. Follow which process is done with each step:

$$3^2 - 5(2 + 3) + 30 \div 2$$
$$3^2 - 5(5) + 30 \div 2$$
$$9 - 5(5) + 30 \div 2$$
$$9 - 25 + 15$$
$$1$$

Now try one for yourself.

Solve 4 * 4 + (8 * 3) -5^2

First, use PEMDAS to make sure you do the order correct.

$$4 * 4 + (8 * 3) - 5^2$$
$$4 * 4 + (24) - 5^2$$
$$4 * 4 + (24) - 25$$
$$16 + (24) - 25$$
$$15$$

Commutative and Distributive Properties

You should always start with the order of operations, but you may find that you need to apply these properties for some problems.

When dealing with the commutative property, the order you do multiplication and addition does not matter because they result in the same answer. Check the following:

2 * 4 = 4 * 2

2 + 4 = 4 + 2

This offers the kind of flexibility you don't usually have with math.

NOTE: This does not apply to division and subtraction. All you have to do is look at the previous examples to see that you don't get the same answer if you switch the position of the numbers.

Look at the following problem, keeping in mind you cannot use a calculator:

Solve for 3 * 17 * 4

Start by using strategic logic instead of jumping into multiplying with larger numbers. It's easier to start by doing 3 * 4 = 12. Now you only have to work with 17 and 12 instead of working with a large number and a larger number.

12 * 17 = 204

That is the easier of the two properties.

The distributive property is the product of multiplying one number either by the sum or the difference of two numbers, can also be reached by multiplying the individual numbers then adding them.

$$5 * (10 + 2) = 5 * 10 + 5 * 2 = 50 + 10 = 60$$
$$5 * (10 - 2) = 5 * 10 - 5 * 2 = 50 - 10 = 40$$

This can be used for fractions too when multiple terms in the numerator and only one term for the denominator can be subdivided.

$$\frac{x^4 + 4x^3}{x^2} = \frac{x^4}{x^2} + \frac{4x^3}{x^2} = x^2 + 4x$$

If there are multiple terms in the denominator, don't use this method. Fore example, you cannot split the following problem:

$$\frac{x^4}{x^3 + 3x^2}$$

Use the distributive property on the following problem:

Multiply 5 * 36 without a calculator.

You can simplify this by writing 5 * (30 + 6) = 5 * 30 + 5 * 6 = 150 + 30 = 80

Try it for yourself.

$13 * 7 + \frac{36x - 6}{6} = ?$

  (A) 70 * 7 + 3 * 7
  (B) 10x + 90
  (C) 6x + 90
  (D) 60 + 90x

Keeping in mind the order of operations, simplify the first part of the problem:

(10 + 3) * 7 = 70 * 7 + 3 * 7 = 70 + 21 = 91

Now look at the fraction to simplify it:

$$\frac{36x - 6}{6} = \frac{36x}{6} - \frac{6}{6} = 6x - 1$$

Write what is left together: 91+ 6x − 1.

Simplify what is left: 90 + 6x

Ratios, Proportions, and Rates

These three terms are similar, but they aren't exactly the same. That means you will have a similar approach, but you need to understand how they differ.

Ratios show the relationship between two or more quantities. You've probably heard people say the ratio of boys to girls in class is 2 to 3, meaning there are more girls in the class. Usually ratios are shown with colons, 2:3.

When you use ratio in math, you set them as a fraction or $\frac{2}{3}$. Ratios don't usually show actual number. For example, you probably don't have just 2 boys and 3 girls in the class. More than likely, you have 8 boy and 12 girls. You can have a part to part ration, such as boys to girls, or a part to whole such as boys to total number of students.

Here's an example:

A training course has 12 labradors and 15 beagles.

  -    The part to part ration of labs to beagles is $\frac{12}{15} = \frac{4}{5}$.
  -    The part to whole of labs to total dogs is $\frac{12}{12+15} = \frac{12}{27} = \frac{4}{9}$.

When working with ratios, you must simplify the fraction.

Let's try one.

Luke loves playing frisbee golf. He has 21 discs. The ration of blue discs to yellow discs to red discs is 4:2:1. How many red discs does he have?

  (A) 6
  (B) 7
  (C) 3
  (D) 1

Start by writing the part to whole ration of red discs:

$$\frac{1}{4+2+1} = \frac{1}{7}$$

Remember, there were 21 games though, so that's 7 *3. There were three times more of all discs. So there are 3 * 1 or 3 red discs. The answer is C, 3.

A proportion compares two rations by using a equation. They can represent number of items, size, degrees, weight, pretty much anything that can be countered or measured.

The best technique to use when working with proportions is the cross-multiplying method. If you are working with two fractions, you just multiply the numerator of the first fraction by the denominator of the other, and then then denominator of the first with the numerator of the other. Let's take a look:

$$\frac{4x}{5} = \frac{3}{4}$$

Use the cross-multiplying method:

4x * 4 = 5 * 3

Now let's look at a problem with proportions.

A blanket is 4 feet wide and 6 feet long is modified to add another $9\frac{1}{2}$ feet long. What is the width of the modified blanket?

Turn this into a proportion to see how the blanket has changed:

$$\frac{4}{6} = \frac{x}{9}$$

4 * 9 = 6x

36 = 6x

x = 6

The width is not 6 feet.

Here is a problem you can try:

Sam skates 12 miles in 2 hours. Assuming he travels at the same rate of speed, how long will it take Sam to go 42 miles?

    (A) 7 hours
    (B) 5.25 hours
    (C) 42 hours
    (D) 7.5 hours

You know that the answer assumes a single speed, and you know the total distance of 42 miles. You need to solve how long it will take. Turn that into fractions:

$$\frac{2 hours}{12 miles} = \frac{x}{42 miles}$$

Now use the cross-multiplication technique to solve the problem.

2 * 42 = 12x

84 = 12 x

X = 7

The correct answer is A. You can then double check it by plugging 7 into the fraction.

The final term is rate, and all that means is a ration comparing two similar quantities. You just did a rate with the last word problem. When you hear the term miles per hour, that is the rate of speed. There are many types of rates, such as rate for an apartment using cost/units.

If you think of a rate in terms of fractions, you can see that the changes to the denominator requires the same change to the numerator. If you are paid $25 for one hour, you will get another $25 for the second hour.

When you have to work with rates, treat them like proportions, and then it is easier to solve for the value you don't know.

A bartender serves 7 drinks every 5 minutes. At this rate, how many drinks will he serve in an hour?

Set it up as a proportion:

$$\frac{7 drinks}{5 minutes} = \frac{x drinks}{1 hour}$$

First, you need to make sure you have the same units in the denominator:

$$\frac{7\,drinks}{5\,minutes} = \frac{x\,drinks}{60\,minutes}$$

Now you can cross-multiply and solve with the cross-multiply method.

7 * 60 = 5x

420 = 5x

x = 84

You should go back and double check by plugging it into the fraction you made to verify your work.

NOTE: Make sure you pay attention to units, especially if the question is about time or a unit that you have converted in the denominator.

Here's one for you to try on your own:

The volume of a liquid is proportionate to its weight. If 32 cubic inches of the substance weigh 112 ounces, what is the volume of 63 ounces of the liquid?

    (A) 20
    (B) 18
    (C) 63
    (D) 32

You have the details to create the ration in the form of a fraction.

$$16 \times 2 = \quad \frac{32}{112} = \frac{x}{63}$$

As you start to do the calculations, you'll notice that you are working with big numbers, which will take time. You can use logic to help reduce the numbers first. Both 32 and 112 are divisible by 16. By reducing the fraction, you are now working with the following:

$$\frac{2}{7} = \frac{x}{63}$$

7x = 2 * 63

7x = 126

x = 18

The answer is B, 18. You can then verify your work.

Combining Problems

The ASVAB often includes a problem or two that want you to calculate how long something takes when effort is combined. You've done one such problem in a previous chapter, but let's look at this problem on its own. For example, if a problems if you have two pipes used to fill tank, you are dealing with one problem with two pipes that work at different rates.

Here's a combined effort problem.

Jake tiles a room in 3 hours. Hannah tiles a room in 2 hours. How long will it take them to tile a room if they work together?

You have two different rates, so write those down:

$$\frac{1}{3} + \frac{1}{2}$$

Find a common denominator to add the rates:

$$\frac{2}{6} + \frac{3}{6} = \frac{5}{6}$$

So they finish 5 rooms in 6 hours, but that's not the answer the problem wants. Now invert the fraction: $\frac{6}{5}$ is how long it takes to paint a room. Simplify that to get the answer = 1.2 hours. Double check your work.

Here's another one.

Three pipes can be used to fill one tank. One of the pipes can fill the tank in 2 hour, another takes 3 hours, and the third takes 4 hours. If you use all three pipes at the same time, how long will it take to fill the tank?

First, you know that it will take less than 1 hour since one pipe alone can finish it that quickly. Create three rates that compare tanks to hours.

$$\frac{1}{2} + \frac{1}{3} + \frac{1}{6}$$

Now you need to find a common denominator:

$$\frac{3}{6} + \frac{2}{6} + \frac{1}{6} = \frac{6}{6} = 1$$

Together it takes 1 hour to fill the tank using all three pipes.

Percents

Percents are something that you probably know, but they boil down to ratios against 100. This gives you a known quantity to work with when dealing with these proportions.

Here's a precent problem that is similar to something you may see on the ASVAB.

A jar has 12 marbles and 6 are the blue. What percent are blue?

Set up the problem as a proportion:

$$\frac{6}{12} = \frac{m}{100}$$

Now use cross-multiply to solve for the percent.

12m = 800

M = 50 or 50%

You could have used logic for this one two. If there are 12 marbles, and 6 are blue, that's exactly half. You didn't have to do the math if you stop and use logic first.

When you are working with percentages, logic is pretty easy. There is also some conversions you can do to better deal with them:

- % means percent, and it is always out of 100
- Of always means to multiple
- What is signified by with a variable of your choice (you can use x or the first letter of the item you are trying to solve.
- Is always means equals

Keep this in mind when you have a question that asks you about percents.

The problems get trickier when the problem asks you to either increase or decrease something by a percentage. In these instances, take the percent of the original amount and add or subtract it from that number.

If you are asked to increase 25 by 60%, calculate 60% of 25:

25 * 0.6 = 15

Now add it to the original number, 25.

25 + 15 = 40

If you are asked to decrease 25 by 60, the steps are the same until the end, then you subtract:

25 − 15 = 10

It's also possible you will be asked to calculate the percentage of change between the original value and a new one. Here's the formula you need for that problem:

$$\frac{new - old}{original}$$

If the percent is increased, the answer will be positive. If the percent is decreased, the answer will be negative. Here's an example.

A game console originally cost $1,000, but was on sale for $750. The sale price was what percent less?

Use the formula:

$$\frac{750 - 1000}{1000} = \frac{-250}{1000} = \frac{-1}{4} = -20\%$$

Let's do another one.

Sean weighed 75 pounds when he finished elementary school, and 80 pounds when he started middle school. What was the percentage of his weight gain between the end of elementary school and beginning of middle school.

    (A) 10% increase
    (B) 10% decrease
    (C) 15% increase
    (D) 15% decrease

Use the formula:

$$100\frac{\% * 80 - 75}{75} = 100\frac{\% * 5}{75} = 100\frac{\% * 1}{15} = 15\%$$

Since the number is positive, it's an increase. C is the right answer. Check your math to verify your answer.

Let's do one more.

Greg invested in the stock market, but the value dropped by 20% in the first month. In the second month it increased by 50%. What was the overall percentage change between her initial investment and the value after two months?

    (A) 20% increase
    (B) 30% increase
    (C) 20% decrease
    (D) No change

There are no specific values given, just percentages. That means, picking numbers could work very well. As it is a percent, you know that the base should be 100.

After the first month of dropping 20%, the stock would be worth 80% of the original value. The value then increased by 50%, or 80% * ½ = 40% + 80% = 120% more. This is a 20% increase from the original 100% she invested. The answer is A.

Make sure to verify your math.

Statistics

Statistics provide averages and are similar to ratio, so you can use fractions to solve them. Here is the formula to determine an average:

$$\frac{\sum of\ Items}{Number of Items}$$

Here's a problem that could be similar to what you see on the ASVAB:

What is the average of 3, 5, and 10?

Average $= \frac{3+5+10}{3} = \frac{18}{3} = 6$

It is more likely you will get problems more like the following.

Six people took a cooking cases. They baked an average of 9 cakes during the first lesson. How many total cakes did they bake?

Plug the numbers they gave you into the formula.

$$9 = \frac{total}{6}$$

Multiple both sides by 6 (the number of people who participated).

$$9 * 6 = 54$$

Go back and verify your answer.

Here's one to try on your own.

Dennis was thrilled to learn that her finally math grade, based on 5 tests was, 90. She scored 83, 87, 90, and 92 on four tests. What did she score on the fifth test?

    (A) 100
    (B) 98

(C) 88
(D) 90

You have most of the information you need.

$$90 = \frac{83 + 87 + 90 + 92 + x}{5}$$

$$450 = 83 + 87 + 90 + 92 + x$$

$$450 = 352 + x$$

$$x = 98$$

The answer is B. make sure to verify your answer.

The ASVAB isn't going to require you to solve the kinds of statistics that go into science reports, but you will be expected to know some basic terms.

- Mean is another term for average, so you solve for the mean the same way.
- Range is how wide the group is, so you subtract the smallest number they provide form the largest.
- The mode is the number that appears the most often
- The median refers to the middle number of the range.

Let's look at these terms when used with a number set:

1, 3, 7, -5, -1, 9, 0

First, put the numbers in order:

-5, -1, 0, 1, 3, 7, 9

The mean is calculated as follows:

$$\frac{-5 + -1 + 0 + 0 + 1 + 3 + 7 + 9}{7} = \frac{14}{7} = 2$$

The range is calculated as follows:

$$9 - (-5) = 14$$

There is no mode because no numbers are repeated or appear more often than any of the others.

The median is 1 because it is the number that falls in the middle of the 7 numbers when they are added in order. If there are an even number of number, add the two middle numbers, then divide them by 2. Here's an example:

-2, 0, 2, 2, 4, 7

The median is 2+2 = 4, then divide it by 2, so 2. The median can have a decimal.

Sequences

A sequence is simply listing numbers in order. When you find the median and range for a set of numbers, you need to sequence them to get the answers.

The ASVAB includes arithmetic sequences, or a sequence that has a pattern in the gaps between each number. The following are arithmetic sequences:

1, 2,3, 4, 5, 6, 7, 8, 9, 10 (each is 1 more than the previous)

2, 4, 6, 8, 10, 12, 14, 16, 18, 20 (each is 2 more than the previous)

1, 3, 5, 7, 9 (each is 2 more than the previous)

3, 6, 9, 12, 15, 18, 21 (each is 3 more than the previous)

All of these are sequences because there is the same number between each of the numbers listed. When you are working with sequences, the median and the mean are equal.

The following is the type of question you may see.

What is the mean of a sequence of multiples of 5, beginning with -5 and ending with 25?

You know the starting and ending points of the sequence, and you know that you need to count by 5.

Start by writing the first number in the sequence: - 10.

Then count by 5s until you reach the ending number, 25.

-5, 0, 5, 10, 15, 20, 25

Since the median and mean are the same, the answer is 10.

Probability

Probability is a type of statistic that calculates the likelihood of a stated outcome. The following is the formula to calculate the probability of an outcome.

$$\frac{Number\,of\,outcomes\,specified}{Number\,of\,possible\,outcomes}$$

Dice and coins are the two most frequently used probability determiners that you are likely to use in real life. Let's look at finding the possibility.

If there are 12 available games, and 9 of them are horror games, what is the probability of randomly selecting a horror game?

Plug the numbers into the formula.

$$\frac{9}{12} = \frac{3}{4}$$

In other words, you have a 3 in 4 chance of choosing a horror game. Go back and check your work.

What if the question asked you the probably of choosing a nonhorror game? You would subtract the answer you got from 1, giving you a probability of $\frac{1}{4}$ of getting a game that wasn't a horror game.

When you working with probabilities, they are a type of percentage, so the range is always between 01 and 1, or 0 and 100%. In the question above, if none of the games are RPGs, there is a 0 probability of choosing an RPG randomly, or there is no chance of selecting that type of game. If you have a probability of 1, it means that something is guaranteed.

If you are asked to find the probability of two events occurring, calculate the probability for the first event, then multiply it by the probability of the second one. This will result in a lower number than either as it is less likely that you will get two events independent of each other.

Let's look at an example.

If there are 12 available games, and 9 of them are horror games, what is the probability of randomly selecting a horror game first, then randomly selecting a nonhorror game second if the games are removed once they are selected?

You've already done the math for the first one, $\frac{3}{4}$. Now you need to calculate the second:

There are 3 nonhorror games, and only 11 games remaining since one horror game has been removed. The probability of getting a nonhorror game is now $\frac{3}{11}$.

Multiply the two answers:

$$\frac{\frac{3}{4} * 3}{11} = \frac{9}{44} \qquad \frac{\frac{2}{4} \times 6}{12} = \frac{12}{48}$$

Verify your work.

Another probability question you may get is the probability of one of two events occurring. In this case, you add the two probabilities, and the result will be larger number than either number on its own.

What is the probability of rolling a 5 or 6 on a single roll of a six-sided die?

You have the same probability of rolling a 5 or 6. You have a 1 in 6 chance, or $\frac{1}{6}$ for rolling both of them. Now add that together:

$$\frac{1}{6} + \frac{1}{6} = \frac{2}{6} = \frac{1}{3}$$

Verify your work. You could also use logic for this one because there are six sides, and they want to know what the probability is for rolling 2 numbers, which is a third of the numbers on the die.

Let's try another.

If a coin is flipped 3 times, what is the probability that none of the flips will result in tails?

    (A) 0

    (B) $\frac{1}{4}$

    (C) $\frac{1}{6}$

(D) $\frac{1}{8}$

You need to multiple the probability for each event together, keeping in mind there are three events, and only two possible results:

$$\frac{\frac{\frac{1}{2} * 1}{2} * 1}{2} = \frac{1}{8}$$

The final answer is D. Verify your work.

WORD PROBLEMS

Math includes a lot of word problems, and you've been dealing with them over most of this chapter, and you've probably groaned when you saw them. Hopefully, the explanations helped you to better manage the problems. What you've hopefully noticed is that the math problems are the focus, you just have to determine how to create an equation so solve the problem. At the same time, the information is added to scenarios that you could actually have in real life. The story isn't what matters, just the numbers and how you use them.

When you have word problems, remember to focus on the math.

This section focuses on how to interpret word problems when you don't know what it is looking for prior to starting. It's a lot easier to anticipate a problem when it is in a section in a guidebook because you know what the focus of that section is – you don't get that on a test.

Translating to Math

The problem with word problems is that it can be tricky to figure out what the problem wants you to find. The following to help you to start assessing problems and determining what operations to use based on the wording:

+	more than, added to, combined with, total, sum, plus
-	difference between, decreased by, less than, minus
x	product, times, multiplied by, of
÷	quotient, per, out of, over, into, divided by
=	equals, is, was, adds up, is the same
Use a variable	what, how many, how much

All of those phrases indicate the type of operation you should use, which can help you to determine how to approach the problem.

When Formulas Are Included

The reason people tend to dislike word problems is that some of them require you to plug the details into a formula. Since the problem doesn't give you the formula, you have to figure out what formulate to use, and hope that you remembered it correctly.

Fortunately, many of the formulas in this part of the test are going to be more straightforward because they have application in your life. That means you are exposed to them more often. The following three are the formulas you are most likely to need:

$$\text{Average} = \frac{\sum of items e}{total number of items}$$

$$\text{Rate} = \frac{Distance}{Time}$$

$$\text{Probability} = \frac{Number of outcomes specified}{Number of possible outcomes}$$

Let's try a couple of problems to give you some practice at figuring out what formula to use.

If an RV drives at 45 miles an hour for 5 hours, how far will it travel during that time?

Assess the problem to determine what it wants. In this case, it is looking for a rate. We can start plugging the details that we do have into the formula.

$$40 = \frac{D}{5}$$

$$40 * 5 = D$$

$$200 = D$$

The RV travels 200 miles during those 5 miles.

Abby biked 40 miles from her home in 2 hours. Unfortunately, she got a flat time, and it took 3 hours to go return home. What was her overall speed for the trip?

- (A) 10 miles an hour
- (B) 16 miles an hour
- (C) 20 miles an hour
- (D) 80 miles an hour

You are given all of the details you need, now you need to determine what formula to use. Since it is providing distance and speed, you know you are working with a rate question. You can add the distance (she travelled 40 miles out and 40 miles back home), and you can add the time it took

$$Rate = \frac{40+40}{2+3} = \frac{80}{5} = 16 \text{ miles an hour}$$

The answer is B, 16 miles an hour.

PRACTICING WHAT YOU'VE LEARNED

Now that you've got the basics, you can test what you've studied.

Here are two practice sets with 15 questions each. This is similar to how the CAT is setup, although obviously your questions aren't going to change the better you do.

The last section of this chapter details the right answers to the questions.

Practice 1

1. If a car travels at a speed of 60 miles per hour, how many miles will it travel in 4.5 hours?

   (A) 240 miles
   (B) 270 miles
   (C) 300 miles
   (D) 330 miles

2. John had $150. He spent $45 on a book and $32 on a shirt. How much money does he have left?

   (A) $68
   (B) $73
   (C) $85
   (D) $98

3. A rectangle has a length of 12 cm and a width of 8 cm. What is its perimeter?

   (A) 24 cm
   (B) 32 cm
   (C) 40 cm
   (D) 48 cm

4. A box contains 15 red balls, 10 blue balls, and 5 green balls. What fraction of the balls are red?

   (A) 1/3
   (B) 3/10
   (C) ½
   (D) 3/8

5. If 5 pencils cost $3.75, what is the cost of one pencil?

   (A) $0.65
   (B) $0.70
   (C) $0.75
   (D) $0.80

6. A school has 450 students. If 55% are girls, how many girls are there in the school?

   (A) 225
   (B) 248
   (C) 248.5
   (D) 247.5

7. A piece of ribbon is 2 meters long. It is cut into 8 equal parts. How long is each part?

   (A) 0.2 meters
   (B) 0.25 meters
   (C) 0.3 meters
   (D) 0.5 meters

8. How much sugar is needed if a recipe calls for 3/4 cup and you want to make half of the recipe?

   (A) 3/8 cup
   (B) 1/2 cup
   (C) 1/3 cup
   (D) 2/5 cup

9. The sum of two numbers is 90. If one number is 57, what is the other number?

   (A) 23
   (B) 33
   (C) 37
   (D) 43

10. A car's fuel tank holds 16 gallons of gasoline. If the car uses 0.05 gallons per mile, how many miles can it travel on a full tank?

    (A) 200 miles
    (B) 240 miles
    (C) 300 miles
    (D) 320 miles

11. The population of a town increased from 20,000 to 25,000 in 5 years. What was the average annual increase in population?

    (A) 800
    (B) 900
    (C) 1,000
    (D) 1,200

12. A store is offering a 20% discount on all items. If a jacket costs $80 before the discount, what is the price after the discount?

    (A) $64
    (B) $66
    (C) $68
    (D) $70

13. If you divide a number by 4 and then subtract 7, you get 15. What is the number?

    (A) 68
    (B) 78
    (C) 88
    (D) 98

14. Three friends share a pizza. One eats 1/3, the second eats 2/5, and the third eats the rest. What fraction of the pizza does the third friend eat?

    (A) 1/15
    (B) 2/15
    (C) 4/15
    (D) 7/15

15. The area of a triangle is 36 square inches, and its base is 9 inches. What is the height of the triangle?

    (A) 6 inches
    (B) 7 inches
    (C) 8 inches
    (D) 9 inches

Practice 2

1. A company sold 200 products in January and 260 products in February. What was the percentage increase in sales?

   (A) 20%
   (B) 25%
   (C) 30%
   (D) 35%

2. A school has 30 teachers. If 40% of them teach math, how many math teachers are there?

   (A) 10
   (B) 12
   (C) 14
   (D) 16

3. If the price of an item is increased by 15%, what is the new price of an item that originally costs $120?

   (A) $130
   (B) $135
   (C) $138
   (D) $140

4. A train travels 300 miles in 5 hours. What is its average speed?

   (A) 50 miles per hour
   (B) 55 miles per hour
   (C) 60 miles per hour
   (D) 65 miles per hour

5. If a worker earns $18.50 per hour and works 35 hours a week, what is their weekly wage?

   (A) $625.50
   (B) $647.50
   (C) $662.50
   (D) $682.50

6. A bag contains 8 red marbles and 12 blue marbles. What is the probability of randomly selecting a blue marble?

   (A) 1/3
   (B) 3/5
   (C) 2/5
   (D) 4/5

7. If a number is increased by 20% and the result is 72, what was the original number?

   (A) 50
   (B) 55
   (C) 60
   (D) 65

8. The ratio of the length to the width of a rectangle is 3:2. If the perimeter of the rectangle is 50 cm, what are its dimensions?

    (A) 12 cm by 8 cm
    (B) 15 cm by 10 cm
    (C) 18 cm by 12 cm
    (D) 21 cm by 14 cm

9. A student scores 85, 90, 78, and 92 on four tests. What is the student's average score?

    (A) 85.25
    (B) 86.25
    (C) 86.75
    (D) 87.25

10. If the product of two numbers is 144 and one of the numbers is 12, what is the other number?

    (A) 10
    (B) 12
    (C) 14
    (D) 15

11. The price of a laptop was $800 last year. This year, the price decreased by 12%. What is the new price of the laptop?

    (A) $688
    (B) $704
    (C) $720
    (D) $752

12. If a car travels at a speed of 50 miles per hour, how many miles will it travel in 3 hours?

    (A) 100 miles
    (B) 150 miles
    (C) 200 miles
    (D) 250 miles

13. Sarah had $200. She spent $35 on groceries and $45 on a new book. How much money does she have left?

    (A) $120
    (B) $125
    (C) $130
    (D) $135

14. A rectangle has a length of 10 cm and a width of 5 cm. What is its area?

    (A) 25 cm²
    (B) 30 cm²
    (C) 40 cm²
    (D) 50 cm²

15. A store is offering a 15% discount on all items. If a shirt costs $40 before the discount, what is the price after the discount?

    (A) $30
    (B) $32
    (C) $34
    (D) $36

CHECKING YOUR ANSWERS

Now it's time to see how well you did.

Practice 1

1. B. 270 miles
2. C. $73
3. C. 40 cm
4. A. 1/3
5. C. $0.75
6. B. 248
7. B. 0.25 meters
8. A. 3/8 cup

9. C. 33
10. D. 320 miles
11. C. 1,000
12. A. $64
13. C. 88
14. D. 7/15
15. A. 8 inches

Practice 2

1. B. 30%
2. A. 12
3. C. $138
4. C. 60 miles per hour
5. C. $647.50
6. B. 3/5
7. C. 60
8. B. 15 cm by 10 cm

9. B. 86.25
10. B. 12
11. B. $704
12. B. 150 miles
13. C. $120
14. D. 50 cm²
15. C. $34

CHAPTER 7

MATHEMATICS KNOWLEDGE

The MK section is the second math portion of the ASVAB, and it includes the kinds of math that you generally associate with problems you are less likely to encounter in daily life. However, it is important to do well on this section in the ASVAB since it is part of your total score. You want to provide the greatest possible answer to each question. The better your performance on the two math sections, the more opportunities you will have in the military.

A BIT MORE ABOUT THIS SECTION OF THE ASVAB

The following is what to expect based on which type of test you take.

-   For the traditional test, you will have 20 problems and 16 minutes to finish, so that's just a little over a minute for each problem. You won't have much time to complete the math because of this. That's why the shortcuts are necessary to finish the full section.
-   For the CAT version, you have 25 problems and 24 minutes to complete it. That means you have less time – less than a minute – to complete each problem. Using shortcuts is going to important to get through this section.

Math isn't as clear cut as the two sections make it seem. When you are doing the different types of math, you are going to need to use applied math, and this is particularly true with the algebra portions of the test.

When you take the ASVAB, the two types of math that you are going to need to know for the MK section are :

-   Algebra
-   Geometry

This chapter covers the concepts that are important for each of these areas of math and works through problems to see how those concepts apply to the kinds of questions you will likely have on the test. You will need to work on the following algebraic concepts as a part of the MK portion of the ASVAB.

-   Algebraic expressions
-   Binomials and monomials
-   Quadratic equations
-   Solving variables
-   Inequalities

You will need to work on the following gemoetry concepts as a part of the MK portion of the ASVAB.

-   Triangles
-   Quadrilaterals

- Circles
- Surface and volume calculations
- Combined geometry

Don't worry if these terms don't immediately seem clear to you – this chapter will help you to understand the concepts and how to use them.

REVIEWING ALGEBRA

So far, we've covered a lot of arithmetic, which has included some algebra. Algebra relies on the use of variables to create generalized relationships.

The ASVAB has a good bit of algebra in this section, and they generally fall into one of two categories:

- Straightforward problems
- Word problems

These are ideal problems for both picking a number to fill in the equation and for working backwards. If you are comfortable with algebra, you may not need to rely on these methods as much since you will be able to do the problems just as quickly for this branch of math. Algebra is like puzzle solving, so the way you best solve puzzles is going to be the best way to work on these problems.

Understanding with Algebra Vocabulary

The following table provides the vocabulary that you need to understand how to work through some of the problems on this portion of the test.

Vocabulary Term	Definition	Examples
Coefficient	The number that comes before a variable	**3**b $5x^2$ **12**y
Variable	A value represented by a letter	3**b** $5x^2$ 12**y**
Equation	Two expressions that equal each other	x + y = y + y x * x = x²
Monomial	An expression with one term.	13 $5x^2 5x^5 yz^2$
Binomial	An expression with two terms	1+ x $2x^2 - 15$
Polynomial	An expression with three or more terms	3x - $5x^2 z + 6z$
Term	In Algebra, term refers to a number, a variable, or a combination of the two that is multiplied by at least one variable	$13x^2 4x 101xy \dfrac{6a}{7}$
Like terms	Two or more terms that have the same variables and exponents.	1x + 7x $5x^2 + 13x^2$
Algebraic express	An equation that has an operation, numbers, and variables.	4x + 12 (9 + 3x) − (9 − 3y)
Linear expression	An equation with at least one variable squared	6x + 8
Quadradic expression		x² *+ 6x + 8

When you are adding and subtracting, you can only do these operations when they have like terms. For example, you can simplify the first example, but not the second:

$$5x^2 + 10x^2 = 10x^2$$
$$5x^2 + 10y^2$$

If you need to combine like terms, you start by adding or subtracting the coefficients. The variables and exponents are not changed, just the coefficient.

$$4x^2 + 2x^2 = 6x^2$$
$$4x^2 - 2x^2 = 2x^2$$
$$3y^2 + 2z^2 - y^2 = 3y^2 - y^2 + 2z^2 = 2y^2 + 2z^2$$

You can't simplify (combine) unlike terms:

$$-y^2 + 2z^2$$
$$-y^2 + 2yz - 5z^{2^2}$$
$$a + b$$
$$x + x^2$$

Now let's go through a problem.

Simplify

$(9 + 5x) - (10 - 4x)$

Remove the parentheses by distributing the subtraction across the second set of parentheses.

9 +5x −10 + 4x

Group like terms

5x + 4x −10 +9

Do the math for like terms

9x - 1

When you multiply and divide monomials, you can do the math for them, even if they are unlike terms.

- Multiple and divide the coefficients
- Add the exponents for variables
- Multiply the variables

Here's an example:

Simplify

(3a)(5b)

Multiply the coefficients and variables:

15ab

Let's try a problem that includes all of these.

Simplify

$(5x)(3xy) + (7x^2)(9y)$

Start by multiplying the terms on each side of the addition sign:

$(5 * 3)(x * x * y) + (7 * 9)(x^2 * y)$

$15\ x^2y + 63\ x^2y = 78\ x^2y$

The answer is $78\ x^2y$.

Verify your work.

If you are working with binomials, there is a specific order of operations known as FOIL:

- First -work with the first terms in parentheses
- Outer – work with the terms at the end of both parentheses
- Inner – work with the terms on the inner sides of the parentheses
- Last – work with the last terms in the parentheses

You need to multiple every term by every other term, and FOIL ensures that you don't miss

Here's how to do that:

Simplify (x+2)(x+ 4)

Multiply each of the terms with the others in this order:

- First terms: (x * x)

- Outer terms: (x * 4)
- Inner terms: (2 * x)
- Last terms: (2 * 4)

This is the result: x^2 *+ 4x + 2x + 8.

Now simply what is left. In this case, you have two like terms– those can be added together.

The final answer is x^2 *+ 6x + 8.

Note this is a quadradic equation. The ASVAB will almost certainly have questions that have you factor a quadratic equation. In this instance you need to go through FOIL in reverse. Here's how you do that.

Factor $x^2 - 3x - 18^-$

Notice that they don't include parentheses. That is what you will do when you do the FOIL steps in reverse.

- The first two terms that go in parentheses are the two xs since x * x = x^2
  (x)(x)
- Determine the last terms by looking at the coefficient with x^2 and the last number:
  (1)(-18)
- Determine the factors that could equal -18:
    o   1, -18 or -1, 18
    o   2, -9 or -2, 9
    o   3, -6 or -3, 6
- Which one of these pairs will result in the coefficient in the middle expression, -3?
  3, -6
- Now combine what you've found in each step:
  (x)(x)
  (x +3)(x − 6)

Now verify that you get the right answers when you simplify it.

Working with Classic Quadratics

There are three quadratics that you are very likely to see, so if you do, you won't need to use FOIL. The following are the three common quadratics that will make problems a lot faster for you to process:

$(a + b)(a − b) = a^2 − b^2$ $(x + 5)(x − 5) = x^2$ - 25

$(a + b)^2 = a^2 + 2ab + b^2$ $(x + 9)^2 = x^2 + 18x + 81$

$(a − b)^2 = a^2 − 2ab + b^2$ $(x − 7)^2 = x^2 − 14x + 49$

If one of the equations ends with a perfect square, such as 16, 25, or 81, as well as the variable being squared, it is probably one of these three equations.

Let's take a look at an example.

Which of the following is equivalent to $x^2 - y^2$?

- (A) (x + y)2
- (B) (x + y)(x + y)
- (C) (x − y)(x − y)
- (D) (x + y)(x - y)

Notice that it is a quadratic equation. You need to find the answer that is the same thing, just expressed differently. Consider the three common types to determine if one of those fits for this problem.

The answer is D because that one is the common equation that would result in the initial expression. You can verify it by working the problem from the answer to the beginning expression.

Working with One Variable

To solve an equation, you need to complete the same tasks for both expressions. Remember, you need to do the same thing to both sides to get the right answer. Here's one for practice.

10x + 5 = 31 − 3x Add 3 x to both sides.

10x + 3x +5 = 31 Subtract 5 from both sides.

10x + 3x = 31 − 5 Now do the math.

13x = 26   Now solve for x by dividing both sides by 13.

X = 2

Go back and verify your answer. You can plug in the answer to the equation to make sure the answer is correct.

Here's another one to try.

Solve for x: $x + 8 = \frac{2x}{6}$

Multiply both sides by 6:   6x+48 = 2x

Subtract 6x form both sides: 48 = -4x

Divide both sides by -4:   x = -12

You can verify the answer by plugging -12 into the original expression .

Working with Two Variables

Algebraic expressions often include more than one variable. The ASVAB usually includes questions that ask you to solve for the value of an expression that has a couple of variables. You learned PEDMAS in the previous chapter, and you will need to use it for working with two variables.

Here's an example problem.

If x = 10 and y = 9, what is the value of 5x(x-y)?

Plug the values into the equation.   5(10)(10-9)

Now use PEMDAS: 5(9)(1)

Multiply them together: (45)(1)

The answer is 45

You may also be asked to solve for one of the variables in terms of the other. The way to do that is to isolate the one that the question once you to figure out. Here's an example.

Solve 7a + 2b = 3a + 10b - 15 for a in terms of b.

Isolate a.

7a − 3a = 10b − 16 − 2b

4a = 8b − 16

$$a = \frac{8b - 16}{4}$$

a = 2b - 4

Here's another one.

Solve 16x + 18y + 8 = -4x − 2y - 6 for x in terms of y.

16x + 4x = -2y − 6 − 18y − 8

20x = -20y -14

$$x = \frac{-20y - 14}{20}$$

$$x = y - \frac{-7}{10}$$

Working with Systems of Equations

If you are given two equations that have two variables, and then are asked to solve for one variable, don't feel overwhelmed. It's very similar to what you've already done. And you have two ways you can do it.

-    Substitution
-    Combination

Let's start with substitution. Use one equation in the place of the variable in the other. Then you'll substitute the resulting expression for the same variable in the other equation. Let's solve and example.

Find the value of m:

$6m + n = 15$

$2m - 3n = -5$

    (A) $-2\frac{1}{4}$

    (B) 3

    (C) $\frac{-5}{4}$

    (D) 2

Both of these are linear equations that have the same two variables. The question wants you to solve for x. Isolate y in the top equation – then you can substitute the result for y in the bottom equation.

$n = -6m + 15$

$2m - 3(-6m + 15) = -5$

$2m + 18m - 45 = -5$

$20m = 40$

$m = 2$

The answer is D. You can verify the solution by plugging it into the equations.

If you wanted, you could also work backwards to solve it. Choose the one that is easiest to plug into the equations, and that would be D.

$6(2) + n = 15$

$12 + n = 15$

$n = 3$

You now have m = 2 and n=3. Plug those into the next equation.

$2(2) - 3(3) = -5$

$4 - 9 = -5$

Yes, it does check out. So D is the right answer. If you get overwhelmed by inserting an equation into another equation, you can just work backward. It will be a lot more guess work, but you can use the process of elimination if that is a faster method for you.

Now let's look at using combination method. You will want to use this when the problem has two equations and two variables. You combine the equations to solve for the specified variable.

Here's an example:

Find the value of x

$4x + 2y = 44$

$6x - 2y = 46$

    (A) 9

    (B) 4

    (C) 6

    (D) 90

Set the equations up to equal each other:

$$4x + 2y = 44$$
$$+\ 6x - 2y = 46$$
$$10x = 90$$

Since x = 9, the answer is A.

You could also solve the problem by working backwards. You would need to pick any number and plug it into both equations to see if it works. That would be pure guesswork, but if you aren't able to do algebraic problems quickly, this could be a better way to do them.

Here's another one:

Solve for m

5n + 17m = 90

-2m − 5n = -30

Since both equations have the term 5n, you can combine them to remove the n term. Adjust the equations to handle the terms.

17m = 5n = 90

+ [-2m − 5n = -30]

15m = 60

m = 4

The answer is x = 4. You can go back and plug it into both equations to verify your answer.

Working with Quadratic Equations

When you are asked to solve for a variable in a quadratic expression that equals 0, you can factor to solve for the variable. Here's an example.

Solve for x

$x^2 + 4x + 3 = 0$

Start by factoring the equation into parentheses.

$(x + 3)(x + 1) = 0$ Now solve for both expressions within parentheses

x + 3 = 0 x = -3

x + 1 = 0 x = -1

This means that x = -3 or -1. When you are working with quadratic equations, you will often have two answers. The exception is when both of the factors are the same. The following is an example of when you will get one value instead of two.

Solve for a

a2 − 4a - 4 = 0

Factor the expression: (a − 2)(a − 2) = 0

When you solve for both expressions, you get 2. Verify your work by plugging it into the original equation.

When you are asked to solve a quadratic expression that equals something other than 0, you adjust it to set it up to equal 0. All you have to do is to move all of the components to one side, leaving 0 on one side. Let's take a look at an example.

Solve for y

$-4 + 2y = 31 − y^2$

This has all of the elements that mean it can be translated into a quadratic equation, then you can use reverse FOIL.

First, adjust the terms on the left side: $2y -4 = 31-y^2$

Now move the 31 over to the other side by subtracting it: $2y − 4 - 31 = -y^2$

$$2y - 35 = -y^2$$

Now subtract y^2 from both sides: $y^2 + 2y -35 = 0$

Now solve by factoring the equation: (y -7)(y + 5) = 0

The answer is y = -7 or 5

Working with Inequalities

You may be asked to plot inequalities along a number line. Compared to a lot of the previous work, this one is easy. However, you need to be careful because it is also easy to make mistakes if you work too quickly.

All inequality can be visually represented on one of these signs.

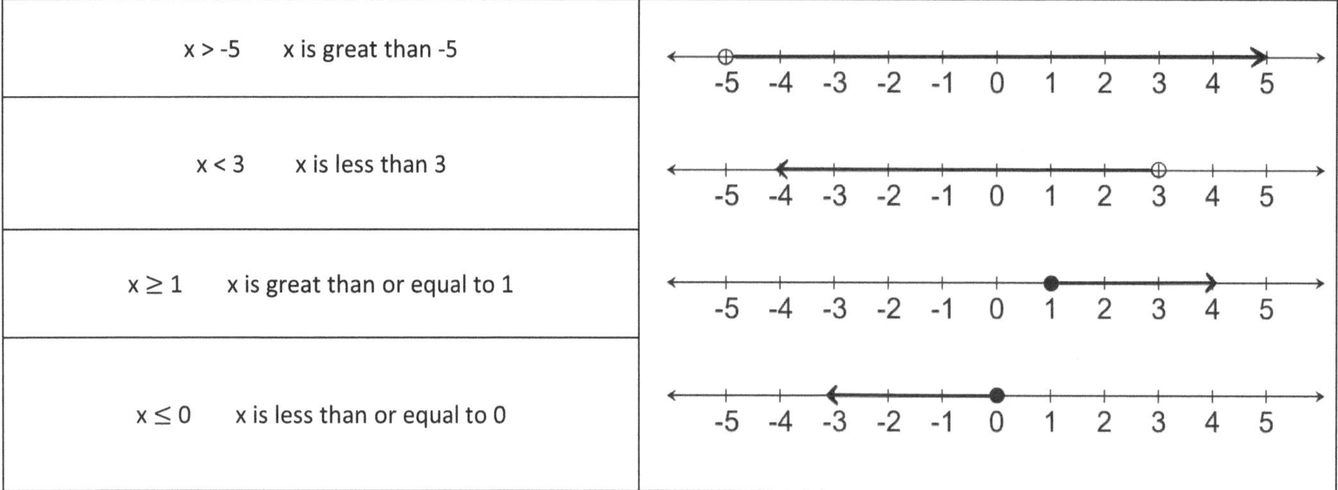

You solve inequalities using the same methods as equations. First isolate for the variable the question wants solved, then do the math. The difference is that you will end up with a range of potential values, not one or two. Here's an example.

Solve for x

$4x + 6 > 2x + 10$

Subtract the coefficient and variable from both sides: $4a - 2a + 6 > 10$

Subtract the 6: $4a - 2a > 10 - 6$

Do the math: $2a > 4$

a > 2 is the answer

This means that the answer can be any number that is larger than 2. So it could be 3, 33, or 333. It cannot equal 2.

When working with inequalities, you treat it just like you were solving an equation. The only different is when you divide with a negative number, it switches the sign of the answer. If you think of it in terms of the number line, it switches the direction the arrow goes.

If you divide -3x > 9 by -3, you get x < -3. Notice that the sign switched from greater than to less than. This also means that you **cannot** multiply or divide by a variable since you can't be sure of if the variable is positive or negative.

Let's look at an example to illustrate this:

$2 < 4$

If you multiple both sides by -2, you get -1 _ -2. -1 is bigger than -2, not smaller, so you flip the direction to -1 > -2.

Let's try an example.

Solve for a

$$\frac{-4a}{2} - 3 > 4a$$

Start by simplifying the fraction (multiplying it by 2): $-2a - 3 > 4a$

Isolate a: $-3 > 6a$

Divide by 6: $\frac{-3}{6} > a = \frac{-1}{2} > a$

NOTE: the answer could be $\frac{-1}{2} > a$ or $a - \frac{1}{2}$. These are the same answers because they both indicate that a is less than $\frac{-1}{2}$. Verify your work.

REVIEWING GEOMETRY

The geometry portion of the ASVAB will include the expected diagrams and shapes that you've seen in class. If the question doesn't indicate that the drawings are to scale, then you should **not** assume that they are. Use only the information that they give in the diagram.

Working with Line Segments

Line segments provide a piece of a number line with a measurable length. Questions might present a line segment divided into several parts, provide the measurements of some of these parts, and ask you to determine the length of another part.

Let's look at an example.

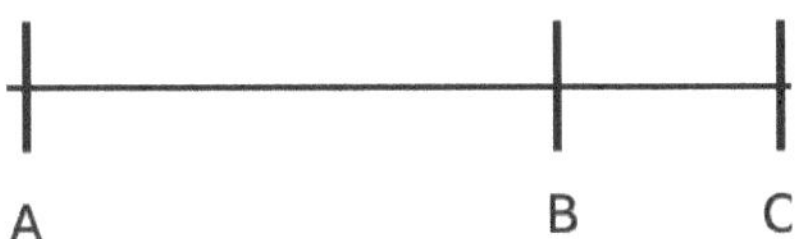

If AC = 12 and BC = 4, AB = ?

AB = AC − BC

AB = 12 − 4

AB = 8

When you have a line segment, the midpoint is the exact middle between both ends. If you bisect the line, that means you cut it in half, so you would bisect the line at the midpoint.

If B is the midpoint, then AB = BC.

Working with Angles

There are three types of angles.

Angle	Definition	Example
Acute	Less than 90 degrees	
Obtuse	Greater than 90 degrees	
Right	Exactly 90 degrees; they are indicated by the square in the middle of the angle	

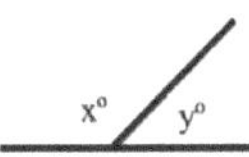

When you have two angles on a straight line, they add up to 180 degrees. That means for the figure x + y = 180.

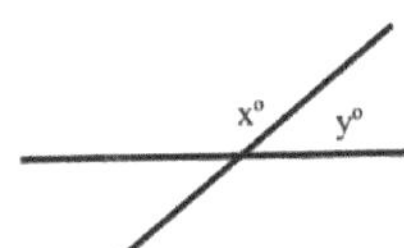

When you have two intersecting lines, the adjacent angles are considered supplementary, or together they equal 180 degrees. For the figure x + y = 180.

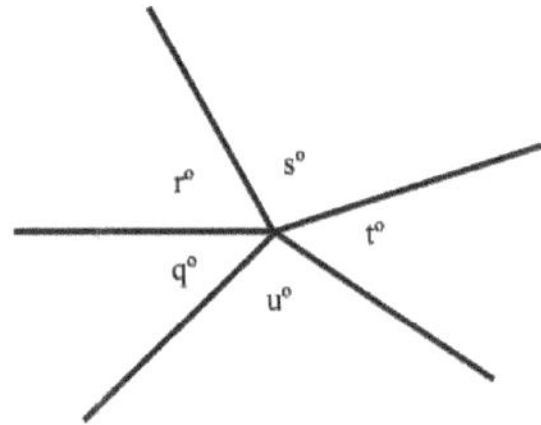

When you have angles around a single point, they add up to 360 degrees. That means for the figure q + r + s + t +u = 360 degrees.

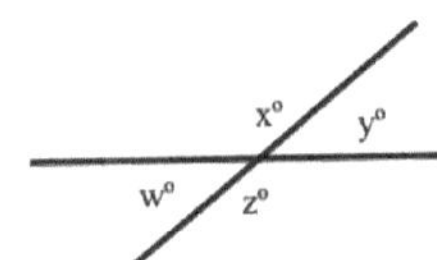

When you have intersecting lines, the angles across the vertex are known as vertical angles, and those angels equal each other. In the figure w = y and x = z.

Let's try a problem.

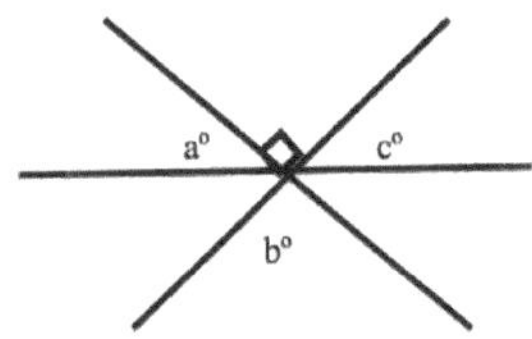

What is the value of a + b + c?

- (A) 90
- (B) 180
- (C) 270
- (D) 360

Really look at the figure. You'll see that there is a right angle on a line (indicated by the rectangle in the center of the angle). This is an unlabeled angle.

The question is looking for the sum of three other angles.

- Since a and c are supplementary to the right angle, together they must equal 90 degrees.
- Since b is a vertical angle to the right angle, it is also a right angle, so it also equals 90 degrees.

So a + c = 90, and b = 90. So a + b + c = 180.

The answer is 180 degrees.

Working with Transversal Crossings of Parallel Lines

The following are parallel lines. That means that they have the same slope and they will never intersect, similar to rail road tracks that are meant to run next to each other. The slope is the steepness of the line.

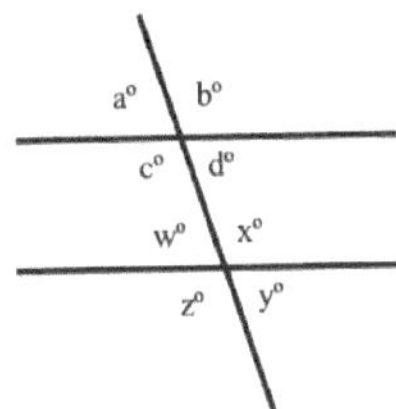

When two parallel lines have a single line intersecting them, the intersecting line is called a transversal. All of the resulting angles are interior angles, or they have the same relationship to the corresponding angle with the other parallel line.

- All corresponding angles are equal, so a = w, b = x, c = z, and d = y.
- All alternate interior angles are equal, so a = y, b = z, c = x, and d = w.
- Angles on the same side of the intersecting line equals 180 degrees, so a + c = 180.
- There are four acute angles, and they are equal.
- There are four obtuse angles, and they are equal.

Let's look at a problem.

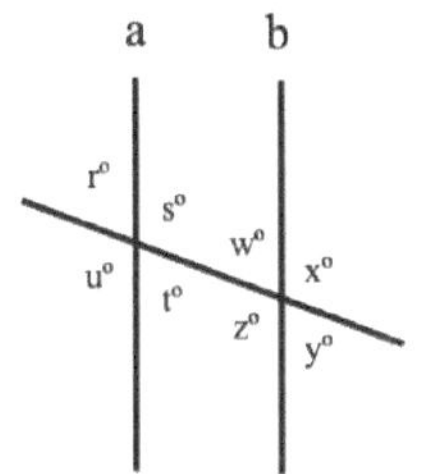

a ||b

What is the value of r + t + x + z?

(A) 45
(B) 90
(C) 270
(D) 360

The image tells you that lines a and b are parallel, so you can apply all of the properties about the transversal crossing. You know that two of the angles are acute, and two are acute. Since the lines are supplemental, you know that when they are paired, they equal 180 (r + x = 180 and t + z = 180). So you have 180 + 180, which equals 360. The answer is D.

Working with Polygons

There are a number of different types of polygons, and you will likely see several on the ASVAB. The following table explains the shapes covered in this section.

Shape	Definition	Examples
Triangle	A shape with three sides and three interior angles	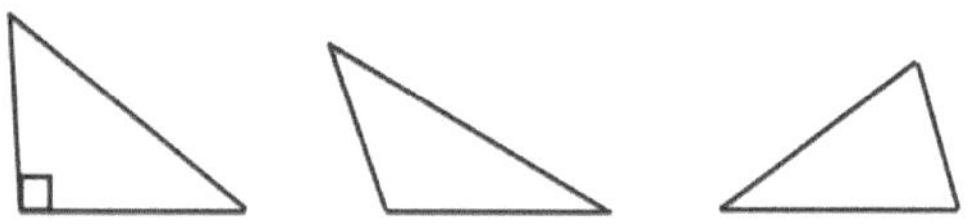
Quadrilateral	A shape with four sides and four interior angles.	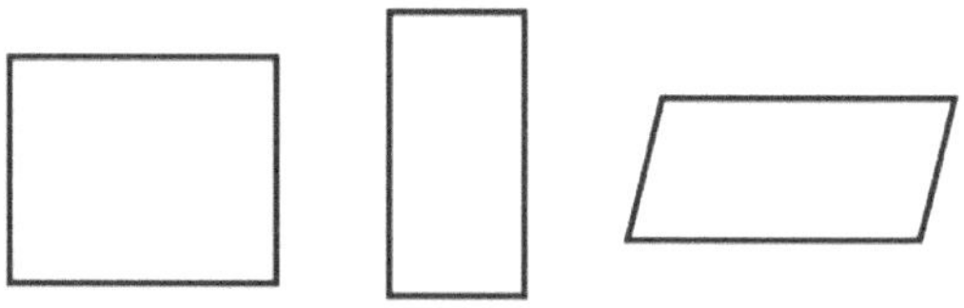

Let's look at triangles first. There are three types of triangles.

Angle	Definition	Example
Isosceles	Has two equal sides and the angles opposite to those sides are also equal; they are called base angles Lines on the sides indicate they are the same length.	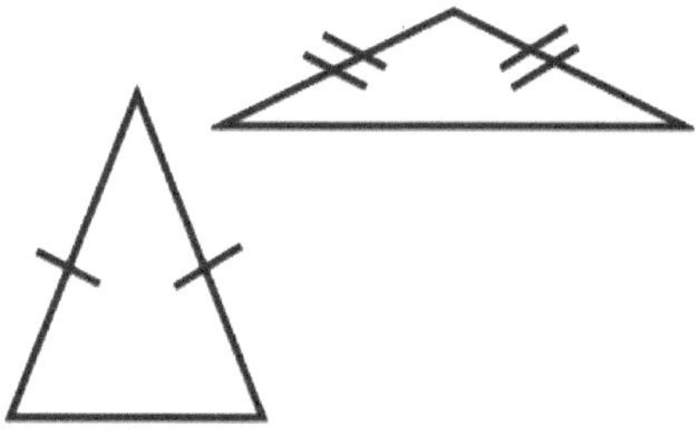
Obtuse	All three sides are the same length, and all three angles are 60 degrees.	
Right	Has one right angle All right triangles have two acute angles and the sides opposite to them are called legs. The side opposite the right angle is called the hypotenuse, and it is longer than the other two sides.	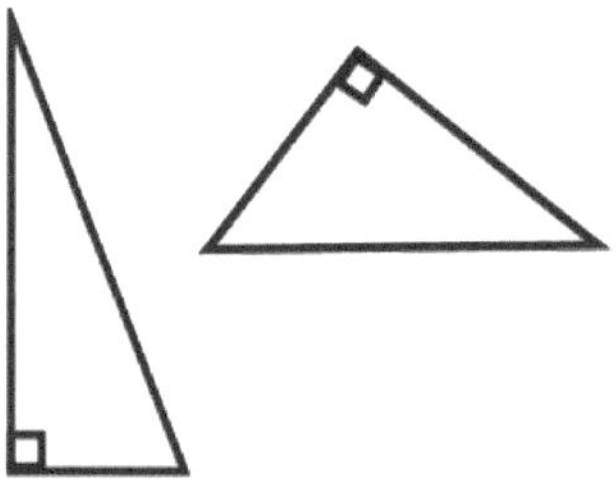

Now let's learn a bit more about working with the sides and angles

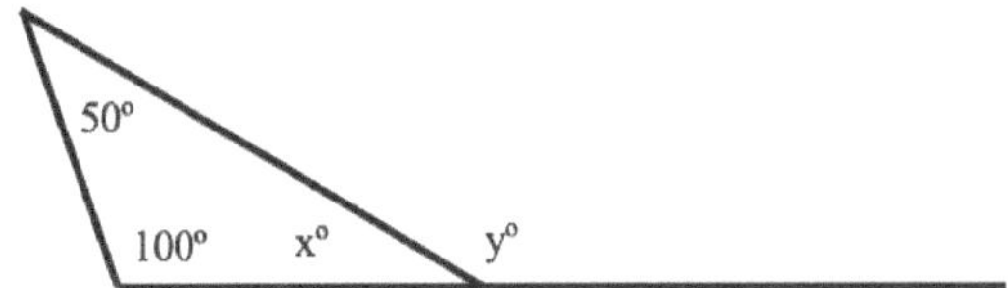

When working with triangles, it's important to remember the following:

- The three interior angles equal 180.
- The exterior angles equal the two nonadjacent interior angles.
- The perimeter is the total of the three sides.
- The area of a triangle equals half of the base times the height, $\frac{1}{2}bh$.

Using this, you can solve for x, y, and a few other data points for the triangle.

- 100 + 50 + x = 180, so x = 30.
- y = 180 − x, so y = 180-30, or y = 150.

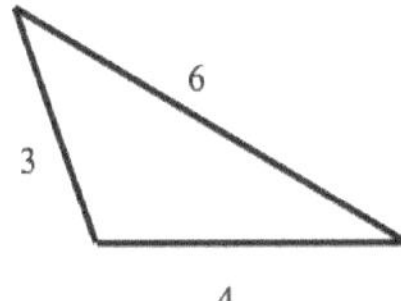

- You can determine the perimeter of the triangle by adding the three sides together 3 + 4 + 6 = 13.

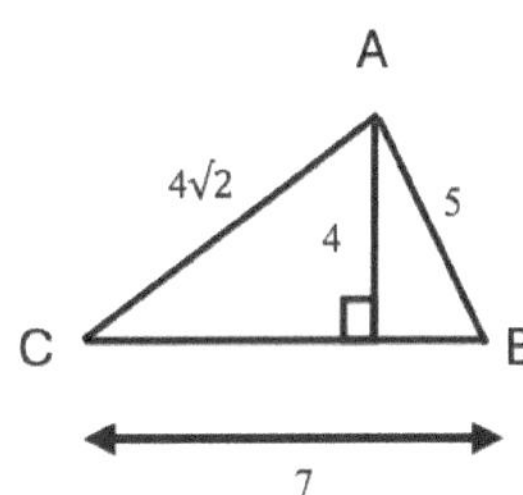

- You can determine the area of the triangle by using the area formula. $\frac{1}{2}(7)(4) = 14$.

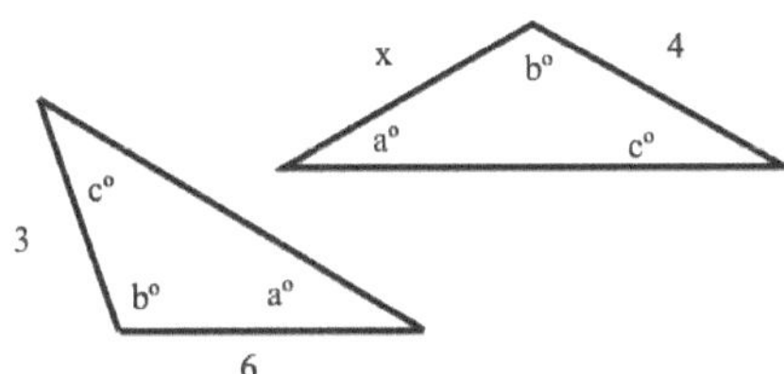

You can't make assumptions with triangles. The above triangles have a similar shape, but they clearly are different sizes. However, they are proportional because the angles are the same between the two triangles. That means that you aren't making assumptions.

Let's see how that works with a problem.

Find the value of s.

You know that they are proportional, so you can set them up as a fraction:

$$\frac{3}{4} = \frac{6}{s}$$

Now you can solve for s. Multiply both sides by 4 and s.

$3s = 24$

$s = 8$

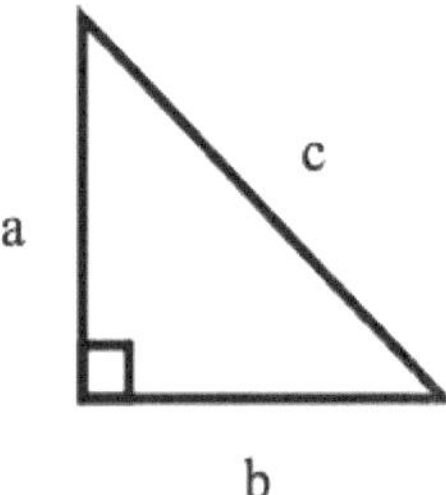

When you have a right triangle and you know the measurements for two of the three sides, you can find the third side with the Pythagorean theorem, $a^2 + b^2 = c^2$. In this equation, a and b are legs of the triangle.

Now let's do a problem for right triangles.

One leg is 2 and the other is 3. How long is the hypotenuse?

Plug the sides into the Pythagorean theorem.

$2^2 + 3^2 = c^2$

$4 + 9 = 13^2$

$C = \sqrt{13}$

Now let's look at quadrilaterals, or shapes with four sides.

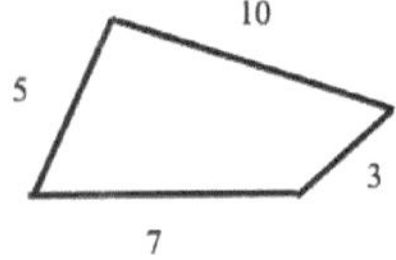

The perimeter of a quadrilateral is the total of all sides. For the above quadrilateral, the perimeter is $5 + 10 + 7 + 3 = 25$.

This equation works for all quadrilaterals.

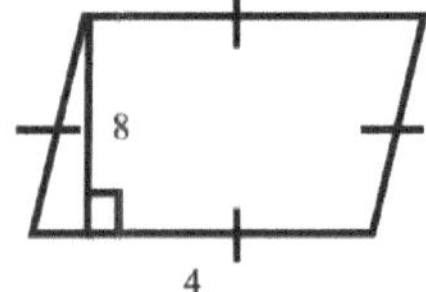

The above shape is a special type of quadrilateral called a parallelogram. It has two sets of parallel sides, and the length of the stets is the same.

You can calculate the area with the following equation for parallelograms; area = base * height. For the parallelogram above, you can determine the area: $4 * 8 = 32$.

Rectangles are a type of parallelogram, and it has four right angles. Sides that are opposite are equal. To find the area of a rectangle, you use a similar equation; area = length * width. You find the perimeter by adding all the sides to get the total, or you can use the formula 2(length * width).

You can use the formula to determine the area for the rectangle: 9 * 4 = 36.

You can also find the perimeter too: 2(9+4) = 2(13) = 26.

Another special type of parallelogram is the rhombus. It has four equal sides. Squares are a type of rhombus.

You find the perimeter for a rhombus by multiplying the length of a side by 4 (or you can total them up, it gives the same result).
Perimeter = 4s

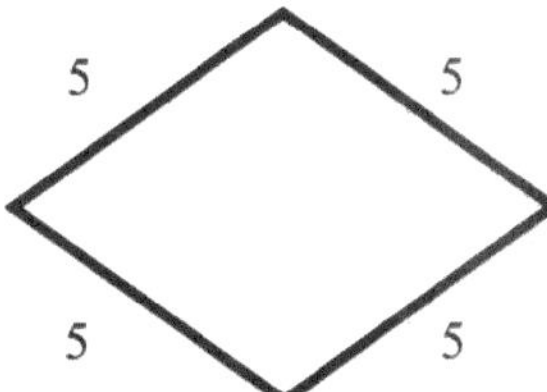

The perimeter for the above rhombus is 4 * 5 = 20.

You can find the area of a square by squaring one of the sides.

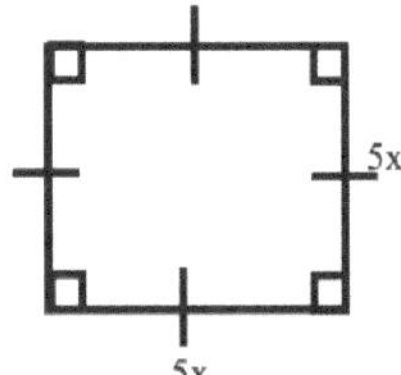

If you use the equation, the area for the above square is 5^2 = 25.

The perimeter is 4 * 5 = 20.

The last shape for this section is the trapezoid. It has one set of parallel sides, and they are called the bases.

You find the area by adding the lengths of the bases times the height, then all of it divided by half. Area = $\frac{1}{2}$(bases)(height)

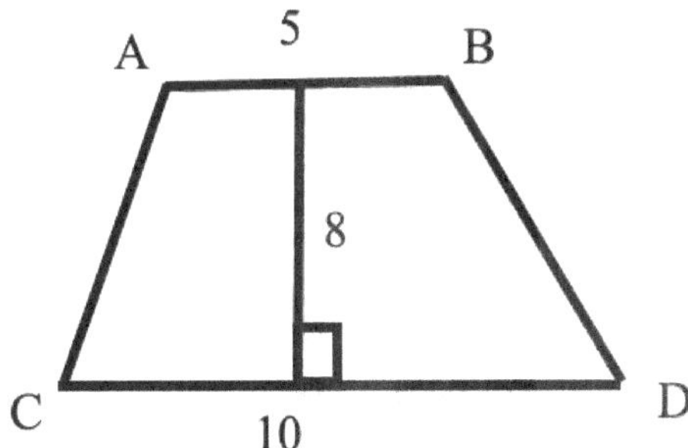

For the trapezoid above, you can find the area:

$\frac{1}{2}$(5+10)(8) = $\frac{1}{2}$(15)(8)= $\frac{1}{2}$(120) = 60

Let's do a problem to get some practice.

The area of the trapezoid is equal to more than twice the area of the square. Which of the following could be a value for x?

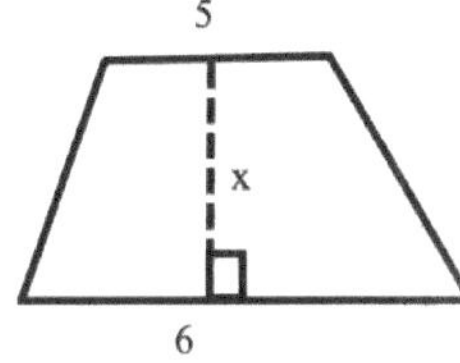

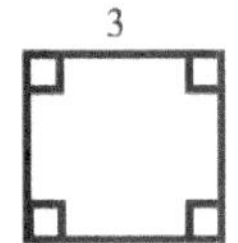

(A) 1

   (B) 3
   (C) 4
   (D) 8

You can calculate the area of the square = 3 * 3 = 9.

You can then calculate the area of the trapezoid with the information given:

A > 2 * A of square

A > 2 * 9

A > 18

Now work backward with the answers given. Let's try 4.

$$A = \frac{1}{2}(5 + 6)4 = 11$$

That area is smaller than the square, so it can't be the right answer. That leaves D. Verify the answer works for the problem.

$$A = \frac{1}{2}(5 + 6)8 = 44$$

44 is more than twice 18, so it is the right answer.

Let's do one more.

The value of the trapezoid's area is divisible by 5 with no remainder. Which of the following could be a value for a?

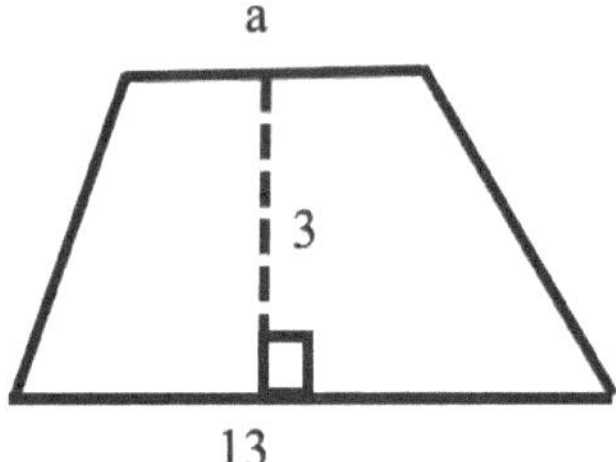

   (A) 11
   (B) 12
   (C) 13
   (D) 14

You know that the area is a multiple of 5.

$$A = \frac{1}{2}(a + 13)3 = (y + 13) \text{ is a multiple of 5.}$$

Work backwards to find the answer. You know that everything that is a multiple of 5 ends with either 0 or 5. The only way to get a number that ends with one of these, y must end with 2 or 7. The only answer that works is B.

Verify the answer. $= \frac{1}{2}(12 + 13)2 = (12 + 13) = 25$. That is a possible value for y that works with the information given.

Working with Circles

In geometry, circles are defined as a figure where each point is an equal distance from the circle's center. This section will define the center as A.

A circle's radius is a straight-line from the center to the perimeter, and all radii are the same length.

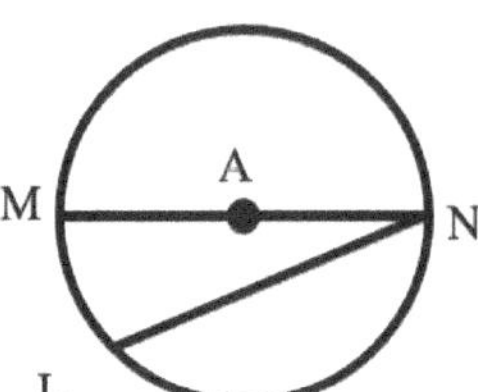

The radius is MA. AN is also a radius.

The above circle also has a chord. This is a segment that connects to point on the perimeter. LN and MN are both chords.

MN is also the diameter of the circle. The diameter always passes through the middle of the circle. All of the diameters of a circle are the same length.

The circumference of a circle is the distance around the circle (called the perimeter for most other shapes). The circumference = π * diameter or πd. You can find the circumference with a radius too, $2\pi r$.

If MN of the circle above = 6, then the circumference = 6π.

You can calculate the area of a circle with the following formula. πr^2. You can find the radius by dividing 6 by 2, so 3, then put that in the area formula:

$\pi r^2 = \pi 3^2 = \pi 9 = 9\pi$

Let's do a couple of problems.

What is the ratio of the area of A to B for the following circles?

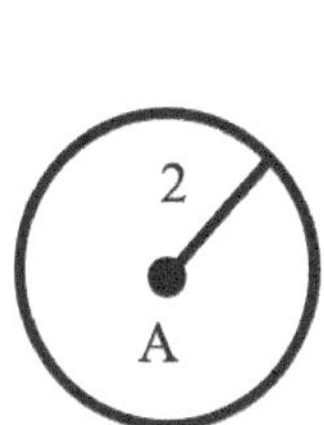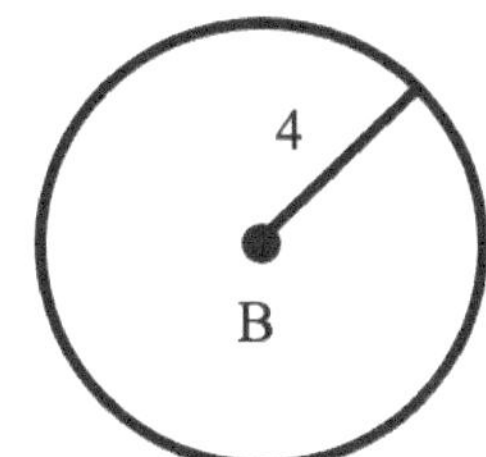

The area of A is half the area of B: $\frac{AreaA}{AreaB} = \frac{\square}{8} = \frac{1}{8}$.

Area B = 8π

Verify your work.

Let's try another one.

The diameter is x. The area of the circle is 81π. What is the value of x?

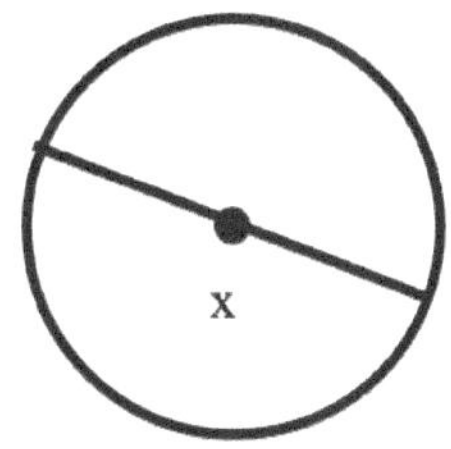

    (A) 20
    (B) 18
    (C) 12
    (D) 2π

A = 81π

d = x

Work backwards to find the one that could be x.

A = $81\pi = r^2\pi$

$r^2 = 81$

r = 9

x = diameter = 2r = 18

The answer is B. Verify your work.

Working with Solid Shapes

Solid shaper are 3D versions, so they rely on the same formulas with on extra aspect to consider.

If you have a rectangular box, you work with the length, width, and height.

- Calculate the volume by multiplying these aspects = l * w * h
- Surface area is calculated by multiplying the thee areas and adding them together, or 2lx + 2wh + 2lh.

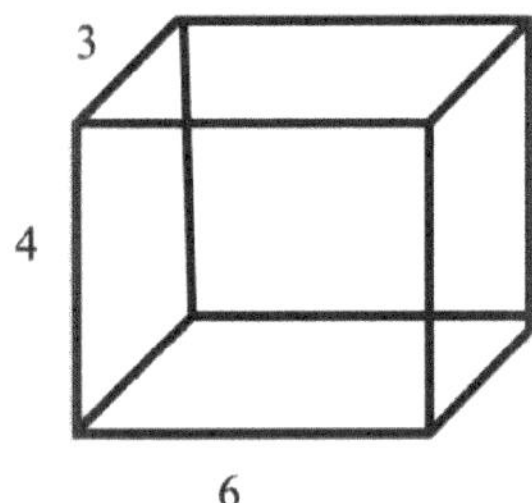

What is the volume of the cube?

V = l * w * h

V = 3 * 4 * 6

V = 72

Let's try another one.

What is the surface of the cube?

S = 2lw + 2wh + 2lh

S = (2 * 6 * 3) + (2 * 3 * 4) + (2* 6 * 4)

S = 36 + 24 + 48

S = 108

Verify your work.

When you have a cylinder, find the volume by finding the area of the base times the cylinder's height.

$V = \pi r^2 h$

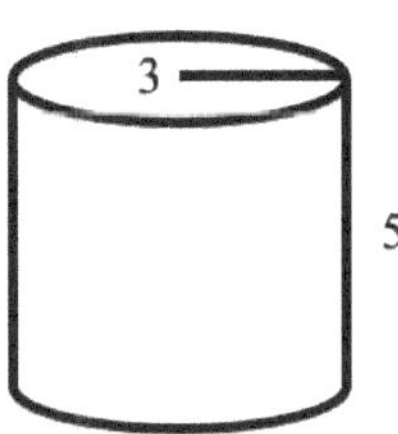

Find the cylinder's volume.

$V = \pi 3^2 5$

$V = \pi(9)(5)$

$V = 45\pi$

When you have a 3D figure, you find the surface by adding the areas of the sides of the figure. When working with a cylinder, add the area of the circular ends with the curved sides. If you were to unroll a cylinder, you will have a rectangle. The top of the rectangle is the circumference of the circle and the side is the height.

You can use the following formula: $S = 2(\pi r^2) + 2\pi rh$

You can also combine figures. The following is an example of two shapes that have been meshed. You can split it into the two shapes to simplify the process of solving the question.

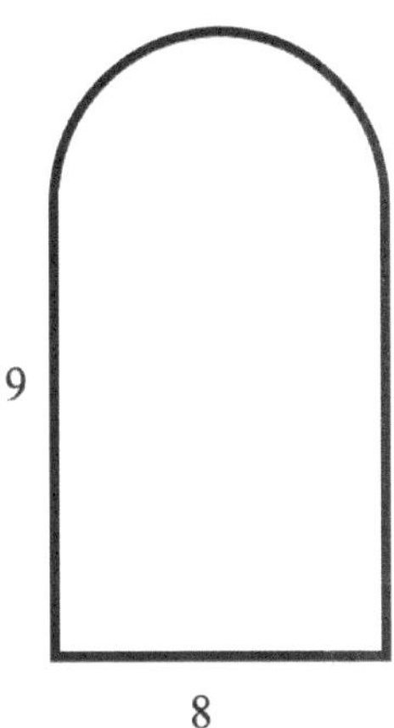

Find the area.

The figure is a combination of a rectangle and a circle. Split it up to do the math.

Rectangle $\qquad$ A = l * w A = 9 * 8 $\qquad$ A = 72

Circle $\qquad$ $A = = r^2\pi$ $\qquad$ $= 4^2\pi$ $\qquad$ $A = 16\pi$

Half circle $\qquad$ $16\pi \div 2 = 8\pi$

Combine $\qquad$ $72 + 8\pi$

Working with Coordinates

The last thing we'll review are coordinates, something you may actually use. Coordinate grids have two axes:

- The x axis runs horizontally.
- The y axis runs vertically.

Every point on a grid has a coordinate pair to represent its location. The first prat of the coordinate is the point on the x axis, and the second part is the point on the y axis. There are three point on the following grid, and you can see what their coordinates are in parentheses.

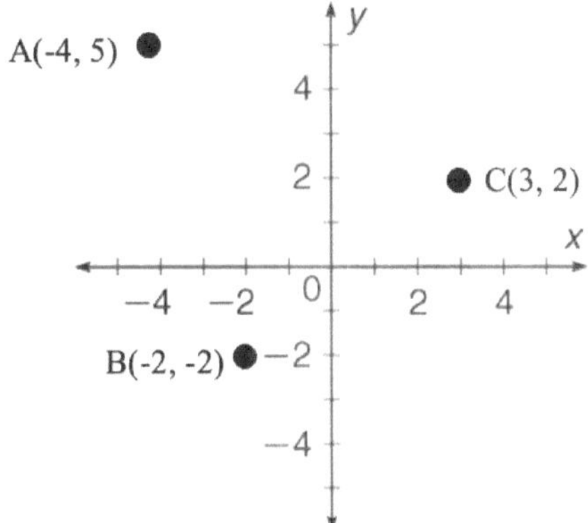

A grid may have a line that represents a slope-intercept relationship, which is what a linear equation looks like when it is graphed. This is usually represented in an equation: y = mx + b, where x and y are the two coordinate points, m is the slope, and b is the y-intercept.

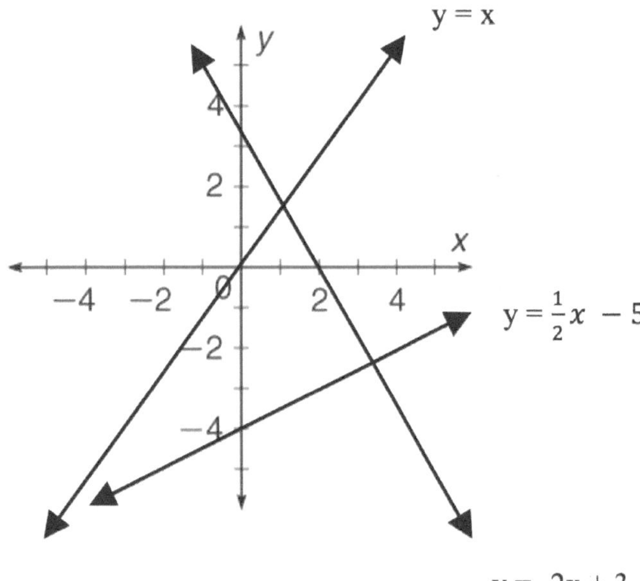

The y-intercept is where the line crosses the y axis. The x coordinate is 0 at the y-intercept.

A line's slope goes from left to right, but it can go up (represents a positive value for x) or down (represents a negative value for x).

Slope $= \frac{ychange}{xchange}$

When a line has a positive slope, that means y both x and y are increasing in value. If slope is negative, y is decreasing while x is increasing. The steeper a line, the higher its absolute value. For example, if you have two lines with slopes of -5 or 5, the slope is steeper than a line with a slope of -2 or 2.

Let's try a problem.

A line has a slope of 2 and a y-intercept of -5. What is the equation of the line?

y = mx + b

y = 2x -5

Let's try another one.

The following points is on a grid: (1, 2) and (4, 5). What is the slope?

Slope $= \frac{ychange}{xchange}$

Add the coordinates from the second point, then the first point, then the x-coordinates.

Slope $= \frac{5-2}{4-1} = \frac{3}{3} = 1$

Verify your answer.

If you have the slope of one line and a single point for another, you can determine the missing coordinate point. Let's keep working with the last example to solve the y-intercept.

Start with the slope formula: y = mx + b.

Use the slope you found in the previous problem: y = 1x + b

Add one of the x and y coordinates: 2 = 1(1) + b

Do the math: 2 = 1 +b b =1

You can also find the distance between two points on a line using the Pythagorean theorem. Using the same points from the last two problems, let's do one more problem.

What is the distance between (1, 2) and (4, 5)?

Map it out on a grid:

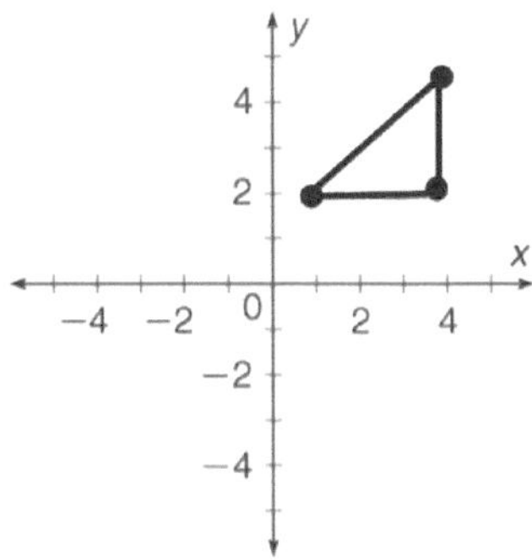

The third dot is a right angle, making the triangle a right triangle. You can count the distance between the points that are a part of the right triangle:

- Along the x axis is 3.
- Along the y axis is 3.

Use the Pythagorean theorem to calculate the distance:

$$\sqrt{3^2 + 3^2} = \sqrt{9 + 9} = \sqrt{18}$$

You could also draw the triangle with the right angle at the top instead of the bottom. Use that to verify your answer.

PRACTICING WHAT YOU'VE LEARNED

Now that you've got the basics, you can see test your newly acquired information and put it to the test.

Here are two practice sets with 15 questions. This is similar to how the CAT is setup, although obviously your questions aren't going to change the better you do.

The last section of this chapter details the right answer, and how you could have reached that answer.

Practice 1

1. Solve for x: 2x + 3 = 11

    (A) 2
    9B) 3
    (C) 4
    (D) 5

2. What is the solution to the equation $x^2 - 4 = 0$?

    (A) x = 2
    (B) x = -2
    (C) x = ±2
    (D) x = 0

3. If y = 3x + 7 and y = 13, what is the value of x?

    (A) 2
    (B) 3
    (C) 4
    (D) 5

4. What is the value of $3^2 - 4 \times 2$?

    (A) 1
    (B) 5
    (C) 9
    (D) 16

5. What is the area of a rectangle with a length of 5 units and a width of 3 units?

    (A) 8 square units
    (B) 15 square units
    (C) 20 square units
    (D) 25 square units

6. Simplify the expression: 2x + 3x - 4

    (A) 5x - 4
    (B) 5x + 4
    (C) x - 4
    (D) 5x

7. Solve for y: 4y - 7 = 5y + 2

    (A) -9
    (B) 9
    (C) -7
    (D) 7

8. What is the value of x if 2x + 5 = 3x - 7?

    (A) 7
    (B) -7
    (C) 12
    (D) -12

9. Factor the expression $x^2 - 9$

  (A) (x - 3)(x - 3)
  (B) (x + 3)(x - 3)
  (C) (x + 9)(x - 9)
  (D) (x - 3)(x + 3)

10. Simplify: $(2x^2)(3xy^2)$

  (A) $6x^2y^3$
  (B) $6x^3y^3$
  (C) $6x^3y^2$
  (D) $6x^2y^2$

11. What is the sum of the interior angles of a triangle?

  (A) 90 degrees
  (B) 180 degrees
  (C) 360 degrees
  (D) 270 degrees

12. Simplify the expression: 4(3x + 2)

  (A) 12x + 2
  (B) 12x + 8
  (C) 7x + 2
  (D) 7x + 8

13. What is the volume of a cube with a side length of 4 units?

  (A) 8 cubic units
  (B) 16 cubic units
  (C) 32 cubic units
  (D) 64 cubic units

14. Solve for x: 5x - 3 = 2x + 9

  (A) 2
  (B) 3
  (C) 4
  (D) 5

15. What is the length of the hypotenuse of a right triangle with legs of 3 units and 4 units?

  (A) 5 units
  (B) 6 units
  (C) 7 units
  (D) 8 units

Practice 2

1. Solve for x:

$$\frac{x}{3} = 4$$

  (A) 1
  (B) 7
  (C) 12
  (D) 13

2. If 2a + 3b = 12 and b = 2, what is the value of a?

  (A) 2
  (B) 3
  (C) 4
  (D) 5

3. What is the perimeter of a square with a side length of 7 units?

  (A) 14 units
  (B) 21 units
  (C) 28 units
  (D) 49 units

4. What is the value of x in the equation $x^2 = 16$?

  (A) 4
  (B) -4
  (C) ±4
  (D) 0

5. Solve for y: 3y + 2 = 14

  (A) 4
  (B) 5
  (C) 6
  (D) 7

6. If a triangle has sides of lengths 5, 12, and 13 units, what type of triangle is it?

  (A) Isosceles
  (B) Equilateral
  (C) Scalene
  (D) Right

7. Simplify: $2x^2 + 4x - 2x^2 + x$

  (A) 4x + x
  (B) 4x
  (C) 5x
  (D) x

8. What is the solution to the inequality 3x - 5 > 1?

    (A) x > 1
    (B) x < 1
    (C) x > 2
    (D) x < 2

9. If $y = x^2 + 3x - 4$, find y when x = 2

    (A) 4
    (B) 2
    (C) 0
    (D) -2

10. What is the measure of each interior angle of a regular hexagon?

    (A) 60 degrees
    (B) 90 degrees
    (C) 120 degrees
    (D) 150 degrees

11. What is the slope of the line passing through the points (2, 3) and (5, 11)?

    (A) 2
    (B) 3
    (C) 4
    (D) 5

12. Solve for x: $\frac{2x - 4}{3} = 2$

    (A) 5
    (B) 6
    (C) 7
    (D) 8

13. What is the circumference of a circle with a radius of 3 units?

    (A) 6π units
    (B) 3π units
    (C) 9π units
    (D) 12π units

14. What is the area of a circle with a diameter of 10 units?

    (A) 10π square units
    (B) 25π square units
    (C) 50π square units
    (D) 100π square units

15. What is the slope of the line given by the equation $y = 2x + 5$?

    (A) 2
    (B) 5
    (C) -2
    (D) -5

CHECKING YOUR ANSWERS

Now it's time to see how well you did.

Practice 1

1. (C) 4

2. (C) x = ±2

3. (A) 2

4. (A) 1

5. (B) 15 square units

6. (A) 5x - 4

7. (A) -9

8. (D) -12

9. (B) (x + 3)(x - 3)

10. (B) 6x^3 y^3

11. (B) 180 degrees

12. (B) 12x + 8

13. (D) 64 cubic units

14. (C) 4

15. (A) 5 units

Practice 2

1. (C) 12

2. (B) 3

3. (C) 28 units

4. (C) ±4

5. (A) 4

6. (D) Right

7. (D) x

8. (C) x > 2

9. (B) 2

10. (C) 120

11. (B) 3

12. (B) 6

13. (A) 6π units

14. (B) 25π square units

15. (A) 2

CHAPTER 8

WORKING THROUGH THE TECHNICAL SUBTESTS

Nicely done! You've made it through the roughest parts that will determine a lot about your military future. With the AFQT complete, it's time for the less critical, but more specialized sections. This portion of the test can help you qualify for specific roles that would use this specialized knowledge. By performing well on the other sections, you may be able to start planning for a position in the armed forces that isn't open to everyone, which can give you a head start when you get out of the military.

WHAT TO KNOW ABOUT THE REMAINING CHAPTERS

Being able to build a career in the military can give you a life path, whether or not you remain in the military. The roles that you may qualify for through these tests are ones that tend to be well-paid in civilian life. While you need to do with on the AFQT, the other subtests can help determine your position in the miliary.

Each of the next six chapters relate to the technical aspects and subtests:

Chapter 9: General Science
Chapter 10: Electronics Information
Chapter 11: Auto Information
Chapter 12: Shop Information
Chapter 13: Mechanical Comprehension
Chapter 14: Assembling Objects

Unlike chapter 3 to 7, you won't need to spend as much times studying for these chapters, or at least not all of them. You want to spend more time with the chapter that is relevant to the role you want to have in the military. This doesn't mean you should ignore them, but you won't need to study for them quite like you do with the reading and math chapters. The exception is if there is a particular job you want.

Before you take the test, talk to a recruiter to see which subtests affect the specialist roles. If there is a particular role you would like or if you are interested in a particular area, find out which tests can help you move into that role. As you go through the specialist subtest sections, if you find one that you enjoy, you may want to spend additional time studying it. This will help you to focus on something that you may enjoy as a career. That's really what you are looking for when you go into the military – a career that you find fulfilling.

Use the results of the diagnostic test to help you determine where you currently have an aptitude in these specialized areas. If you didn't do particularly well in an area where you are interested, take the time to practice the subject. These are the kinds of things where the more you learn, the better you will do.

Remember the point of these is to get into positions that have more obvious career paths. If there are a couple of areas that you enjoy, you can keep your options open. Doing well on several specialized sections can give you even more opportunities in the military. It's possible that you end up taking one role, but decide to take a different path later. You may be able to take the test again later and switch fields if you do well.

APPROACHING THE SPECIALIZED SECTIONS

Just like with the other sections, you need to have a process for how you approach each section.

- Always make sure you understand the question. You need to know what the question is asking before you look for the answer. You can check the answers to see if they guide you, especially if you don't have a clue what the question is asking. For example, if all of the answers are in mHz, you know that you need to calculate the frequency. If you recognize the answers are all the names of bodies of water, you know that the question wants you to identify them. You can treat these similar to how you did the reding comprehension. The answers can be a guide.
- Determine if you are solving for sampling or if you are simplifying something if the problem involves math. There will be plenty of questions that test your knowledge of things like basic science and shop vocabulary, in which case, you aren't really solving. That means you don't need to take the time to deal with the math aspect. Otherwise, there are several types of solving and simplifying you may need to do.
  - o Diagrams are a common question type for some of these subtests. When working with a diagram, determine the following:
    - What is the diagram showing?
    - What relationships is the diagram depicting?
    - How can I use the diagram to better answer the question?
  - o These sections will contain variables and formulas. You can approach them with the methods you used in Chapters 6 and 7.
  - o Charts, tables, and graphs are common in these sections, and they show you details about the data being discussed. You should ask yourself the following questions.
    - How does the graph help answer the question?
    - What units are used in the graph?
    - What do the two axes represent?
    - Is there a pattern shown in the graph?
  - o Many of the questions will provide some introductory information, then provide some details. Take the time to make sure you understand all of the written information so that you answer the correct question.
- Try to predict what kind of answer the question wants. Approach these subtests in a way that is similar to how you approached the reading and math sections. Having a prediction can help you see the right answer faster.
- Be strategic in how you approach the answers. If you can make a good prediction, look for an answer that fits. If not, you can use the process of elimination to remove answers that don't make sense in terms of what the question wants. Look for the following kinds of details that can help you eliminate answers.
  - o The result is too small or large.
  - o The wrong unit is provided.
  - o Is the answer from the wrong category, such as a naming an organ instead of a bodily system.
  - o Are any two answer synonyms? Since only one answer can be correct, if there are two answers that have the same meaning, then you know that they are both wrong.

Keep these methods in mind so that you have a better chance of getting more questions right, even if you aren't particularly familiar with the subject. The more you review the content within these sections, the more likely you will have at least a basic understanding. This will help you to better narrow down the answers.

REMAINING CHAPTERS

The remaining chapters, Chapters 14 and 15, require a different approach. Chapter 14 is essentially logic puzzles. You'll need to go through that chapter to better understand how to answer those questions. You will have puzzles that you need to figure out how they go together, or how to assemble parts into a whole.

Chapter 15 provides practice tests to help you study. That way you will have a chance to test your knowledge and ensure you are able to apply a process reliably to do better on your test.

CHAPTER 9

GENERAL SCIENCES

General Science is just that – a lot of questions about many different scientific topics. Most of what it covers are things that you will have studied in high school. There are three primary areas covered in the General Science section:

- Life science
- Earth and space
- Physics

This chapter covers these three areas in their own sections. The final section gives you a chance to test what you've learned and your test taking process when it comes to science questions.

LIFE SCIENCE

Life science does change the more we learn about different aspects of life. Things like nutrition seem to have significant shifts every decade or so, and today, what is considered nutritious varies by different scientific studies and groups. That's because life science is an area where we still have a lot to learn, even something as apparently simple as what we should eat. That human body is incredibly complicated, and not everyone's body is the same. When you consider just how much life science covers – plants and animals – it's easy to see how much it changes.

Fortunately, the basics of life science are pretty well-known. Things like the basics of genetics, cells, and ecology are predictable. This chapter covers the areas that make up the basics of life and are things that have been true for decades or longer.

Classifications

All living things fall into a classification, whether it's plant or animal. There is a third classification, mineral, that is also robust, but minerals are not living, like plants and animals. Minerals will be discussed a bit more in the Earth and space chapter as that is where you find minerals – in the Earth and in celestial objects.

When you are working with animals, they all fall into very specific schools for their classifications. The following tables provides the layers of classification, from the broadest to the smallest.

Classification	Description	Examples
Domain	There are three biological domains.	Archaea – single-celled organism without a nucleus Bacteria – single-celled organism without a nucleus Eucarya – living creatures with a nucleus in the cells, and it includes most animals, plants, and fungi

		The difference between Archaea and Bacteira is largely metabolic and chemical, but they are both much simpler than Eucrya
Kingdom	There five kingdoms and they include all living beings	Animalia – all animals are in this kingdom Plantae – a plants fall into this kingdom Fungi. – all fungus belong in this kingdom Monera – single-celled organisms, like bacteria Protista – single-celled organisms that have chloroplasts, like algae and protozoans
Phylum	Most animals fall into one of seven phylum	Porifera – simple marine animals, like sponges Cnidaria – more complex marine animals, such as jellyfish and coral Platyhelminthes -mostly flatworms that are parasites, like tapeworms Annelida – segmented worms, like earthworms and leeches Mollusca – large, complex marine animals, such as clams and mussels Arthropoda – invertebrate creatures with segmented bodies and exoskeletons, like insects, arachnids, and crustacean, like butterflies and shrimp Chordata – vertebrates, or animals with a spine, which includes most animals you know, like fish, dogs, and people
Class	Each phylum is subdivided into classes. For the Chordata Phylum, there are seven classes, most of which you've probably heard of.	Agnetha – jawless fish, such as lampreys, many are extinct Chondrichthyes – fish with cartilage, such as stingrays and sharks Osteichthyes – fish with bones, like clown fish, catfish, and trout Amphibia – better known as amphibians, including newts, frogs, and salamanders Reptilia – better known as reptiles, this includes turtles, lizards, snakes, and alligators Aves better known as birds, this includes all animals with two legs, two wings, and feathers Mammalia -better known as mammals, these are any warm-blooded creatures that have lungs and give birth to live babies
Order	Animals within each of the different classes are further divided into orders.	In the Mammalian class, there are roughly 25 orders (it varies as scientists reorganize based on new understanding of creatures. Rodentia is an order, and it is better known as a rodents. Primates is the order that includes chimps, gorillas, and

		humans.
Family	Each order has a different number of families with in it.	In the Carnivora (better known as carnivore) order, there are 12 families, including Ursidae (bears), Canidae (wolves, coyotes, and dogs), and Felidae (all types of cats).
Genus	By this point, the animals in this classification are much more closely related.	For Felidae, it breaks down cats into more specific types, with there being dozens. The following are a just a few of the genus in the Felidae family. - Felis (domestic cats) - Panthera (Leopards, Lions, Jaguars, Tigers) - Saber-toothed cats - Lynxes
Species	The smallest group, there are thousands of species because every animal has its own unique species.	The species Delphinus Delphis refers to dolphins, and there is no other animal in that species. People fall into homo sapiens, and there are no other animals with that classification

There is a mnemonic to help you remember the order of classifications: Dear King Philip Came Over For Good Soup. Domain is a more recent addition, so it may not appear on the ASVAB, but the rest have been established for a long time, so you will almost certainly see them. Here's how it looks when you exclude Domain:

- Kingdom
- Phylum
- Class
- Order
- Family
- Genus
- Species

To better understand these classifications, spend time studying the different areas and learning what falls into the first four or five classifications. Once you get to the families, there are so many that it will be nearly impossible to learn all of them for one order, let alone all of them. It will be more important to understand how the classification are broken down.

The following is the kind of question you may see on the test.

Snakes are a part of what kingdom?

- (A) Animalia
- (B) Mammalia
- (C) Amphibia
- (D) Reptilia

Snakes do belong to the Reptilia class, but the question asks about the kingdom. Like all animals, they are a part of the Animalia kingdom, so the right answer is A.

Here's another example question.

What is the order of categories of taxonomy, from specific to general?

- (A) Species, Genus, Family, Order, Class, Phylum, Kingdom
- (B) Kingdom, Phylum, Family, Order, Class, Genus, Species
- (C) Kingdom, Phylum, Class, Order, Family, Genus, Species
- (D) Species, Genus, Order, Family, Class, Phylum, Kingdom

Kingdom is the most generic, so you know that it needs to be at the end. Species is the most specific, so it needs to be at the start. That rules out B and C .When you go through the mnemonic, you will see that A because D has Order and Family switched.

Nutrition

All organisms require basic nutrition to survive. As mentioned, nutrition for people continues to change and evolve as we learn more. While what an individual may have different needs, there are a few universal requirements for the human body.

Nutritional Aspect	Basic Function	How the Body Uses It
Carbohydrates	Energy	Sugars and starches provide carbs, and they are one of the primary sources for energy. Sugar is naturally found in fruits, beets, and cane sugar. Starches are found in processed foods, such as breads and pastas.
Fats	Energy	This is another significant source of energy, and there are three types: - Monounsaturated tends to decrease bad cholesterol. Things like almonds, cashews, avocados, and olives have monounsaturated fats. - Polyunsaturated tends to decrease bad cholesterol. Things like flaxseed, pumpkin seed, soybean oil, and sunflower oil have polyunsaturated fats. - Saturated tends to contribute to bad cholesterol and increase the risk of stroke and heart disease. Eggs, dairy, shellfish, and meats are the most common items with saturated fats.
Fiber	For the body to function	Fiber helps make the large intestine more robust to better process all of the wastes that goes through it. Things like carrots, greens, beans, potatoes, fruits, and whole grains are good sources of fiber.
Minerals	For the body to function	Isn't something you need a lot of since it includes things like iron, calcium, magnesium, and sodium, but they are essential for building bones, blood, and teeth. Minerals play a variety of different roles in helping your body to function.
Proteins	Energy	Provide energy to grow, maintain, and repair the body. You can get proteins from eating meats, eggs, and cheeses, or form eating some vegetables, such as beans, peans, and nuts.
Vitamins	For the body to function	This is probably the one you've been hearing about the longest because it includes things like Vitamin C and D. Like minerals, vitamins play a lot of different roles in how your body functions. Most of these come from fruits and vegetables, but you can also get things like Vitamin D from sun light.

Not included in the table is the most basic requirement for sustaining life – water. Since the human body loses about four pints of water daily, and that's why there is a recommendation to drink a lot of water every day. You can get water form most foots, but it isn't enough to sustain life without drinking. Without enough water, you will become dehydrated. Initially, this will cause cramps, lightheadedness , and dizziness. When not fixed, it will cause death.

To be healthy, you need the right balance of nutrient's in your diet. Different people need different things, and you have different needs when you are young compared to when you are an adult. If you are pregnant, your needs change again.

It's important to maintain a balance that goes with your current situation. There are many ailments that you can get from a lack of different nutrients. For example, when you don't get enough iron, it can cause anemia, which ahs symptoms like dizziness, headaches, and weakness. If you don't get enough Vitamin C, you could get scurvy, which has symptoms like gum disease and tooth lose, irritability, depression, muscle pains, and spotty skin. While the recommendations change, you can tell when you aren't meeting your own requirements because you will fall ill.

The following table provides details about the kinds of nutrients that you need to ensure you have in your regular diet to avoid some of the most common ailments associated with insufficient nutrients.

Dairy products like milk, yogurt, and cheese; spinach

Nutrient	What It Does	What It's In
Calcium	Growing and repairing bones, functioning muscles	Largely found in dairy productions, but also in spinach
Iron	Functions in the red blood cells to move oxygen across the body.	Meat, some beans, whole grains
Magnesium	Developing bones and muscles, and ensures the functioning of enzymes and nerves	Fortified foods, leafy greens, nuts, whole grains
Potassium	Ensures a balance with fluids	Bananas, nuts, seeds, sweet potatoes
Vitamin A	Helps with cell growth, the immune system, and vision	Carrots, eggs, liver, milk, spinach
Vitamin C	Helps with the functioning of the immune system, cell repair, and formation of collagen	Broccoli, citrus, peppers
Vitamin D	Strengthening bones (it helps absorb calcium), muscles, and improves immune system and nerves	Cereals (some), fortified milk, juice; exposure to the sun

The following is an example of the type of nutrition question you may see on the ASVAB.

An expecting mother experiences dizziness and headache. A doctor told her that she was anemic. What food would help her?

- (A) Peppers
- (B) Meats
- (C) Cereals
- (D) Bananas

The question identifies the ailment as anemia, and that means that the person has an iron deficiency. Meat is one of the foods recommended to help improve the amount of iron a person has in their blood. The right answer is B.

Let's take a look at another one.

If a doctor diagnoses someone as having high cholesterol, what food should the person avoid to lower their cholesterol numbers?

- (A) almonds
- (B) avocados
- (C) flaxseed
- (D) red meat

The answer is D red meats because they contain saturated fats. All of the others can help lower the person's cholesterol because they are either monounsaturated or polyunsaturated fats.

Bodily Systems and Diseases

Besides nutrition, there are a lot of things that can cause a person to become ill. That's because there are a lot of different components and systems. These are all things that could appear on the ASVAB. This section covers some of the basics that are likely to be covered on the test.

First, we'll look at the basics of most of the human's systems.

Muscles and Skeletal System

The skeleton and muscles work together to support the body and enable movement. Without a skeleton, the body would be a motionless collection of organs, veins, and skin. Some organisms, like arthropods (including insects, spiders, and crustaceans), possess exoskeletons, or external skeletons. Vertebrate animals, like humans, have internal skeletons, which are also called endoskeletons.

The human skeleton is made of bone and cartilage.

- Bones re the primary structural support, and it is hard.
- Cartilage is more flexible, and it is located at the ends of all bones, within joints, and in areas like the nose and ears.

Bones support the body, protect vital organs, produce blood cells, and store essential minerals, such as calcium. There are two types of connective tissue:

- Tendons are tough fibrous cords of connective tissue that attach muscles to the skeleton.
- Ligaments connect bones to each other at joints such as the elbow, knee, fingers, and vertebral column.

The following is an image that shows some of the parts of the skeleton:

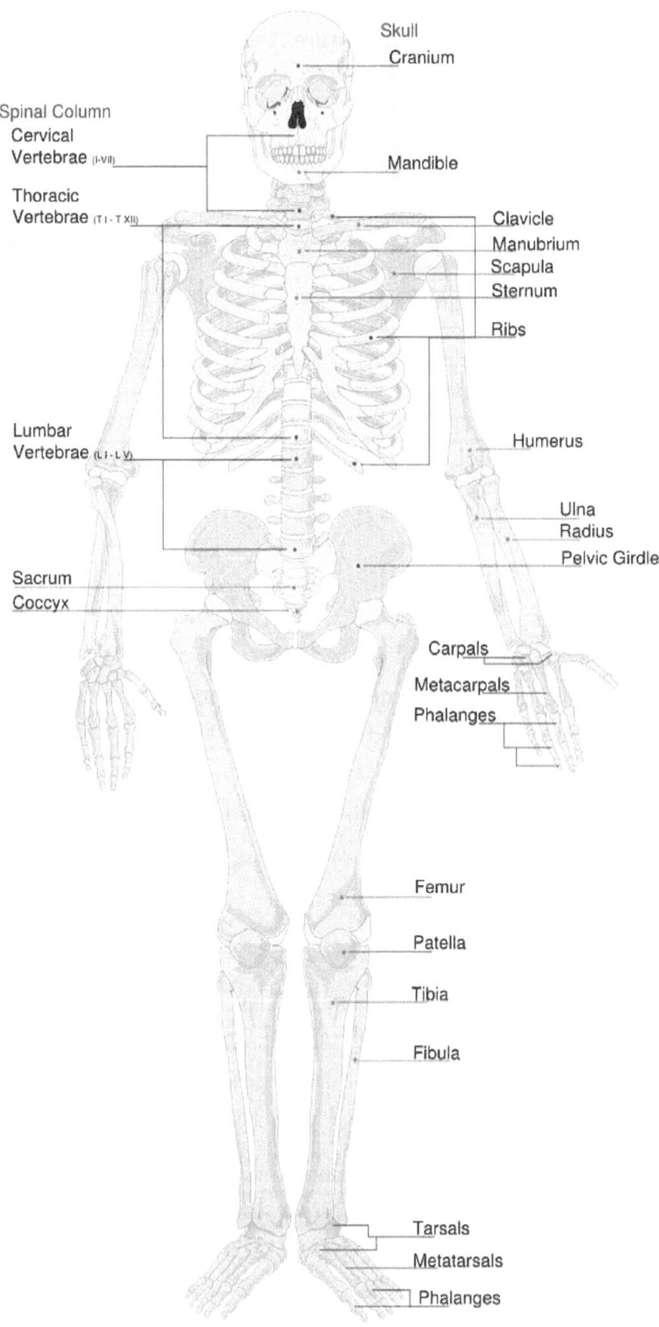

The following are the kinds of questions that could be asked about skeletons and muscles.

What type of tissue connects the ulna and humorous?

- (A) cartilage
- (B) ligament
- (C) muscle
- (D) bone

The only one of these that is a connective tissue is B, ligament, so that is the right answer. Make sure to know the difference between ligaments and tendons though. Tendons connect muscles, but ligaments connect bones to each other, just like in the question.

Respiratory System

You probably know that the respiratory system is how you take air into your body and it is then processed into the blood and moved around your body.

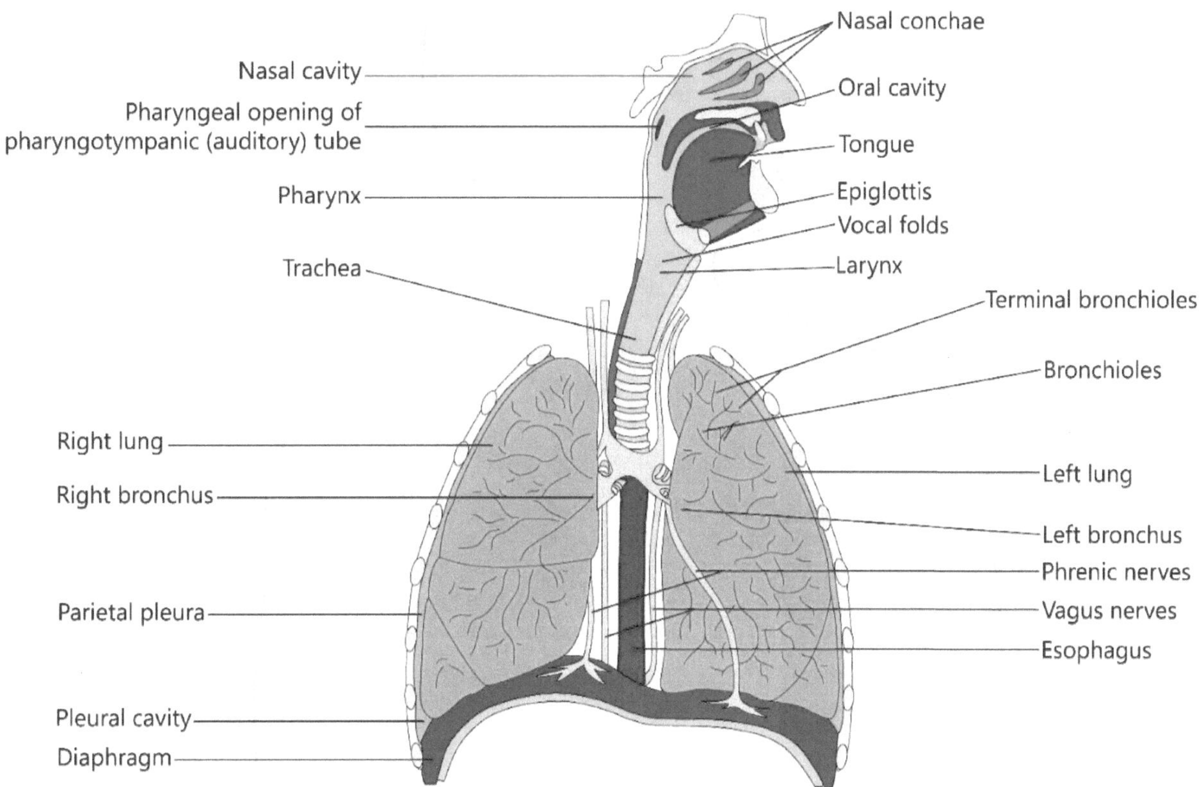

When you breath in through your nose, the air goes through the nasal cavity, where it is filtered, moistened, and warmed. Next, the air moves through the pharynx where it is filtered to protect against infection. It then goes through the open epiglottis. This is the thing at the back of the throat that closes when you swallowing to prevent food from entering the airway. From there, the air enters the trachea for additional filtering. The trachea splits into the left and right bronchi, leading into the left and right lungs. Once in the lungs, the bronchi divide into bronchioles that end in alveolus. Here, the oxygen moves through capillaries (tiny blood vessels) and enters the bloodstream.

While air comes through the nose, it is the diaphragm facilitates breathing. The diaphragm is a system of muscles that contracts to expand the lungs and drawing air in the body. When you exhale, the diaphragm relaxes, the lungs contract, and the air is pushed out of your body

Circulatory System

Once the air is in the blood, it is circulated around the body, which is why the circulatory system has that name. The primary purpose of blood is to take oxygen and nutrients to all parts of the body, then to remove carbon dioxide and wastes.

The heart is the primary organ that moves the blood through the body. It has four chambers:

- Each of the atrium collects the blood. .
- The ventricles pump the blood through both of the venae cavae. The vena cava is the largest vein in your body, and you have two of them.

Blood runs through the vena cava and through the pulmonary artery and lungs. Here, oxygen is taken up by the blood and distributed throughout the body. From the system pumps the blood through the left atrium and based it to the left ventricle to go through the aorta. This is the largest artery in your body. From here, the blood is pumped through several smaller arteries and out to the rest of the body.

The vales are what makes a heart efficient. They close once the blood leaves the heart so that the blood doesn't flow back into the heart.

The right side is typically responsible for handling deoxygenated blood since it is pumping the blood into the lungs to pick up oxygen. The left side handles the oxygenated blood since it has just returned from the lungs.

The most common cause of death in the US is heart disease. Contributors to this problem include high blood pressures, high cholesterol, smoking, and a sedentary lifestyle.

CIRCULATION OF BLOOD THROUGH THE HEART

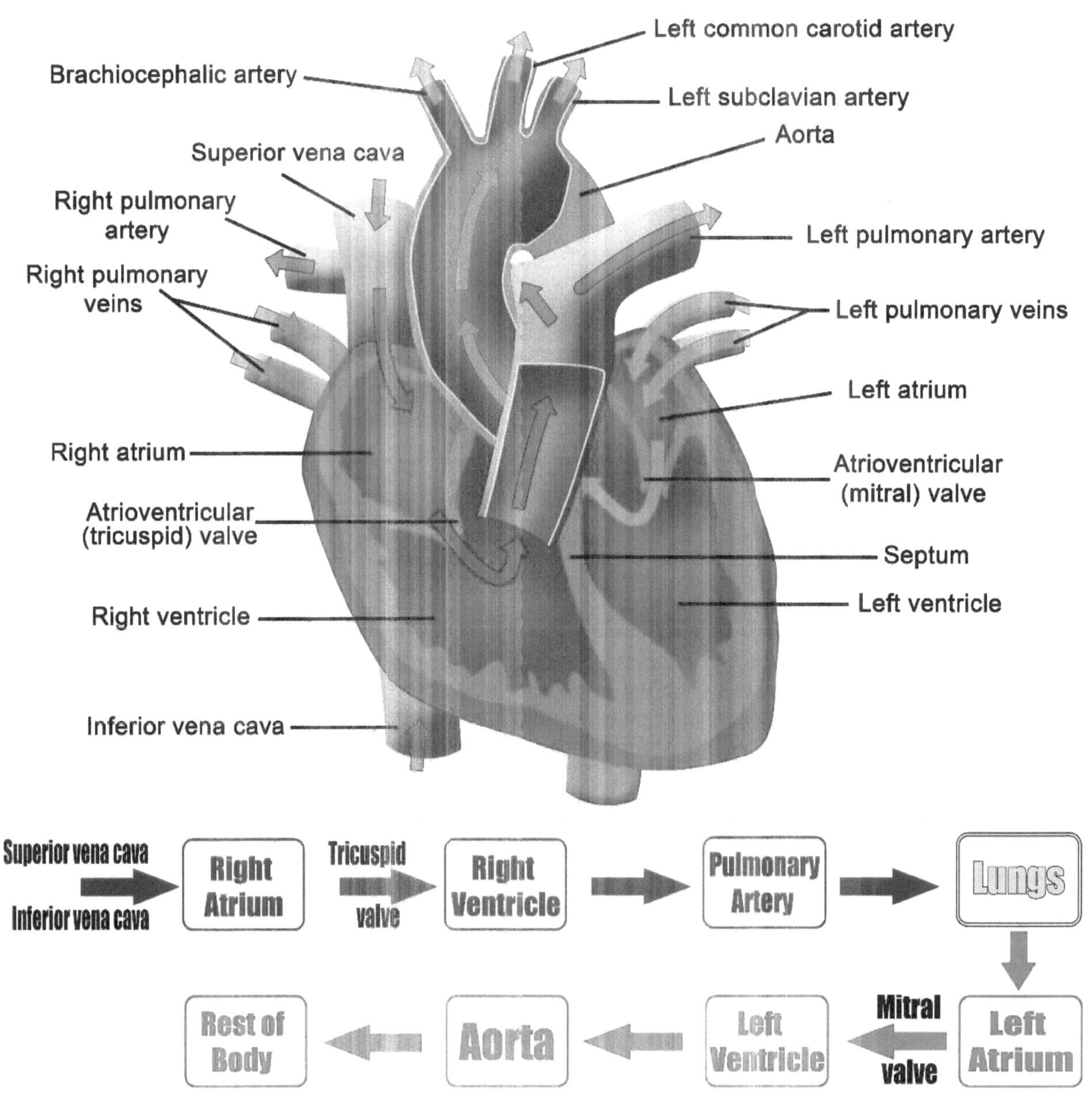

The following image shows the system of veins, arteries, and capillaries.

Human circulatory system

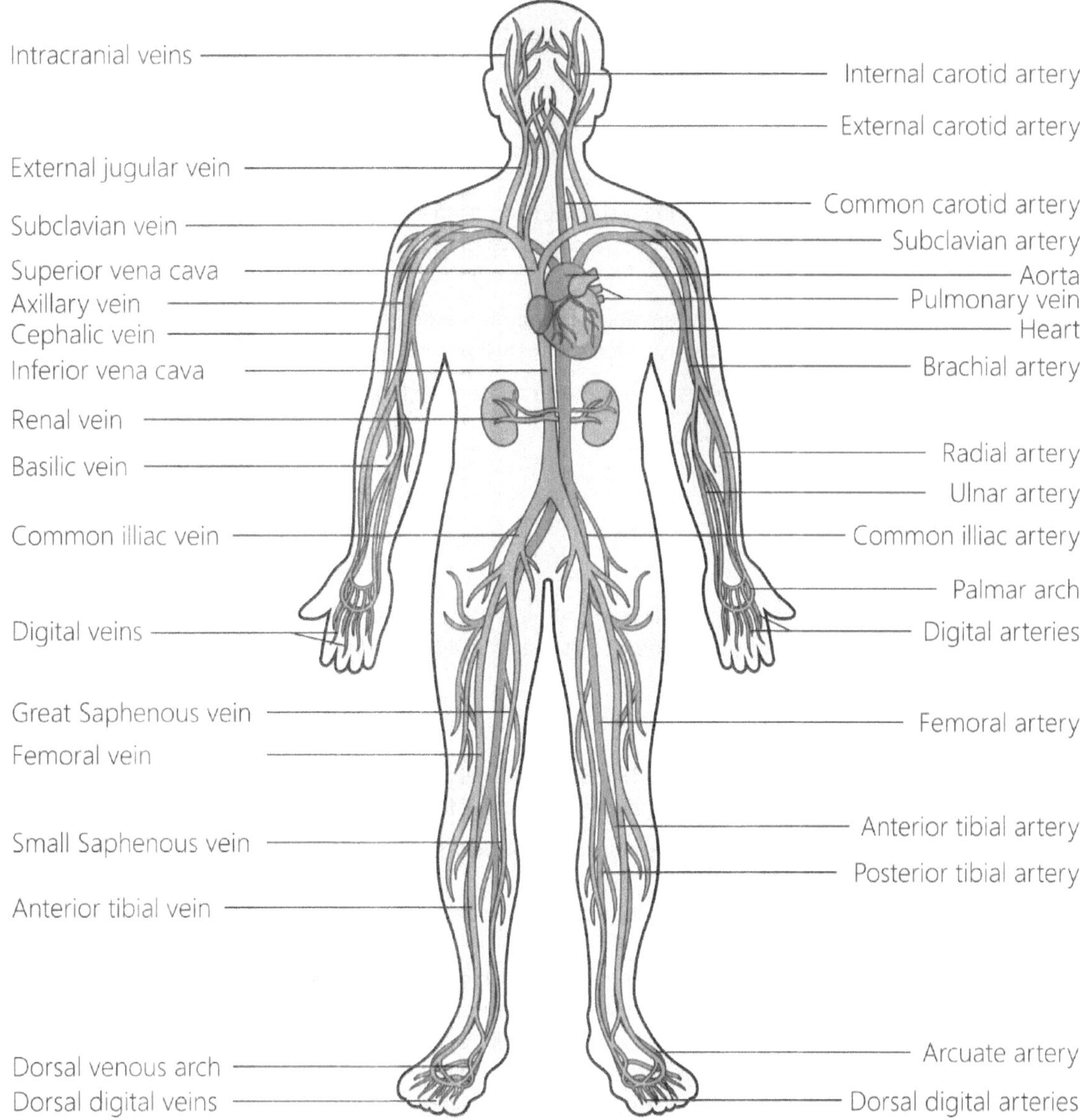

Arteries take blood away from the heart. They become increasingly smaller the farther from the heart as they go into smaller parts of the body, including the extremities. They have thick walls because the blood moves at a high pressure with the oxygen. When you take a pulse, you are feeling for the arteries.

Veins return blood to the heart from all over the body. They have thinner walls as the blood returns as a lower pressure. They do have more valves to keep the blood from flowing back to other places. You cannot feel a pulse in a vein, except for the pulmonary vein, which is the only one that contains oxygenated blood as it has just come from the lungs.

Capillaries are the small vessels with thin walls that allow for the exchange of different materials, including carbon dioxide, nutrients, oxygen, and waste.

Capillaries can be adversely affected by heart disease and other problems. High blood pressure in particular can cause problem as it damages blood vessels, in addition to other body parts like kidneys. The best way to prevent high blood pressure is to keep a moderate salt intake and to exercise daily.

Blood is a liquid that has a number of different components. Three types of blood cells are suspended in plasma to make blood:

- Red blood cells carry things like oxygen and nutrients and remove waste.
- While blood cells combat illness by cooperating with the immune system.
- Platelets provide a way of clotting the blood to for scabs.

All three of these are made in bone marrow. To determine a person's health, all three of these parts of blood have to be examined. When white blood cells are elevated, this is usually a sign that someone is fighting an infection or other ailment. Red blood cells help to indicate other kinds of illness, such as anemia because it will not have adequate iron.

There are four different blood types, all of which can be either positive or negative:

- A
- B
- AB
- O

The Letter indicates the types of antigens are in the plasma outside of the red blood cells. Whether it is positive or negative indicates the Rh factor of the blood. People with a positive blood type may be able to have a blood transfusion with someone who has a negative type, but someone with a negative Rh factor cannot.

Blood type O negative is a universal donor type as anyone who needs a blood transfusion can accept this time. If a person has blood type AB positive, they can receive any blood type for a transfusion.

Here's the type of question you may see on the test about the circulatory system.

What type of vessel does not have oxygen rich blood?

(A) Aorta
(B) Right atrium
(C) Left ventricle
(D) Pulmonary vein

The aorta is a primary carrier of oxygen as it is moving it from the left ventricle of the heart, ruling both of these out. Since the pulmonary vein carries oxygen from the lungs to the heart (it's the only vein that has oxygen-rich blood), the answer must be A, the right atrium.

Digestive System

The digestive system breaks food into something that the body can use for energy and to keep it functioning. This system starts with the mouth and ends with the anus. The following is the process for food from start to finish.

- The tongue and teeth help with the majority of breaking down food in your mouth. Your saliva helps to break down starch.
- When you swallow your food or liquid, it travels down the esophagus where muscles contract to move it down to the stomach.
- The stomach contains gastric juices (acids and pepsin) to further break down the food into proteins.
- Food the moves into the small intestine where the majority of the digestion occurs. Despite the name, this organ is actually roughly 23 feet long. It contains enzymes to break down the food. The enzymes are produced in the small intestine walls, in the pancreas, and the liver. The pancreas provides lipase to turn fat into glycerol and acids. Pancreatic amylase turns complex carbohydrates into sugars. Trypsin converts polypeptides into amino acids that cells use. The liver produces bile, which breaks down fats into separate molecules. Most of these substances are then absorbed by blood in the capillaries that then goes to the liver, and from there across the body. Fatty acids and glycerol are the only substances not moved through this last part of the process.
- Once most of the food is broken down, it moves into the large intestine (also called the colon), any remaining water or minerals are reabsorbed into the body.
- Waste produced during this process then move to the kidneys for filtering and release as urine. This waste byproduct moves through the ureters to the bladder. When enough is stored, it is released through the urethra.
- Any remaining solid waste moves to the rectum for storage until your body has enough ready to release through the anus.

The following image shows each of the primary organs in the digestive system.

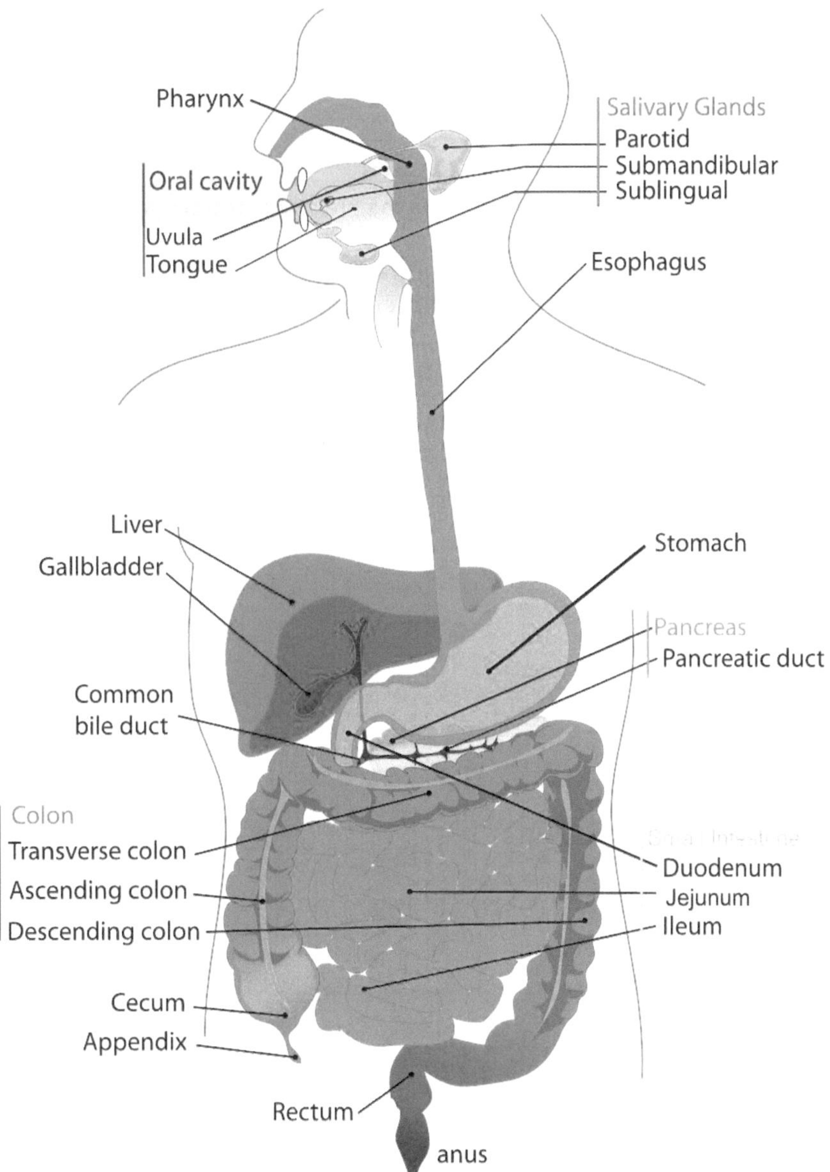

The following is an example of the kind of question you may have on the ASVAB.

Where does food go once the proteins start to be digested?

(A) Esophagus
(B) Large intestine
(C) Small intestine
(D) Stomach

The stomach starts to really break down the food into small units, but the small intestine is where most of the digestion happens after the process begins. C is the correct answer.

Nervous System

This is the stems that includes the brain and all of the components of the body that relay messages from the brain to the rest of the body. The following are the components of the nervous system:

- The brain
- The spinal column, which is encased in bone going from the brain to the tail bone at the buttock.
- Billions of neurons, which includes billions of nerves to send out signals all over the body

This is the system that both tells the rest of the body how to act, but also receives signals go process what is happening around the body. The central part of the system are the brain and the spinal column, but the neurons are required to process everything.

The brain is often divided into four different parts to process different aspects of the world.

- The cerebrum is where scientists think intelligence resides, and it is responsible for processing the signals that return from the body, particularly the different senses.
- The cerebellum is a group of nerve tissues located at the brain's base. It is thought to be associated with basic functions, especially movement like balance and muscle coordination.
- The medulla is part of the brain stem and it connects the brain with the spinal cord. It is responsible for involuntary functions, like breathing and your heart beat.
- The spinal cord provides the connection between the brain and nerves. Impulses run up and down the spine from the brain to the rest of the body and bringing data back tot eh brain from everywhere else. It also manages reflexes.

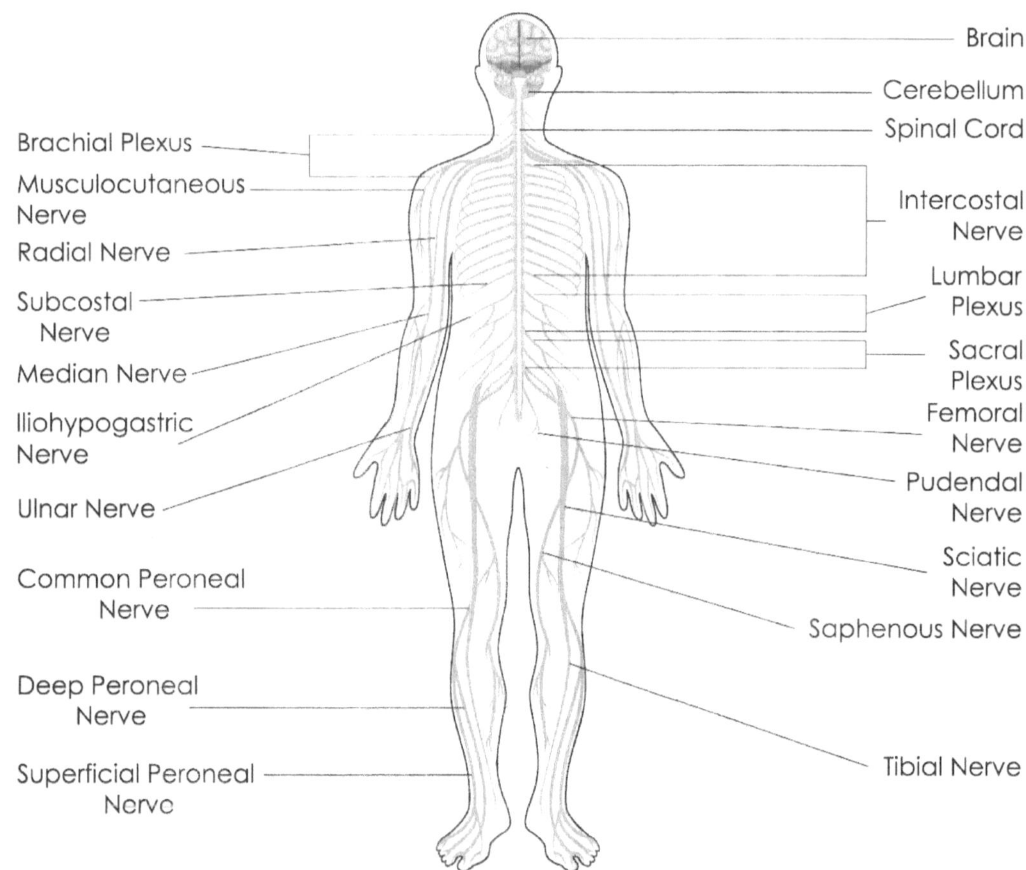

NERVOUS SYSTEM

The other elements of the system are called the peripheral parts, and they are divided into two groups.

- The somatic system includes the nerve fibers that manage sensory data across the central nervous system, so it manages voluntary actions.
- The automatic system manages the involuntary activities.

The following is the kind of question you may see about this system.

Damage to which part of your brain will cause you to have reduced balance?

- (A) Cerebrum
- (B) Cerebellum
- (C) Medulla
- (D) Spinal cord

The answer is the cerebellum, B.

Reproductive System

When people reproduce, the male fertilizes the egg in the woman's body (if you'll recall, this is something that is unique to mammals). Roughly every 28 days, one of two ovaries releases an egg (called an ovum), which travels through the fallopian tube into the uterus. In preparation for the egg to be fertilized, the uterus wall begins to line with blood so that the egg can attach to it and start to grow.

A male and a female have to have intercourse and the male must ejaculate to release sperm. On average, more than 250 million sperm are released from the testes, through the penis and into the woman through her vagina. From this point, the sperm travels to the uterus to fertilize the egg. If one of the sperm is able to find and move into the egg, the egg is said to be fertilized, forming a zygote. If the fertilized egg attaches to the uterine wall, it will begin to form a fetus.

When a woman is pregnant, the pituitary gland begins producing prolactin, which then helps to start the lactation process to produce milk. If the woman does not become pregnant after an egg is released, the blood that lines the uterus then flows out of the body, along with some lining.

The cycle of the release of the egg through the blood being expelled is called the menstrual cycle. A woman begins menstruating during puberty. When her body stops menstruating, she goes through menopause. Between when she begins puberty and when she finishes menopause, she will have roughly one period a month. This goes on for several decades.

Diseases and Pathogens

Diet and lifestyle factors contribute to diseases like deficiency disorders, hypertension, and heart disease. Pathogens are disease-causing agents like bacteria and viruses, and they cause a different type of ailment to the host. Bacteria cause well-known diseases like staph infections, pneumonia, and strep throat. Antibiotics are the most common treatment for bacterial infections.

Viruses are very different because they aren't living organisms. They are not able to reproduce on their own – they require the cells of a host to replicate themselves. The most common viruses are the flu and the cold. Herpes and AIDS are more serious viral disease. Treating this kind of ailment requires the use of antiviral drugs.

Both of these types of ailments are communicable, or they are spread from person to person. Many are contagious through the respiratory system, so you breath them in, with the flu and cold being the most common diseases to be spread through the air. If someone has contradicted the disease sneezes or coughs, they release the disease into the air where others may breath it in and contract it. Other disease or contracted through fluids, with AIDS being the most notable disease that moves through bodily fluid exchange. Herpes can be transferred between people through skin contact.

Vaccination is the best way of preventing most types of viral infections, and it can be used for a few bacterial diseases. Vaccination introduces the disease into your body in small doses that allow your body to build up antibodies against the disease. If you have been vaccinated against a disease, then you are far less likely to suffer serious side effects if you do come into contact with the disease. One of the most prominent examples of successful vaccinations was against polio. The disease caused serious illness, sometimes permanently disabling a person, and in the worst cases, it killed people.

Here is the kind of question you may get about it on the ASVAB.

You can use antibiotics to treat which of the following diseases?

- (A) Heart disease
- (B) Strep throat
- (C) Anemia
- (D) Influenza

Antibiotics are used to treat bacterial infections. Strep throat is the only bacterial disease listed, so that is the correct choice.

Genetics

Genetics are a fairly new area of science because you cannot see the basic components of genetics, just the outcome. All hereditary traits that a parent passes to their offspring are a part of their genetics. Scientists began to study genetics back in the 18th century, so the foundations of the science have been around since the late 1700s. Gregor Mendel is the person who discovered this area of science. He studied the hereditary traits passed on through reproduction and discovered that individual traits separate so that gametes (reproductive cells) have half of the chromosomes as a normal cell. Where a normal cell is a diploid, gametes are haploids (you can think of them as half a diploid to remember the name).

In humans, women have the female gamete (the egg) and it mixes with the male gamete (the sperm). Both have 23 chromosomes, so when they combine, there are 46 chromosomes (or 23 pairs of chromosomes). Sexual intercourse is the means by which the process begins, so it is the meiosis, or the process that creates gametes. There is a huge variety of potential results because the process has many potential outcomes.

Deoxyribonucleic acid, better known as DNA, is a molecule with genetic information. Each strand of DNA contains genetic code that is a combination of nucleotides that provide instructions to the cell for how the cell should present and behave.

DNA is made of individual units called genes. There are several types of genes called alleles. You have a gene for your eye color, but you may have alleles for both green and blue eyes. You inherit one allele from each parent, so you have two alleles. In some cases the alleles may be the same, in which case there is no question of how that trait shows up, and that gene is homozygous. If the alleles are not the same, the gene is heterozygous. A babies gender is determine by the sex chromosome; when there are two X chromosomes, the child is a female, when there is one Y and one X chromosome, the child is a male.

Traits are inherited on independent of other traits, and often there is one dominant and one recessive trait.

- A dominant trait is likely to present, even if only one copy is present.
- A recessive trait typically only presents when someone has two copies of it.

Huntington's disease is a degenerative disease that effects nerve cells and the muscles, and it is inherited through genetics from the parents. As a dominant trait, it only takes one parent to have this gene for their children to be at higher risk of having Huntington's disease. Cystic fibrosis is another genetic disease, and it causes problems with the lungs. However, it is recessive, so both parents have to contribute this gene for their children to have the disease.

Not all genetic diseases are caused by dominant and recessive genes. Down syndrome is caused by having three chromosomes (instead of two) for the 21st chromosome pair. Color blindness is passed down through an X chromosome, and it causes some of the cones to be missing from the eyes.

Your genetic makeup is your genotype, and it includes your dominant and recessive alleles. The term phenotype means how your genetics present themselves in your physical traits. The color of your hair and eyes are two of the most often notices parts of your phenotype.

The following is the kind of question you may see on the ASVAB.

What if you have two dogs, one the dominant brown fur trait and one with a recessive red fur trait. If they had a red coated puppy, which of the following could be true?

- (A) The next puppy will be brown.
- (B) The red dog has a recessive brown fur allele.
- (C) The puppy's fur will change colors to be brown later.
- (D) The brown dog has a recessive red fur allele.

The only one that can be inferred from the information given is D. While it's possible that a puppy's coat color will change, there is nothing in the text to support that prediction, meaning D is the correct answer.

Cells

Since we are still learning about cells, but we do know enough to have established a reliable theory about them.

- Everything alive is made of cells.
- Cells are the basic life unit.
- Existing cells make new cells.

There are two different types of cells based on whether the cells has a nucleolus or not.

- A cell without a nucleus is prokaryotic. Bacteria are prokaryotic.
- A cell with a nucleus is eukaryotic.

Most living organisms are made of eukaryotic cells, including animals, fungi, plants, and protists. Compared to prokaryotic cells, all of them exhibit greater complexity. The cell's nucleus has all of the necessary genetic cell material. Much of the area around the nucleus is cytoplasm, a substance that holds most of the cell contents in place. The other contents in the cytoplasm are called organelles (like organs in animals). Like organs, each organelle has a specific function. The following are the other elements within eukaryotic cells:

- Ribosomes make proteins.
- Mitochondria make energy.
- The Endoplasmic reticulum synthesis fats and proteins.

- The Golgi apparatus makes proteins usable.
- Lysosomes manage cell waste.
- Centrosomes provide the guide for cell reproduction.

The following provides a look at the eukaryotic cell structure.

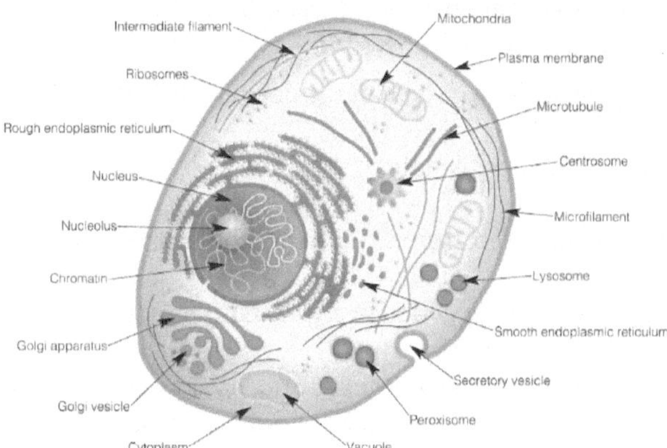

Plant cells are different from animal cells. One of the most noticeable differences is the cell wall, which is rigid. Plants are able to make their own energy through photosynthesis. The process requires carbon dioxide, sunlight, and water, then it uses these with sugar and oxygen to generate energy.

Animal cells just have the semipermeable membrane, which lets oxygen and water into and out of the cell. Unlike plants, animal cells cannot create their own energy, so it relies on taking in materials form outside of the cell to make energy. Cellular respiration is how it does that, and it requires the mitochondria to process oxygen and sugar, carbon dioxide, and water. Without oxygen, the process results in fermentation, which is not something that cells can live on. However, it does result in lactic acid or alcohol.

The replication of genetic materials is called cell division, and it occurs in the nucleus. The process starts with interphase, which is DNA replication. This also replicates the chromosomes that then divide. The interphase process takes the longest to complete, and it includes both DNA replication and cell growth. During this time the cell grows enough to hold the larger number of chromosomes. Once it ends, prophase starts, during which the chromatids stat to pair up with each other. As this happens, the metaphase begins as the sister chromatids migrate to opposite sides. Anaphase comes next, and the chromatids begin to separate into two poles. The cell becomes longer. The next phase is called the telophase, and during this time two nuclei separate. The last phase is cytokinesis, where the cells completely separate, with the last parts to split being the cell membranes and cytoplasm.

The entire reproductive process is regulated through genetic signals that identify when reproduction should end. Cancer is an example of when the signals aren't working as they should because they are mutated, resulting in cells growing unchecked. There are a number of factors that contribute to mutating cells to the point that it causes cancer:

- Exposure to UV light (usually from the sun), which results in skin cancer, the most common type of cancer
- Smoking
- Genetics

The following is the kind of question you may have on the ASVAB.

When does cellular fermentation occur?

(A) When there is no oxygen during the cellular respiration process
(B) When there is yeast during the cellular respiration process
(C) When there is lactic acid during the cellular respiration process
(D) When there is oxygen during the cellular respiration process

The answer is A, when oxygen is absent during cellular respiration.

Here's another one.

Which organelle is responsible for DNA replication?

(A) Endoplasmic reticulum
(B) Mitochondria
(C) Nucleus

(D) Cell wall

The answer is the nucleus.

Ecology

The last thing to cover in this section is Ecology, which is the study of the relationships between living things and their environment. The following are some of the most well-known and common elements in ecosystems.

- Biosphere is the places around the Earth where life occurs on its own, including some of the deeper parts of the Earth's crust and up to the air.
- Biome is a major zone where many species life together with similar life and climate.
- Community a collection of species within a set region.
- Consumers are creatures that eat other organisms, and are also called heterotrophs. There are three classifications of consumers:
    o Primary consumers, better known as herbivores, largely live on eating producers (plants). Rabbits, cows, and grasshoppers are primary consumers.
    o Secondary producers, better known as carnivores or predators, and they mostly eat primary consumers. Foxes, owls, and snakes are secondary producers. Omnivores are also considered secondary producers, such as bears, rats, and otters.
    o Tertiary consumers, often called top carnivores, can eat secondary consumers. Often tertiary consumers are omnivores, such as bears, wolves, sharks, lions, and people.
- Decomposers are mostly fungi and bacteria, and they help to break down matter, like dead plants and animals, and then release minerals to encourage future growth.
- Ecosystem is a system of animals, plants, organisms, and non-living things making up a community. They may be small or large, clean or developed.
- Population is a collection many of the same species living within a set area.
- Producers are largely plants as it refers to organisms that can make their own energy through photosynthesis.
- Scavengers are creatures that are similar to decomposers because they help with breaking down organic matter. Many insects are scavengers, but some larger animals are, with vultures being the most well-known for it.

Food chains are an example of how ecosystems are structures, particularly with the different types of consumers.

The following is the type of question you may see on ASVAB about ecology.

What is the term for the places where life exists on Earth?

(A) Population
(B) Biome
(C) Producers
(D) Biosphere

The answer is biosphere.

Practicing What You've Learned

Here are 10 questions about life sciences.

1. What is the fundamental unit of heredity?

(A) Chromosome
(B) Gene
(C) Nucleotide
(D) Protein

2. Which organelle is responsible for energy production in the cell?

(A) Nucleus
(B) Ribosome
(C) Mitochondrion
(D) Lysosome

3. Which of the following is an example of a biotic factor in an ecosystem?

(A) Temperature
(B) Water
(C) Soil pH
(D) Plants

4. What term describes the process by which organisms better adapted to their environment tend to survive and produce more offspring?

(A) Genetic Drift
(B) Gene Flow
(C) Mutation
(D) Natural Selection

5. In Mendelian genetics, what term describes an allele that masks the effect of another allele?

(A) Recessive
(B) Dominant
(C) Co-dominant
(D) Incomplete dominant

6. What is the scientific name for humans?

(A) Homo sapiens
(B) Pan troglodytes
(C) Gorilla gorilla
(D) Canis lupus

7. Which taxonomic rank is directly below "Class" in the hierarchy of biological classification?

(A) Order
(B) Family
(C) Genus
(D) Species

8. Which kingdom includes multicellular, photosynthetic organisms?

(A) Animalia
(B) Fungi
(C) Plantae
(D) Protista

9. What is the main pigment involved in photosynthesis?

(A) Carotene
(B) Xanthophyll
(C) Chlorophyll
(D) Anthocyanin

10. Which system in the human body is responsible for transporting nutrients, gases, and wastes to and from cells?

(A) Digestive System
(B) Respiratory System
(C) Circulatory System
(D) Nervous System

Checking Your Answers

1. (B) Gene

2. (C) Mitochondrion

3. (D) Plants

4. (D) Natural Selection

5. (B) Dominant

6. (A) Homo sapiens

7. (A) Order

8. (C) Plantae

9. (C) Chlorophyll

10. (C) Circulatory System

EARTH AND SPACE

The next general science section that you are likely to see on the ASVAB is about Earth and space. As the name suggests, it covers everything about the Earth and what we know about the universe. They won't expect you to have a lot of deep information, but you should know the basics of these sciences. Fortunately, the kind of stuff you are likely to be asked are the kinds of questions you asked when you were young or were taught in school.

Geology

Geology is the study of the Earth, particularly its composition and history. Rocks are the best way to determine the history of the planet. Our planet has a very rigid structure that makes it so that scientists can study different parts to learn more about what has happened over Earth's history. There are three layers:

- The crust is only about 1% of the Earth's solid layers. It is much thicker in some places than others, with the thinnest being about 10 kilometers, and the thickest parts being around 100 kilometers.
- The mantle is the second layer from the surface. It makes up the majority of the Earth at about 75%. It is about 3,000 kilometers thick and is largely made of metals like calcium, iron, and magnesium. It is considerably denser and hotter than the crust.
- The core is at the center of the Earth, and it is about twice as dense as the second layer. It is made mostly of iron-nickel alloy. It is roughly 2,200 kilometers thick and is largely liquid across the outer core and 1,300 kilometers thick in the inner core.

The middle of the Earth is hot, between 3,000 and 4,000 degrees Celsius. The reason that the heat doesn't escape is because of the layers of mantal and crust. Most of the crust and the top part of the mantle is made of plates, roughly 30 of them. When they move, it causes earthquakes. The plates actually move all of the time, and this is why the continents have formed and literally drifted apart from each other. It is thought that the land was all one large mass called Pangaea.

The plates meet at fault lines. When these run into each other, it causes a lot of action on the surface, with earthquakes and tidal waves being some of the most devastating. Earthquakes are measured by the Richter scale, which starts at 1 and reached

10. The higher the number, the stronger the effects on the surface. Each new number is 10 times stronger than the previous level.

On the surface, there are three primary types of rocks classed by their formation.

- Igneous rocks are mad of molten lave cooling; it is molten rock, better known as magma, until it reaches the surface when it is called lava.
  Examples of igneous rocks include basalt, granite, obsidian, and pumice
- Sedimentary rocks are made of small deposits of small pieces of other materials, known as sediments. These materials are usually made of a number of different things like clay and rock. The small pieces of material form into a single substance over long periods of time. The most well-known types of sedimentary rocks surround fossils because the rocks formed around the deceased organisms.
  Examples of the sedimentary rock include coal, dolomite, gypsum, sandstone, and shale.
- Metamorphic rocks are made of rocks that are changed through chemicals, pressure, or temperatures.
  Examples of metamorphic rocks include gneiss, marble, quartzite, and slate.

Scientists have speculated about the planet's history based on studying rocks and fossils, mostly found in sedimentary rock. According to scientific calculation based on this type of rock, the Earth has been around for about 4.6 billion years. Much of the rock lacks fossils, indicating that there wasn't large life on the Earth's surface, including plants. It wasn't until about 570 million years ago that fossils began to show up, and this part of the Earth's history is called the Precambrian eon, or the time before fossils kept a record of Earth's history. It's similar to how most of human history is undocumented because humans did not record history over much of its existence. These layers are not entirely lacking in fossils though. Scientists who have studied this area have find life that is over 3.5 billion years old. However, scientists aren't able to identify what the animals and plants are.

The following are the kinds of questions you will probably have on the test.

Scientists have identified that life appeared roughly how long ago on Earth?

- (A) 4.6 billion years ago
- (B) 5 billion years ago
- (C) 3.5 years ago
- (D) 570 million years ago

The answer is 3.5 billion years ago, even if they aren't sure how to classify life that far back in history.

Which is a type of metamorphic rock?

- (A) Shale
- (B) Granite
- (C) Coal
- (D) Slate

The answer is slate.

Cycles

Earth has a lot of cycles, including biogeochemical cycles. Probably the most well-known is the water cycle (hydrologic cycle), but less well-known is the carbon cycle.

The water cycle explains how water moves through multiple states (gas, liquid, solid) as it goes from the atmosphere to the Earth as rain, then into bodies of water. The exception is when water is frozen, like in the ice caps and icebergs. This water can melt, releasing liquid water into the surrounding waters or landscape. When heated up by the environment, the water becomes a gas and returns to the atmosphere. The process of water turning into gas is known as evaporation. Water can evaporate from living creatures, and often does from plants through photosynthesis. It can also evaporate from animals, as you've probably experienced when you sweat – the water evaporates from your skin and returns to the atmosphere, leaving salt on your skin and clothing.

Once back in the atmosphere, water creates clouds. As more water accumulates in the clouds, it becomes too heavy, so it releases the water as precipitation, whether that is rain, sleet, or snow. The process begins again as the snow is absorbed by the Earth or runs into other bodies of water. Snow collects on mountains and land. Over time, it melts, and the water flows underground or into streams and creeks until it reaches rivers and other bodies of water. When water is absorbed into the Earth is called infiltration. It collects underground in an underground water table. In some places with little water on the surface, water can be found underground. Places that have wells drill into the ground, they are tapping into these underground water tables.

Carbon follows a similar cycle, but is also far more prevalent across the Earth since this gas is what helps balance most of our ecosystems. The exchange of carbon is what keeps our atmosphere in a state where organisms can breathe. Most living creatures release carbon, with humans being particularly efficient at producing the gas. However, most living creatures generate carbon through respiration, or breathing. When an animal exhales, they release carbon dioxide into the air. Dead animals and plants also release carbon dioxide. Once released into the air, carbon moves into the atmosphere, where plants then take it in to complete the photosynthesis process. The soil also absorbs some of the carbon.

The following are a couple of examples of the kinds of questions you may have about these cycles.

Which is one of the ways carbon can be released into the air?

- (A) When oak trees photosynthesize
- (B) Decomposing ferns
- (C) A growing sapling
- (D) Sediment settling in a river

Plants produce oxygen, not carbon, so that rules out A and Ce. Sediment doesn't create carbon, leaving just B. Everything that decomposes creates carbon, even plants.

What is t called when water is absorbed into the soil?

- (A) Evaporation
- (B) Respiration
- (C) Condensation
- (D) Infiltration

The answer is infiltration, D.

Meteorology

Meteorology is the study of the atmosphere, and it includes understanding the weather. The Earth's atmosphere is divided into several layers.

- The troposphere is the first layer is the one that we experience on the surface. This is where weather occurs, largely because of the air pockets that rise and fall over time. Different seasons and regions (based on latitude) affect the thickness of the atmosphere, with the thinnest being 6 kilometers and the thickest being 17 kilometers. About 21% of the troposphere is made of oxygen, and most of the remaining 79% is nitrogen. As a result, the temperature varies wildly from the frozen polar regions that are well below freezing to the tropical regions that have a pretty steady temperature to the deserts where the temperatures can be too hot to sustain life.
- The stratosphere is the next layer, and air moves largely horizontally in this region. At the top of the stratosphere is the ozone, and it is particularly reactive type of oxygen. The ozone absorbs a lot of the ultraviolet radiation that reaches the planet from the sun. The temperature in this layer is around -60 degrees Celsius.
- The mesosphere is the next layer and it is about 90 kilometers above the surface. From the surface, we see this layer when space debris enters the atmosphere and burns up, a phenomenon we call falling stars. As the debris enters Earth's atmosphere, it burns up. The temperature is significantly lower at about -90 degrees Celsius.
- The thermosphere is the top layer. Unlike the rest of the layers, it is extremely hot because there is no protection from the radiation from the sun. As a result, temperatures in this layer have been recorded at up to 2,000 degree Celsius.

The image on the right shows the layers of the Earth's atmosphere.

When there is a different in air pressure in the atmosphere, it results in wind. This moves the air masses with different temperatures together. Warm fronts are made when cold air are overtaken by warm air. The warmer air moves over the cold air because it is lighter. With the rise of the warm air, vapor condenses, making the lower clouds release the water. This can create one of four types of precipitation, freezing rain, rain, sleet, or snow. In some cases, it can release several types of precipitation, and in many cases, it will release all types of precipitation.

Cold front are created when cold air overtakes a warm mass of air. As the cold start to arrive, it often makes the clouds release precipitation, so rain usually happens as a cold front is coming. Once the cold air arrives though,

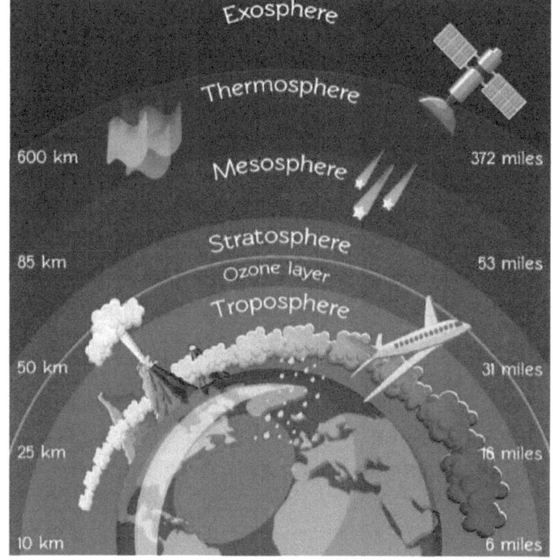

it usually produces a drop in temperature and wing, not any kind of precipitation.

If two air masses meet without displacing each other, they push together to create a stationary front. This usually is accompanied by clouds and precipitation that will last for about a week, but can go for longer as the two masses are at a stalemate.

Although clouds can come in many different shapes and sizes, there are three primary types of clouds.

- Stratus clouds are the clouds that occur closer to the ground. They are flat and broad, creating a kind of blanket. In certain environments, they can get so low that they are on the ground, creating fog. When stratus clouds are dark, it is an indication that they will be releasing their water soon, so you can expect rain.
- Cumulous clous are the large clouds that look very fluffy on top, like cotton candy, but they have a flat bottom. When they appear dark, you know that it will almost certainly bring heavy rains with it.
- Cirrus clouds are the highest, and they appear very wispy, almost see through. They can go as high as 20,000 feet over the surface.

The following are the kinds of questions you will likely see meteorology.

If a weather report says that a warm from is coming, what is the most likely result when it arrives?

(A) A drastic temperature drop
(B) The formation of a hurricane
(C) A week of rain
(D) Freezing rain

The most likely result is freezing rain as the warm mass moves over the cold mass.

You look up to see dark, puffy clouds and realize a thunderstorm is coming. What type of clouds do you see?

(A) Stratus
(B) Cumulous
(C) Air mass
(D) Cirrus

The answer is cumulus, or B, and they are often called thunderclouds because they are what you see when there is a thunderstorm coming.

The Solar System

We live in a solar system that has one start, eight planets, thousands of asteroids, and thousands of comets.

The start in our solar system is called the sun, and scientists have classified it as a G2V star, also known as a yellow dwarf (as indicated by the V in the name). On the surface, they are about 6,000 degrees Celsius. They are yellow and are usually made of calcium, iron, magnesium and other metals. Our sun in an estimated 4.7 billion years old, which is not much older than the estimated age of Earth.

Our sun may be smaller than a lot of stars, but it makes up about 99.9% of our solar system's mass. It is a large superheated plasma ball that is hot because of continuous atomic reactions at its center. At the core, scientists estimate it to be about 15,000,000 degrees Celsius. On the surface it is between 4,000 and 15,000 degree Celsius. The sun's diameter is about 1.4 kilometers, which is over 100 times larger than the Earth's diameter, and the sun's surface is roughly 12,000 times larger than Earth's surface.

The eight planets can be divided into two groups.

- The four planets nearest the sun are made largely of rocks. Mercury, Venus, Earth, and Mars are those planets, arranged from closest to furthest from the sun. They are called terrestrial because they have mostly the same make up. At their core, they have metal, and on the surface they have rocks. Only Mars and Earth have their own moons. Earth's moon is larger than any of Mars's moon. Earth is the largest of these four planets.
- The other four planets are called gas giants and they are out past Mars, which is why they are also called the outer planets. In order from Mars, they are Jupiter, Saturn, Uranus, and Neptune. These plants have rings and moons, with Saturn having the most number of rings that are made largely of ice crystals.

Pluto is out past the Neptune is Pluto, which was once classified as a planet, but has been recategorized as it does not meet the criteria to be either type of planet.

There is a mnemonic to remember the order of the planets, My Very Educated Mother Just Serve Us Nachos. Notice that Pluto has been removed because it is not a planet.

Many of the other things that revolve around the sun are either asteroids or comets. When asteroids and comets lose material, the waste becomes meteoroids, and those small pieces may strike other parts of the solar system. When they enter the Earth's atmosphere, the tend to burn up, making falling stars as the material burns up upon entry. If any part of the mater strikes the planet, it is called a meteorite.

There is a group of asteroids located between the last terrestrial planet, Mars, and the first gas giant, Jupiter. This group of asteroids is called a belt. There is a much larger collection of asteroids past Neptune that is called the Kuiper Belt. It is thought that this is the left over material from when the solar system formed.

Comets are like dirty snowballs because they are largely made of ice (a mix of frozen gases and frozen water) and space dust that was not pulled into the formation of any of the planets when they were forming. They are only visible when they get close to the sun because they begin to melting, creating long tails that cane be a few hundred million kilometers in length. They have a predictable trajectory around the sun, just like the planets.

The Earth's moon is probably the other celestial body that humans have studied in depth. The Earth keeps the moon in orbit because of gravity. The gravitation pull of the moon also effects Earth by creating tides. Twice a day, the water goes farther up the banks, and this time is called high tide. When the tide is "out" or farther from the land, it is low tide. High tide is caused by the moon being at the closest and farthest points, so it occurs twice a day. It is thought that life on land occurred because of the tide since it changed the environment around life in the sea, and those creatures adapted to live outside of the water during those times. Over time, life moved away from water.

A year on Earth is one revolution around the sun, and this is called orbiting. Earth is tilted a little, so the north part of the Earth actually points toward or away from the sun during different parts of the orbit. The seasons are different between the northern part of the planet and the southern part, with winter in the north being summer in the south, and winter in the south being summer in the north. This means that the other two season are also opposite to each other since the poles are not pointing toward or away from the sun.

The following images shows how the position of Earth in relation to the sun affects the seasons.

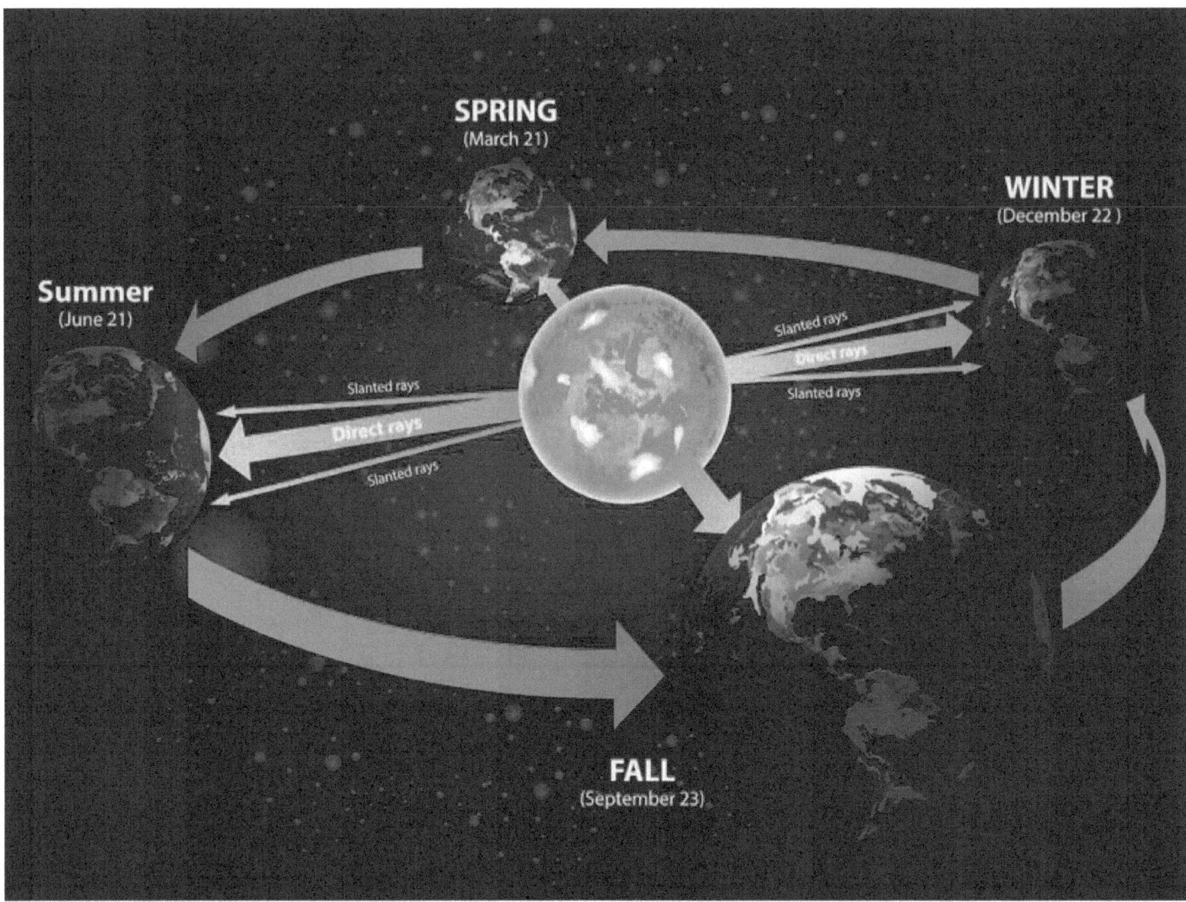

Eclipses occur when the moon passes in front of the sun or the Earth passes in front of the moon. There are two types of eclipses

- A solar eclipse is when the moon comes between the Earth and the sun, obscuring the sun during the day (or when it should be light outside). The moon cases shadows on the Earth during this time. This does not last long, and the sun will appear again soon. Since the Earth is tilted, usually only a small part of the world in one hemisphere will see the eclipse.
- A lunar eclipse is when the Earth comes between the moon and the sun. This can be seen from anywhere as long as the moon is currently over the horizon in the region. They can also be partial or total, meaning only part of the moon or all of the moon is covered by the Earth's shadow as it comes between the moon and the sun. They do last longer since the Earth's shadow is much bigger than the moons.

The following are a couple of examples of questions about the solar system.

What is one of the characteristics of a terrestrial planet?

- (A) It has made of ice crystals.
- (B) It has a temperature like Earth's.
- (C) The surface is made of a mix of gases.
- (D) The core is metal.

The answer is D, they have a metal core.

Here's another one.

What is another name for a meteor that enters Earths' atmosphere?

- (A) Falling star
- (B) Comet
- (C) Planet
- (D) Asteroid

The answer is A, fallen star.

Practicing What You've Learned

Here are 10 questions about Earth and space science.

1. What is the most abundant gas in Earth's atmosphere?

   (A) Oxygen
   (B) Nitrogen
   (C) Carbon Dioxide
   (D) Argon

2. Which layer of the Earth is liquid?

   (A) Crust
   (B) Mantle
   (C) Outer Core
   (D) Inner Core

3. What is the term for the surface of the Earth?

   (A) Crust
   (B) Core
   (C) Mantle
   (D) Atmosphere

4. Which type of rock is formed from cooling magma?

   (A) Sedimentary
   (B) Metamorphic
   (C) Igneous
   (D) Fossiliferous

5. What is the name of the supercontinent that existed around 300 million years ago?

   (A) Gondwana
   (B) Laurasia
   (C) Pangaea
   (D) Rodinia

6. What causes Earth's magnetic field?

   (A) The rotation of the crust
   (B) Movements in the mantle
   (C) Electric currents in the outer core
   (D) Sun's influence

7. Which ocean is the largest?

   (A) Atlantic Ocean
   (B) Indian Ocean
   (C) Arctic Ocean
   (D) Pacific Ocean

8. What is the process by which water vapor turns into liquid water?

   (A) Evaporation
   (B) Condensation
   (C) Sublimation
   (D) Precipitation

9. What is the primary cause of earthquakes?

   (A) Volcanic eruptions
   (B) Tsunamis
   (C) Plate tectonics
   (D) Wind erosion

10. What is the largest planet in our solar system?

    (A) Earth
    (B) Saturn
    (C) Jupiter
    (D) Neptune

Checking Your Answers

(B) Nitrogen

(C) Outer Core

(A) Crust

(C) Igneous

(C) Pangaea

(C) Electric currents in the outer core

(D) Pacific Ocean

(B) Condensation

(C) Plate tectonics

(C) Jupiter.

PHYSICS AND CHEMISTRY

The last general science questions are about physics and chemistry. Don't worry, it's not the advanced stuff that you would get in an AP class. It's more about the basics, including a good bit of math and measurements.

Measurements

All branches of science use the Metric System. You have seen the use of it often in this chapter (with a few exceptions), because scientists don't use feet, yards, miles, inches, or anything from the British measurement system.

The metric system is based on a base unit for any king of measurement. The Metric System has a set unit for all types of measurements (meters, liters, etc.). Everything is measured in relation to that unit, and when the item is substantially larger or smaller, the unit indicates that through a prefix. For example, a centimeter is 100 times smaller than a meter (something you may recognize if you have been studying prefixes from one of the earlier chapters0. The meter is the base unit to measure something length.

The following table shows the relationship between the base unit and the different prefixes that indicate how much larger or smaller something is.

Prefix	Abbreviation	How Much Larger Or Smaller from the Base Unit
mega	M	1,000,000 (or 10^6)
kilo	k	1,000 (or 10^3)
hecto	h	100 (or 10^2)
deka	da	10
base (no prefix)	—	1
deci	d	0.1 (or 10^{-1})
centi	c	0.01 (or 10^{-2})
milli	m	0.001 (or 10^{-3})
micro	μ	0.000001 (or 10^{-6})

The prefixes you will most likely encounter are kilo-, centi-, and milli-, but you could see any of them, so it is best to study all of the prefixes (it will help you in both science and vocabulary, so it is something that could be doubly beneficial).

The following table provides the base units for many of the types of measurements you are likely to see.

Base	Abbreviation	What it Measures
meter	m	Length
grams	g	Mass, such as the amount of nutrients or food packaging
cubic base	cb	Volume of three-dimensional objects; it is measured as cubic of the base measurement. For example, you can have a cubic meter or cubic liter. When abbreviated, the prefix is added to the base measurement, not the term cubic.
seconds	s	Time, with smaller units getting prefixes. Larger units do use minutes and hours instead of the prefixes.
degrees Celsius	C	Temperature; water freezes at 0 degrees Celsius and it boils at 100 degrees Celsius.
Kelvin	K	Temperature, measuring against absolute zero

Since you are likely more familiar with Fahrenheit to measure temperature, you can use the following equation to convert measurements between the two measurement systems.

$$F = \frac{9}{5}C + 32$$

$$C = \frac{5}{9}F - 32$$

When temperatures are measure in Kelvin, it is based on the item's state of rest, so it is not making heat or moving. When something is 0 K, that means that it is -273 degrees C. Kelvin is not measured in degrees like the other two temperature standards.

There are many other measurements, but those are the ones you are most likely to see on the ASVAB.

Here's an example of the kinds of problems you may see on the test.

During an experiment, the procedure calls for the addition of 50 mL of water at room temperature. How many liters is that?

    (A) 50,000 L

    (B) 50 L

(C) 5 L
(D) 0.05 L

Milliliters means that there is less and it is a 1,000 times smaller, or move the decimal 3 places. So 50 mL is 0.o5 L, or D.

When you are working on a chemistry experiment, the instructions ask you to add0.13 kg of sodium chloride. All of your tools measure in grams. Convert the requested measurement into grams.

(A) 130 g
(B) 13 g
(C) 1.3 g
(D) 0.0013 g

Kilograms is larger than grams, so you know the number should be larger, not smaller (eliminates D). The measurement is 1,000 times larger than the base, so move the decimal three places to the right. The answer is A, 130 g.

Physics

Physics is the study of the world based on changes, properties, and how mass and energy interact. There a many areas that fall under this umbrella, including the following:

-   Electricity
-   Magnetism
-   Mechanics
-   Optics
-   Thermodynamics

If you are interested in any military role that falls into one of the categories, you will want to spend some extra time studying physics to do well on the test.

Mass and weight may seem like the same thing when used in everyday speech, but they aren't even remotely same thing when discussed as concepts in physics.

-   Mass refers to how much matter a thing contains.
-   Weight is much force gravity places on an object's mass.

The best way to think about this is when you consider that people are weightless when they are in space, but they still have the same mass.

Understanding Motion

Velocity refers to how fast something is displaced, or changes position. The following is the formula to determine velocity:

$$V = \frac{d}{t}$$

Velocity is a vector quantity, or a term that described direction and magnitude. If a motorcycle travels west, covering 15 m every two seconds is traveling at a velocity of 7.5 m/s. Displacement can also be a vector, and it has an arrow over it to indicate movement. Time, however, is not a vector.

Momentum is the quantity of motion for an object that is measured as mass plus velocity. The faster an object moves, the harder it is to stop. The following is the formula to determine the momentum of an object:

$$p = mv$$

For example, if a bike is moving at 5 km/h it has more momentum than a person jogging at that speed because it has more mass. Momentum is another vector quantity, so if two objects are moving toward each other, they will start to cancel each other out if they hit each other.

Acceleration is the speed of an object in a specific direction. The following is the formula to determine the velocity:

$$a = \frac{v}{t}$$

You can actually see acceleration when you are in a car. As the driver removes their foot from the break and presses the gas, the speedometer shows the acceleration of the car, and velocity increasingly goes up until the driver stops increasing the acceleration. Also, acceleration is another vector quantity.

Let's take a look at a few problems you may see regarding these concepts – and be aware they will be word problems.

A luxury car brakes, changing the velocity from 80 m/s to 40 m/s in 5 seconds. What is the average acceleration rate?

    (A) 13 m/s
    (B) 8 m/s
    (C) 5 m/s
    (D) -8 m/s

Use the velocity equation:

$$V = \frac{d}{t} = 40\text{-}80 = -40 \text{ m/s}$$

Plug this into the acceleration formula:

$$a = \frac{v}{t} = \frac{-40}{5} = -8 \text{ m/s}^2$$

In this case, the car is changing direction because it is breaking. That is why it has a negative value. The answer is D.

A cyclist reaches the top of a hill and rests. She then begins coasting down the hill, accelerating at 7.2 m/s². If the cyclist is halfway down the hill after 6.5 seconds, what is the bike's velocity?

    (A) 82.8 m/s, uphill
    (B) 76.8 m/s, uphill
    (C) 7.2 m/s, uphill
    (D) 76.8 m/s, downhill

Start with the acceleration equation, plugging in the information you have:

$$a = \frac{v}{t} = 7.2 = \frac{v}{6.5} = 7.2 * 6.5 = 76.8 \text{ m/s, downhill}$$

The velocity 76.8 m/s downhill, so D is the correct answer.

Understanding Forces and Energy

Force occurs when an object is either pushed or pulled to change either the objects speed or direction. Weight is a type of force and is a result of gravity. Force is measured in newtons (N), named after the man who first theorized about gravity, Isaac Newton. A newton is how much force is needed to accelerate 1 meter per second squared for each mass of 1 kilogram.

Work is required to move an object, and work is an applied force in the same direction of the movement. There is a formula to calculate work:

$$W = force * displacement$$

Work is measured in units called newton-meter, or joule, represented as J. Work uses energy, something you probably now because when you work, you are tired at the end of it. If you look at food packages, you may see some include details about how much energy per serving is provided, and the numerical value is followed by kJ.

Power is the rate that work is performed, or how much energy is converted. It is expressed by the formula power = work divided by time or power = (force times distance) divided by time:

$$P = \frac{w}{t} \text{ or } P = \frac{w*d}{t}$$

Power is measured in watts (W), and a watt is 1 J per second. Be careful to pay attention to the use of W so that you don't confuse when it is used to represent work and when it represents watts. If W is used to express watts, it will follow a number as it represents units.

The following is an example question of what you could see on the ASVAB.

A space shuttle is outside of the Earth's atmosphere where it's propulsion system has a net force of 10,000 N for 50 m. The craft's potential energy does not change. What is the increase in kinetic energy when the system is engaged?

    (A) 0 J
    (B) 0.01 J
    (C) 200,000 J
    (D) 500,000 J

Use the work formula:

$$W = force * displacement \text{ (10,000 N) * (50 m) = 500,000 J}$$

The answer is D.

Looking at Newton's Laws

As mentioned earlier, Sir Isac Newton was an important figure in Physics because he studied the subject and discovered and theorized much of what we know about it today. As a mathematician, he used a lot of math to find a reliable way of reaching answer about the world around him. He developed a lot of formulas, but he is best known for the laws he wrote in this area. The following are the three laws that are critical to know for the Physics section.

1. The Law of Inertia: The first law states: "An object at rest tends to stay at rest, and an object in motion tends to stay in motion at a constant speed in a straight line (constant velocity), unless acted upon by an unbalanced force." Friction is an example of a force that resists motion and naturally slows items on Earth, so friction can act as an unbalanced force that stops an object from staying in motion.

2. The second law states: "When dealing with an object for which all existing forces are not balanced, the acceleration of that object, as produced by the net force, is in the same direction as the net force and directly proportional to the magnitude of the net force, and is inversely proportional to the object's mass." It is easier to describe this by the formula acceleration equals the net force divided by the mass:

$$a = \frac{F}{m}$$

   The greater the mass, the more force required to move it. The result could be zero acceleration, or no change in motion. It can also be expressed in the formula net force = mass times acceleration:

$$F = ma$$

3. Law of Action and Reaction: Newton's third law states: "For every action, there is an equal and opposite reaction." So if one objects puts force on another object, the second object is exerting the same amount of force back on the first object.

These laws are commonly discussed in Physics, but equally important is the law about gravitation: "All objects in the universe attract each other with an equal force that varies directly as a product of their masses, and inversely as a square of their distance from each other. This force is known as gravity." This can be expressed with the formula:

$$\vec{F}_g = \frac{Gm_1 m_2}{r^2}$$

G is a constant that equals $6.67 * 10^{11}$ and r represents the distance between the objects that are interacting.

There is a gravitational pull between the sun and all of the planets, even Earth. With the law of universal gravitation, you know the following are true:

- If you double Earth's mass, Earth's force would double.
- If the sun's mass doubled, the force it has on the Earth also doubles.
- If there was twice the distance between the Earth and sun, the force would be four times smaller.
- The sun's force is both equal and opposite to the force exerted by the Earth, based on Newton's third law.

Let's look at a couple of problems.

A 600 g squid moves forward using a force of 21 N. The propulsion of the water jet points east. What is the squid's acceleration?

(A) 24.4 m/s2, east
(B) 35 m/s2, west
(C) 32 m/s2, east
(D) .035 m/s2, west

Use the third law.

600 g = 0.6 kg

Acceleration = $\frac{21}{0.6}$ = 35 m/s^2, going west. B is the correct answer.

Here's another one.

A kid plays with a 15 kg shopping cart in the parking lot on a rough surface, using 15 N of force. Friction resists the movement with a magnitude of 15 N. What is the effect on the shopping cart's motion?

(A) The cart speeds up at 3 m/s^2.
(B) The cart slows down at a rate of 1 m/s^2.
(C) The cart will keep a uniform velocity.
(D) The cart will slow down at a rate of 2 m/s^2.

Since friction is acting with the same force in the opposite direction, C is the right answer.

Energy

Energy is the capacity or ability to do work. On the ASVAB, a lot of questions about energy will relate to mechanical energy, which can be potential or kinetic energy.

- Potential energy is when energy is stored, whether because of the object's position, state, or shape.
- Kinetic energy is when energy is active, so the object with that energy is moving.

The law of conservation of energy states "Energy can neither be created nor destroyed – only converted from one from of energy to another." A classic example of this is if you pick up a rock and skip it across the water. That rock is resting, but you put it into motion. You transfer your energy into the rock. Another example is a rock on a ledge. It has potential energy because of the ground and gravity. If something dislodges the rock, causing the rock to fall over the edge, the potential energy becomes kinetic energy.

Light and sound are both types of energy that move in waves. They each deserve a closer look.

Understanding Sound Waves

When an object moves or vibrates, it creates sound waves that ripple away from the object in all directions. Sound waves can go through air (in which case we hear it), liquids (usually shown in actual water ripples), and solids (in which we can feel the waves). Sound does not travel in empty space or a vacuum. Sound travels slowest in air and fastest through solids, particularly metal and wood.

Pitch is directly linked to the vibration's rate (called frequency) of the sound waves. High-frequency sound waves vibrate rapidly and produce a high pitch. The unit of measurement for frequency is the hertz (Hz), and it represents the number of cycles per second. Extremely high frequencies are inaudible to humans, but are audible to other animals like dogs. Most humans hear sounds that have a frequency between 20 Hz to 20 kHz. Occasionally, a person (or other creature) may perceive a sound outside of their usual auditory frequency than its actual frequency because of something called the Doppler effect. The Doppler effect is when the source of a sound, the creature hearing it, or both are moving toward each other, making the pitch of the sound seem higher or lower than it is. The most frequently cited example of this is an emergency vehicle with its sirens going sounding different when it is heading toward you compared to when it reaches you, and then again it sounds different when it is moving away from you.

Understanding Light Waves

Light is much more complex because what people can see is only a small part of the light spectrum, called the electromagnetic spectrum. There is a vast range of radiation wavelengths and frequencies in the electromagnetic spectrum. What we call visible light waves appear in the middle of the spectrum. There are seven areas along the electromagnetic spectrum:

- Radio has the longest wave length
- Microwave
- Infrared
- Visible
- Ultraviolet
- X-ray
- Gamma

The following image shows the different wave lengths and other details about the electromagnetic wavelength.

ELECTROMAGNETIC SPECTRUM

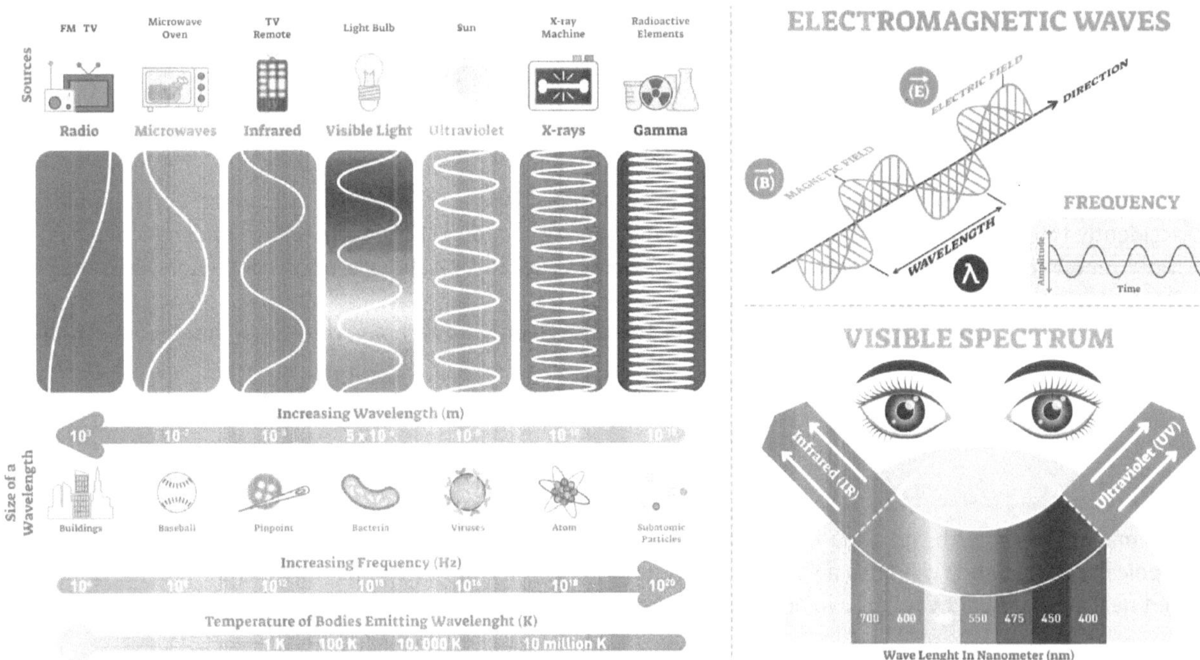

Visible light comes in many colors, which are determined by the frequency of the waves. At the law end of the spectrum is red. As you can see in the image everything to the left of red is absent of color, and that is why it is called infrared. At the other end with the highest frequency is violet, and wavelengths that are shorter and beyond this are called ultraviolet.

Light has wave properties, but it is very different compared to sound. Unlike sound, light can travel through empty space. They also travel much faster than sound, reaching speeds in a vacuum of 299,792,458 m/s.

Refraction is the changing of light waves when it passes through something. The object bens the light at different angles, separating it. The best example of this is a diamond. When you shine light on a diamond, it refracts the light, showing a rainbow of color.

Reflection is caused when light bounces off a surface, usually smooth, flat one that acts as a barrier. The law of reflection sates: "The angle of incidence is equal to the angle of reflection as measured from a line normal to the barrier." The most obvious example of this is a mirror, but there are other things that can reflect light, such as bodies of water.

You use mirrors and lenses regularly, so you know that they aren't necessarily flat.

- Concave surfaces curve in, similar to a cave. Converging mirrors are concave, and their name comes form the many incidental angles of light.
- Convex surfaces curve out, similar to a wave. Diverging mirrors are convex because it spreads light wave out once the light hits it.

If you move the mirror away from the image source, the image blurs as it approaches the mirror's focal point, where the angles of incidence converge. Once the image source moves past the focal point, the image reappears in the mirror but inverted.

Unlike mirrors, lenses function based on refraction. Converging lenses are made of convex lenses, which are thicker in the center than the borders. This is because they combine parallel light waves. This type of lens is used in reading glasses to correct farsightedness, as well as in magnifying glasses, cameras, telescopes, and microscopes.

Conversely, a concave lens, thicker at the edges than in the middle, is known as a diverging lens because it spreads out the light waves that pass through it. In nearsightedness, light waves converge before they reach the retina. A nearsighted person can see nearby objects clearly but not distant ones. Placing a concave lens in front of the eyes of someone who is nearsighted helps fix the problem because the glasses bend light so that it converges further back on the eye reaching the retina.

The following is the kind of question you may see on the ASVAB.

What is the medium where light travels fastest?

(A) air

(B) solids
(C) liquid
(D) A vacuum

The answer is D, a vacuum.

Heat

Heat always moves from warmer to cooler environments. Heat energy can move from one object to another through one of three means:

- Conduction is the simplest way of transferring heat, and it happens through direct contact. For example when you accidently touch a hot eye or the oven right after it's been used, you get burned because of conductive heat transfer. Most metals are good heat conductors. Materials like wood and plastic are poor conductors, but they are good insulators.
- Convection is the transfer of heat through a fluid. The heat given off by a fire is convection heat. Convection heat also affects the wind and ocean currents.
- Radiation is heat transfer through electromagnetic waves. The sun's heat moves through space as radiation.

Magnetism

Every simple magnet has a north and a south pole. Opposite ends of magnets attract, and same sides repulse. If you try to bring two north poles together, they will push away from each other, and you can see this as the magnets will move until the two sides are not near each other. If you put two opposite poles near each other, the magnets will move together, usually until they touch.

The Earth is magnetized with a North Pole and a South Pole, so you can use a magnetic compass to determine direction. A compass has a needle (small, lightweight magnet) that rests on a nearly frictionless, nonmagnetic surface. The end marked "N" points toward the magnetic south of the Earth, which constantly points towards the Earth's North Pole, helping the user find their bearings.

The following is an example of the kind of questions you may see about these subjects on the ASVAB.

When the warm front arrived, a the mass of hot air is an example of which of the following?

(A) The workings of the Earth's magnetic field
(B) Conduction
(C) Convection
(D) Radiation

The answer is convection, C.

Chemistry

Chemistry is the study of properties and composition of matter, as well as the reactions they have to each other. The definition of matter is anything with mass and takes up space. The chemical properties of matter is based on the molecular composition. There are a lot basic terms, but we'll look at the ones that are likely to be important for the test.

- Element: A pure substance that cannot be broken down into different types of matter through ordinary chemical means. Everything is composed of elements, which are listed on the Periodic Table of Elements.
- Molecule: The basic multi-atom particle that exists on its own and still has all of the compounds and characteristics of the element or compound. Elemental molecules have more than one similar atoms.
- Atom: The smallest elemental unit that has the element's properties. Compounds are made by atoms combining with each other. Each atom consists of a complex arrangement of electrons orbiting a positively charged nucleus, which contains protons and neutrons. The exception is hydrogen, which does not have any neutrons.
- Proton: A subatomic component located in every atom's nucleus, and it has a positive charge. There are a different number of protons in different elements.
- Neutron: A subatomic component located in every atom's nucleus, and it has no charge, making it a neutral part of the atom.
- Electron: A subatomic component located outside the nucleus in orbits around it, and it has a negative charge. They are tiny compared to the other parts of an atom.

Atoms usually have the same number of electrons as protons, so that the atoms have no charge to them.

The following is an image of how atoms are structured.

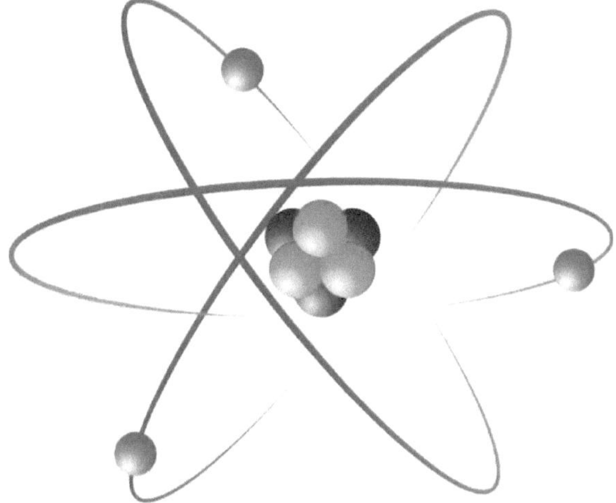

Source: https://img.freepik.com/free-vector/diagram-atom-structure_1308-97176.jpg

Scientists have spent centuries coming up with a chart of all of the known elements and their molecular structure, resulting in the Periodic Table.

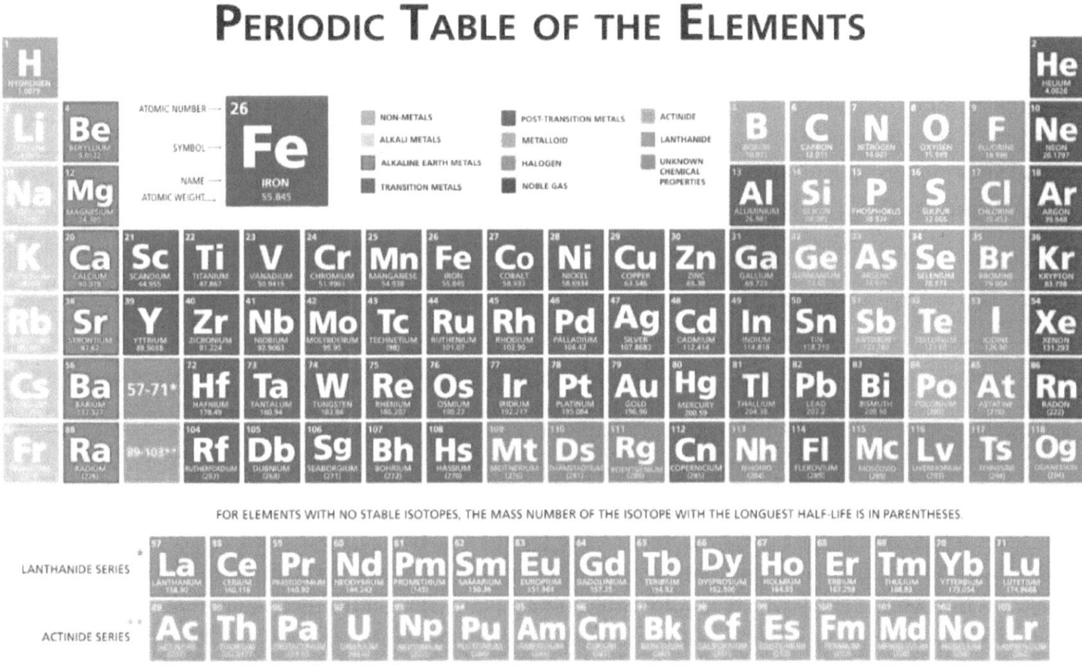

The Periodic Table of Elements classifies all known chemical elements according to their atomic numbers, electron configurations, and recurrent chemical properties. Each element is represented by a unique symbol (usually based on their Latin names) and is placed in a specific location that provides a number of details about the element. It is very regimentally structured to make it easy to understand a lot about the element with just a quick look at its location on the table.

The table is structured in periods (rows) and families (columns).

- Periods make up the seven horizontal rows. Each period represents a new principal energy level of electrons. Moving left to right across a period, the atomic number of the elements increases, meaning each subsequent element has one more proton and is generally heavier than the previous one.
- Families make up the 18 vertical columns. Each family has elements that share similar chemical and physical properties because they have the same number of electrons. For example, Group 1 elements make up the alkali metals and are highly reactive, especially with water. They have one electron on the outer orbit. The next family, Group II, every element has two electrons in the outer orbit. The final family on the right is called the Noble Gas column, and they are the most stable elements, and it is incredibly difficult to get them to react with other elements.

The periodic table also has several specific categories and blocks:

- Main Group Elements: These include the s-block and p-block elements, covering Groups 1, 2, and 13 to 18. They include metals, nonmetals, and metalloids with a wide range of properties.
- Transition Metals: Groups 3 to 12 make up the transitional metals, and they are the d-block of the table. Transition metals are characterized by their ability to form variable oxidation states and colored compounds.
- Lanthanides and Actinides: The detached rows appear below the table, making up a large part of f-block. The include elements with similar properties, such as the lanthanides which are rare earth metals, and the actinides, many of which are radioactive.

In addition to the key chemical properties, an element's atomic mass is related to its position on the table. The atomic mass shown in the element's box is the average mass of a single atom of that element. This number is an average because elements can have different numbers of neutrons, which means that atoms can come in slightly different sizes. These variations are known as isotopes.

A good way to estimate an element's atomic mass is to total the number of protons and neutrons. Each proton and neutron has a mass close to one atomic mass unit, known as an amu. For example, most hydrogen atoms have just one proton, so its atomic mass is very close to one, (it's actually 1.007 amu).

Let's take a look at silicon:

Silicon has an atomic number of 14, indicating that it has 14 protons in the nucleus and 14 electrons orbiting it. It's symbol is Si, which his close to the name, making it easy to remember. It isn't like gold (AU), iron (FE), or tungsten (W) which have symbols that aren't close to their names. The atomic mass of silicon is 28.085.

The following is an example of the kinds of questions you may get on this subject.

How many neutrons are in one sulfur (S) atom?

- (A) 48
- (B) 32
- (C) 16.1
- (D) 16

The answer is A, 16.

Compounds

Compounds are several unstable elements mixed together to create something that is distinctly different from the individual elements. For example, both sodium and chlorine are unstable and noxious elements, but when you combine them, you get sodium chloride, with the symbol NaCl, which you probably eat regularly since it is better known as table salt. This is known as an ionic compound because chorine borrows sodium's electrons, so the atoms stay close together, making a very stable crystalline structure. When you introduce this compound into water, it breaks the bond, leaving the electronically charge ion.

Your table sugar is similar, but it has a covalent compound, meaning it does not ionize in water, but it does dissolve. The atoms have covalent bonds, or they shar electrons pairs, giving each atom half the electrons.

Acids and Bases

An acid is a chemical that, when dissolved in water, releases positively charged hydrogen ions (H+). Acids are known for having a sour taste and can corrode metal. Obviously, not all acids are safe to consume. Vinegar (which contains acetic acid) and lemon juice (which contains citric acid) are both safe acids to consume. The three acids that are frequently found in lead batteries—hydrochloric, nitric, and sulfuric—are quite caustic and need to be handled carefully. It's also obvious that you shouldn't eat them.

When a material dissolves in water, it releases negatively charged hydroxyl ions (OH-), which is why it is called a base. Bases are known for being bitter and may be referred to as alkaline substances. Common basic substances found in the kitchen include baking soda (sodium bicarbonate) and liquid soap (often containing potassium hydroxide). Stronger bases can be just as harmful and caustic as strong acids. Examples of such bases are sodium hydroxide, sometimes known as lye, and sodium hypochlorite, which is the main component of bleach.

When you mix acids and bases, they neutralize each other, producing water and a salt.

How acidic or basic a compound or solution is indicated by the pH, which has a scale from 0 to 14. If something has a pH that is less than 7, it is acidic, with acidity increasing tenfold for each level. For example, black coffee has a pH of 5; vinegar has a pH of 3 (making it 100 times more acidic than coffee); and battery acid, has a pH of 1 (making it 100 times more acidic than vinegar). Anything with a pH over 7 is considered basic.

If you have a solution with a pH of 7, it is considered neutral, such as pure water.

The following image depicts the pH scale:

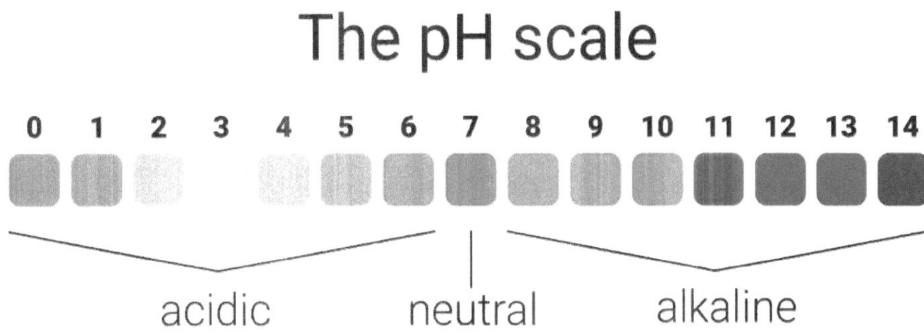

The following is an example of the kind of question you may have on this on the ASVAB>

If you have four solutions and measure the pH for all of them, which solution has the most hydrogen ions?

(A) pH 9
(B) pH 7.1
(C) pH 3.7
(D) pH 2.5

The answer is D, 2.5 because it is the most acidic of the solutions.

Physical and Chemical Changes

Matter can change both physically and chemically.

A physical change includes superficial changes, such as the form, size, or shape of matter. However, it does not undergo any molecular changes. When you build a sandcastle, you are physically changing the sand, but you aren't changing any of the materials in the sane. Making balloon animals is another type of common physical change. You've rearranged the matter, but it still consists of the same types of atoms and molecules.

A phase transition occurs when matter changes from one state, such as solid, liquid, or gas, to another. Water is still water, whether it is ice, rain, or gas.

- Solid states occur at lower temperature, compared to the liquid and gaseous states, and the atoms or molecules are packed very closely together. This means they do not move freely, resulting in solids maintaining a constant volume and shape.
- Liquid states occur at a higher temperatures than solids, but at a lower temperature compared to gases. Molecules able to move freely around each other. While liquids have a constant volume, they do not have a fixed shape. If in a contain, liquids take the same shape; if unconfined, the liquid spills out into puddles.
- Gas states occur when a liquid reaches a critical temperature that evaporates the liquid. In this state, the molecules are most free to move and spread as far as they are able to go. They have no shaper or volume, expanding to fill an enclose space. For example, you can confine oxygen to a tank, but as soon as you open it, the oxygen releases into the air around you.

Under certain conditions, solids and gases can transition directly between each other without passing through the liquid state.

- Sublimation is when a solid changes into a gas without becoming a liquid first.
- Deposition is when a gas changes into a solid without becoming a liquid first.

A chemical change occurs when new molecules are formed that differ from the original molecules. When elements and compounds undergo a chemical change, this process is known as a chemical reaction. Atoms are rearranged to create a new combinations, resulting in different molecules. The starting atoms and molecules are called reactants, and the resulting atoms and molecules are called products.

For example, water (H_2O) has two hydrogen atoms and one oxygen atom. If you have a reaction where the water molecules break apart, making the atoms rearrange to form hydrogen molecules (H_2) and oxygen molecules (O_2), water is the reactant, and pure hydrogen gas and oxygen gas are the products. When iron rusts, that's a chemical reaction, as is burning wood.

Here is an example of a question.

A small sample has a constant volume, but an amorphous shape is probably a

    (A) A compound
    (B) A gas
    (C) A liquid
    (D) A solid

The answer is C, a liquid.

Practicing What You've Learned

Here are 10 questions about Physics and Chemistry

1. What is the smallest unit of an element that retains the properties of that element?

    (A) Molecule
    (B) Proton
    (C) Atom
    (D) Electron

2. Which state of matter has a definite volume but no definite shape?

    (A) Solid
    (B) Liquid
    (C) Gas
    (D) Plasma

3. What type of bond involves the sharing of electron pairs between atoms?

    (A) Ionic bond
    (B) Hydrogen bond
    (C) Covalent bond
    (D) Metallic bond

4. What is the chemical symbol for gold?

    (A) Au
    (B) Ag
    (C) Gd
    (D) Pb

5. Which law states that energy cannot be created or destroyed, only transformed?

    (A) Law of Conservation of Mass
    (B) Law of Conservation of Energy
    (C) Newton's First Law
    (D) Law of Thermodynamics

6. Which of the following is a physical change?

    (A) Rusting of iron
    (B) Burning of wood
    (C) Melting of ice
    (D) Baking a cake

7. What term describes the amount of matter in an object?

    (A) Volume
    (B) Density
    (C) Mass
    (D) Weight

8. What is the pH of a neutral solution?

    (A) 0
    (B) 7
    (C) 14
    (D) 10

9. Which element is the most abundant in the Earth's crust?

    (A) Carbon
    (B) Iron
    (C) Silicon
    (D) Oxygen

10. What is the acceleration due to gravity on Earth's surface?

    (A) 9.8 m/s^2
    (B) 6.2 m/s^2
    (C) 12.4 m/s^2
    (D) 15.0 m/s^2

Checking Your Answers

1. (C) Atom

2. (B) Liquid

3. (C) Covalent bond

4. (A) Au

5. (B) Law of Conservation of Energy

6. (C) Melting of ice

7. (C) Mass

8. (B) 7

9. (D) Oxygen

10. (A) $9.8 \ m/s^2$

PRACTICING WHAT YOU'VE LEARNED

Now that you've got the basics, you can see test your newly acquired information and put it to the test.

Here are two practice sets with 15 questions each. This is similar to how the CAT is setup, although obviously your questions aren't going to change the better you do.

The last section of this chapter details the right answer, and how you could have reached that answer.

Practice 1

1. What is the powerhouse of the cell?

    (A) Nucleus
    (B) Ribosome
    (C) Mitochondrion
    (D) Endoplasmic reticulum

2. What molecule carries genetic information in most living organisms?

    (A) RNA
    (B) DNA
    (C) Lipid
    (D) Protein

3. What process do plants use to convert sunlight into energy?

    (A) Respiration
    (B) Digestion
    (C) Fermentation
    (D) Photosynthesis

4. Which organ is responsible for pumping blood throughout the human body?

    (A) Brain
    (B) Liver
    (C) Heart
    (D) Kidney

5. What is the basic unit of life?

    (A) Cell
    (B) Atom
    (C) Molecule
    (D) Tissue

6. What is the primary gas in Earth's atmosphere?

    (A) Oxygen
    (B) Carbon dioxide
    (C) Hydrogen
    (D) Nitrogen

7. What celestial body is at the center of our solar system?

    (A) Earth
    (B) Sun
    (D) Moon
    (D) Mars

8. What type of rock is formed from cooling lava or magma?

    (A) Igneous
    (B) Sedimentary
    (C) Metamorphic
    (D) Organic

9. What is the term for a natural satellite that orbits a planet?

    (A) Asteroid
    (B) Comet
    (C) Moon
    (D) Star

10. Which layer of the Earth is composed primarily of solid iron and nickel?

    (A) Crust
    (B) Mantle
    (C) Outer core
    (D) Inner core

11. What is the unit of force in the International System of Units (SI)?

    (A) Joule
    (B) Newton
    (C) Watt
    (D) Pascal

12. What is the speed of light in a vacuum?

    (A) 3×10^8 m/s
    (B) 3×10^6 m/s
    (C) 3×10^5 m/s
    (D) 3×10^7 m/s

13. Which law states that for every action, there is an equal and opposite reaction?

    (A) Newton's First Law
    (B) Newton's Second Law
    (C) Newton's Third Law
    (D) Law of Conservation of Energy

14. What type of energy is stored in a compressed spring?

    (A) Kinetic energy
    (B) Thermal energy
    (C) Electrical energy
    (D) Potential energy

15. Which subatomic particle has a negative charge?

    (A) Proton

(B) Neutron
(C) Electron

Practice 2

1. What organelle is responsible for producing energy in eukaryotic cells?

(A) Nucleus
(B) Golgi apparatus
(C) Mitochondria
(D) Endoplasmic reticulum

2. Which of the following is not a type of tissue in the human body?

(A) Epithelial
(B) Connective
(C) Nervous
(D) Contractile

3. What process do plants use to convert sunlight into energy?

(A) Respiration
(B) Digestion
(C) Photosynthesis
(D) Fermentation

4. What is the layer of the Earth directly beneath the crust called?

(A) Mantle
(B) Inner core
(C) Outer core
(D) Lithosphere

5. What is the name of the largest moon in the solar system, orbiting Jupiter?

(A) Titan
(B) Europa
(C) Ganymede
(D) Io

6. What is the name of the process by which water vapor changes into liquid water?

(A) Melting
(B) Condensation
(C) Evaporation
(D) Sublimation

7. What is the SI unit of force?

(A) Watt
(B) Joule
(C) Newton
(D) Volt

8. Which of the following is a measure of the resistance of an object to a change in its state of motion?

(A) Mass
(B) Weight
(C) Acceleration
(D) Velocity

(D) Photon

9. What is the name of the force that opposes the motion of objects as they move through a fluid?

(A) Friction
(B) Tension
(C) Gravity
(D) Drag

10. Which of the following is an example of potential energy?

(A) A moving car
(B) A stretched rubber band
(C) A spinning top
(D) A falling apple

11. What is the law that states that energy cannot be created or destroyed, only transformed from one form to another?

(A) Law of Conservation of Mass
(B) Law of Inertia
(C) Law of Conservation of Energy
(D) Law of Universal Gravitation

12. What is the process by which a liquid changes into a gas at its boiling point called?

(A) Sublimation
(B) Condensation
(C) Evaporation

13. What is the pH of a neutral solution?

(A) 0
(B) 7
(C) 14
(D) 10

14. What is the basic unit of heredity?

(A) Gene
(B) Chromosome
(C) Allele
(D) DNA

15. What is the chemical symbol for water?

    (A) W

(B) H2O
(C) HO
(D) WT

CHECKING YOUR ANSWERS

Now it's time to see how well you did.

Practice 1

1. (C) Mitochondrion
2. (B) DNA
3. (D) Photosynthesis
4. (C) Heart
5. (A) Cell
6. (D) Nitrogen
7. (B) Sun
8. (A) Igneous

9. (C) Moon
10. (D) Inner core
11. (B) Newton
12. (A) 3 x 10^8 m/s
13. (C) Newton's Third Law
14. (D) Potential energy
15. (C) Electron

Practice 2

1. (C) Mitochondria
2. (D) Contractile
3. (C) Photosynthesis
4. (A) Mantle
5. (C) Ganymede
6. (B) Condensation
7. (C) Newton
8. (A) Mass

9. (D) Drag
10. (B) A stretched rubber band
11. (C) Law of Conservation of Energy
12. (D) Vaporization
13. (B) 7
14. (A) Gene
15. (B) H2O

CHAPTER 10

ELECTRONICS INFORMATION

Modern technology is largely based on the discovery of electricity. Although electricity has been around forever, we've only really started to understand it over the last few centuries. Electrical and electronic devices may have different function, but they both operate because of the movement of electrons (covered in the Chemistry section of the previous chapter).

This chapter focuses on the basic principles of electricity and the components you find in many electronics.

A BIT MORE ABOUT THIS SECTION OF THE ASVAB

Circuits are among the most well-known of the components of an electrical system, so it is important to understand the basic components and their functions. In this chapter, you'll learn about those, as well as learning how to calculate details about their properties. Finally, you'll need to understand the safety devices that are used to prevent electrocution in modern-day wiring.

- The ASVAB gives you 10 minutes to finish 15 questions about Electronics Information (EI) if you take the CAT.
- If you take the traditional ASVAB, you will have 20 questions with just 9 minutes to finish. This does mean finishing more than 2 questions a minute.

Let's take a look at the kinds of knowledge you need to do well in this portion of the test.

UNDERSTANDING THE ELECTRON FLOW THEORY

You have to study electricity on an atomic level to understand it. Make sure to understand the chemistry section of the last chapter to understand atoms and their subatomic particles, particularly electrons. As a reminder, protons and neutrons are in the nucleus. Protons are positively charged and neutrons don't have a charge (they are neutral). Since it has two of the three particles, it has most of the atom's mass. The third subatomic particle is the electron (with a negative charge), and they orbit the nucleus. When an atom is neutrally charged, It means that the number of electrons is the same as the number of protons. Every manifestation of electricity, whether it's a static shock or a lightning strike, involves the movement of these electrons.

Electrons are arranged in different energy levels called shells. As one shell fills up with electrons, another shell begins. The valence shell (outermost shell) determines whether an element behaves as a conductor, semiconductor, or insulator. The number of electrons in the valence shell determines how the element behaves. Typically, the valence shell can host a maximum of eight electrons.

- A conductor is an element that has free flowing electrons. All conductors have the presence of one or more mobile valence electrons per atom, enabling them to move freely between atoms. This mobility stems from the fact that the valence shell possesses more vacancies than electrons.
- An insulator possesses a valence shell that is either more than half full or completely filled, hindering the flow of electricity. Electrons in insulators are tightly bound within their respective valence shells and do not readily leave.
- A Semiconductor has a valence shell that is exactly half full. While not excellent conductors or insulators, semiconductors possess unique properties that render them invaluable for electronic components.

A fundamental principle of electricity dictates that similar charges repel each other, while opposite charges attract (just like magnetism). A useful analogy for understanding electricity is to envision electrons as water droplets flowing through a pipe. Just as pressure applied at one end of the pipe propels the water in the opposite direction, the movement of electrons in a conductor is a result of the repulsive pressure exerted by the negative terminal of a battery. This pressure drives the electrons toward the positive terminal. The ensuing flow of electrons through a conductor constitutes an electric current.

Current is the flow rate of electrons through a conductor and you can think of it as the flow of water through a drain and pipe, the movement of electricity also exhibits a current. When you turn on a faucet, the distinction between a trickle and a deluge is based on the varying volumes of water flowing at each moment. Similarly, electrical current is quantified by the quantity of charge passing through per unit of time. Current is quantified in amperes, abbreviated as amps, with one ampere defined as one coulomb (C) of electrical charge traversing a specified point within one second. The symbol denoted for amperes is A. For instance, if five coulombs of charge pass by a point every second, this would represent a current of five amps (5 A).

The following is the kind of question you may see about this section.

Calculate the current if 18 coulombs move through a point in 3 seconds.

- (A) 3 A
- (B) 5 A
- (C) 6 A
- (D) 18 A

If you divide 18 C by 3 s, you get 6 A, so the answer is C.

Voltage and Resistance

To set electrons in motion through a conductor, you must have a force compelling their movement. Just as water flows from an place that has high pressure to somewhere with lower pressure in a pipe, if pressure remains balanced at both ends, no movement occurs. Electricity operates on a similar principle—electrical pressure must be applied to a conductor to induce electron movement.

For an electron, this pressure arises from electromagnetic repulsion generated by a significant negative charge, such as that found at the negative terminal of a battery. This terminal harbors an abundance of electrons, each exerting force against the others. When a conducting wire links the positive and negative terminals of a battery, completing the circuit, electrons are propelled from the negative terminal into the wire. They then push along the existing conducting electrons, propagating down the line until each electron eventually reaches the positive terminal. This is where a dearth of electrons and an excess of attractive protons draw the electrons inward. This process persists until the surplus of electrons at the negative terminal depletes, resulting in the battery's discharge.

Voltage, referred to as electrical pressure, is quantified in volts (denoted by the letter V). The magnitude of voltage applied to a conductor directly correlates with the electrical pressure exerted on electrons. Consequently, higher voltage leads to increased electron flow, resulting in a higher current.

Voltage is defined as the disparity in electric pressure between any two given points. Electrons are compelled to move from regions of higher electric potential (equivalent to higher pressure) to those of lower electric potential (lower pressure). Hence, voltage is also termed electrical potential difference and occasionally referred to as electromotive force. Technicians typically measure voltage using a voltmeter.

Resistance, the opposition to current flow, is quantified in ohms (Ω). An ohm is the measurement of electrical resistance. Resistance can be gauged using an ohmmeter.

As resistance in a conductor rises, current flow diminishes, indicating an inverse relationship between current and resistance. Materials with lower resistance serve as better conductors. Remember that current and voltage are directly proportional. Therefore, an increase in voltage prompts a corresponding rise in current. If a material has a higher resistance, then it needs a higher voltage to keep a steady flow.

All conductors possess some level of resistance, albeit varying. Silver, copper, and aluminum are three commonly utilized conductive materials. Among them, silver offers the lowest resistance, making it the optimal conductor. However, due to its high cost, copper emerges as the next favorable option. While copper's resistance is marginally higher than silver's, it is significantly more cost-effective. Consequently, most electrical cables and wires are crafted from copper. Conversely, aluminum exhibits higher resistance than copper and presents additional drawbacks, rendering it less desirable for electrical applications. Though aluminum was previously prevalent in residential wiring, it is now reserved for select uses.

Examining the Exceptions and Additional Notes

While the analogy between electron flow and water flow in a pipe is often illuminating, it has its limitations. For instance, when a pipe ruptures, water spurts out, yet if a conducting wire within an electronic device breaks, the current simply halts along that route. This disparity arises because electricity courses through the conducting metal of a wire, not within a hollow pipe as water does. Without a conducting medium, electrons cannot advance, and air molecules prove to be highly inadequate conductors. With exceptions such as when the gap is minuscule (a characteristic of spark plugs) or the voltage is exceptionally high (as seen in lightning strikes), an electric current will not traverse through air.

Another intriguing aspect of electrical studies is conventional current, defined by the (hypothetical) flow of positive charge and opposite in direction to actual electron flow. This peculiar notion emerged due to the challenge in directly observing electricity at a microscopic scale.

Observing and sensing the direction of water flow is relatively straightforward, but electricity presents a more intricate puzzle. In the early days of serious electrical inquiry, when knowledge about atoms was still forming, scientists tried to understand electricity moved. They speculated one of two things were true:

-    Negatively charged electron streams moved in a single direction
-    Positively charged proton streams moved in the opposite direction.

Either one of these could cause the charge dynamics inherent in electricity.

Initially, scientists hypothesized that it was the protons that underwent movement, basing the definition of current direction on this assumption. However, subsequent discoveries revealed that protons remain confined within the nucleus and do not participate in movement. Nonetheless, the direction of current had already been established using this framework. Consequently, if a circuit exhibits a conventional current coursing clockwise, it indicates that, in reality, electrons navigate the circuit counterclockwise. Presently, we can interchangeably refer to either conventional current or electron flow, but it's imperative to recalibrate the direction when transitioning between the two.

While it's common to envision electrical current as the movement of individual electrons through a circuit, in reality, a single electron may not journey across the entire circuit. Instead, the process resembles a relay race, where successive electrons pass the charge along. Hence, some sources may characterize it as the flow of a charge.

You now have enough information to answer some basic questions, so let's look at one.

Which lamp lights based on the following diagram?

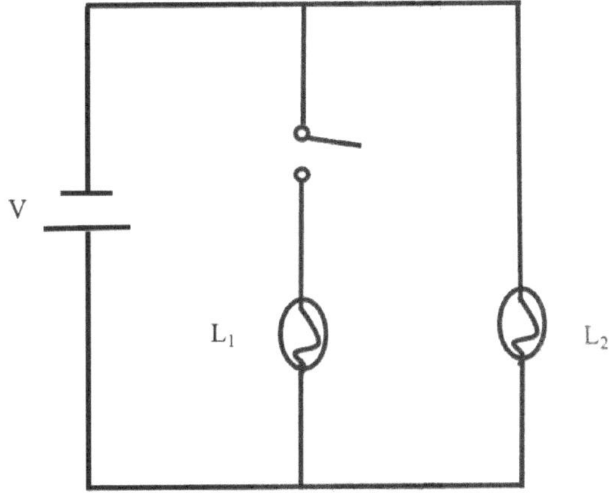

(A) L1 only
(B) L2 only
(C) Both
(D) Neither

The answer is B.

Let's try one more.

If conducting materials decrease in their resistance as the heat in the wire increases, what will happen?

- (A) The amount of current should decrease.
- (B) The amount of current should increase.
- (C) The current will be the same.
- (D) The current's direction will change.

The right answer is B, the current increases as resistance increases.

EXPLAINING CIRCUITS

A circuit is a known pathway that allows for electric current to circulate. A circuit comprises a complete loop or pathway that electricity can traverse. There are three fundamental elements constitute an electrical circuit:

- A voltage source
- A load
- Conductors linking the load to the voltage source.

When these components are interconnected to facilitate current flow, a closed (or completed) circuit is established.

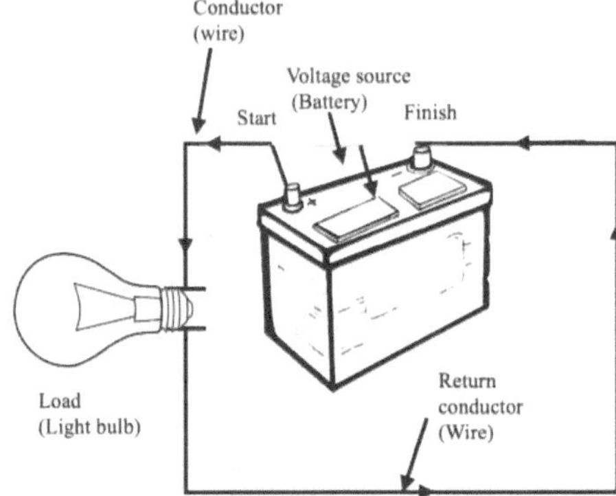

The term load refers to the conversion of electrical energy into some other form by a resistance source. For example, a light bulb acts as a load by possessing resistance and transforming electrical energy into light and heat. Other examples of loads include electric motors, heating elements, and solenoids.

As current increases, the load's resistance goes down. If the circuit is interrupted, such as by a disconnected wire (conductor), no current flows. Open circuits use this kind of setup to work.

When current is flowing, each electron carries the voltage along the conductor, pushing the electrons ahead while being pushed by those behind. Any stop along the line stops it everywhere else, so the absence of a clear path, like a gap, stops the flow of electrons. This is what results in current being zero.

Understanding Ohm's Law

Ohm's Law describes how current, resistance, and voltage relate to each other. Ohm's Law states that the voltage (potential difference) in volts equals the current in amperes multiplied by the resistance in ohms. Ohm's Law can be expressed with the following formula

$V = IR$

- V is voltage, or the potential difference
- I is for the current (think of the intensity of the charge to remember what the I means)
- R is resistance

The voltage source is the circuit's starting and the endpoint. The negative terminal repels electrons, propelling them through the circuit, while the positive terminal attracts them. An electron released from the negative side will eventually travel through the entire circuit and reach the positive side. Voltage can be thought of as the electric pressure exerted on a charged object, making the voltage source responsible for driving the electric flow in a circuit.

If there is voltage in a closed circuit, it will produce a set amount of current, depending on the resistance. Batteries, which have positive and negative terminals, are examples of voltage sources (such as a 12 V battery). A household electrical outlet also provides voltage, originating from a power plant. These sources are also known as power sources because the electric current they generate carries energy.

The conductor is the wire that connects the voltage source and the load. For simplicity, the wire is often assumed to have zero resistance. To t more accurate, you can usually ignore the resistance that goes through a conducting wire when running calculations since it is negligible. If you are working with zero resistance, the voltage low enough to be considered zero, so you can expect no change in electrical pressure between the two ends.

According to Ohm's law, if a load's resistance decreases, the current flowing in the circuit increases. Conversely, if the circuit is broken, such as when a wire (conductor) is disconnected, no current will flow. This situation is called an open circuit. According to Ohm's Law, if a circuit has a break, that is like having infinite resistance.

Looking at Series Circuits

A series circuit is an electrical circuit that has only one path. Any break (opening) at any point in the circuit will stop current flow throughout the entire circuit. The simplest possible series circuit consists of a single voltage source connected to a single load by conductors.

The following image is of a series circuit.

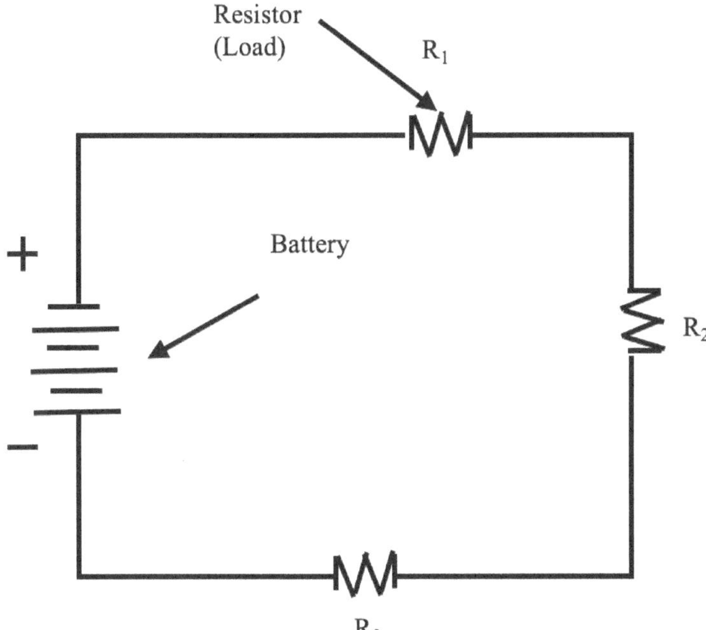

Series circuits have a single path for the current, ensuring a consistent current flow across the circuit. For example, you don't need one million electrons leaving the negative terminal every second, but two million reaching the positive terminal. It would be akin to a plane taking off from New York and landing in Miami with twice as many passengers. This ensures the current passing through each individual load in a series circuit doesn't change.

On the other hand, what happens if multiple loads are present in a series circuit? When two or more loads are added to a series circuit, their resistances combine, resulting in the total or effective resistance across the circuit.

$R_{total} = R_1 + R_2 + R_3 + \ldots$

You can use a single load to calculate the voltage. In a series circuit, the sum of all voltage decreases over every load, and the drop equals is the same as the circuit's total voltage. This matches the source's voltage.

$V_{total} = V_1 + V_2 + V_3 + \ldots$

Let's look at an example question.

If three different loads all have different resistance rates and they are wired in a series, which will remain consistent across all of them?

(A) The resistance
(B) The voltage drop
(C) The heat created
(D) The current passing through them

The answer is C.

Looking at Parallel Circuits

Parallel loads are far more common. In this arrangement, loads are connected in a separate path. If any one of these paths has a break, the current will still flow through the other paths, maintaining a closed circuit.

The following image is of a parallel circuit.

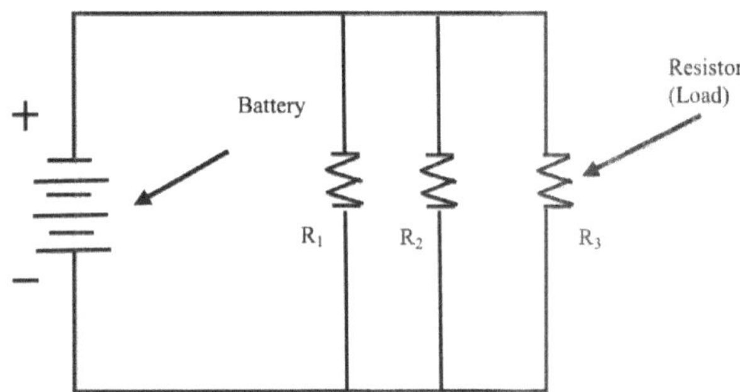

Parallel circuits differ from series circuits in that the voltage remains consistent across each parallel branch, while the current flow varies. Since voltage measures the difference in electric potential between two points, the voltage across different paths connecting the same two points in a circuit must be equal.

$V_1 = V_2 = V_3 = ...$

However, when there are parallel paths, the current flow becomes divided. It's often incorrectly stated that electricity follows the path of least resistance. In reality, electricity follows every available path, although a greater number of electrons will choose a path with low resistance while fewer will opt for a path with higher resistance. In a parallel circuit, you can calculate the total current by adding all of the currents that goes through each of the branches.

$I_{total} = I_1 + I_2 + I_3 + ...$

What about the effective resistance of a circuit when multiple loads are connected in parallel? It may seem intuitive to assume that adding more loads will increase resistance, but this isn't necessarily the case. Consider a busy retail store that is having a major sales event. One aisle is crowded with customers vying for the best deals, creating resistance to movement and slowing down the flow of people through the aisle. In this scenario, the customers represent electrons, and the rate at which they move symbolizes the current.

Suddenly, a store clerk clears away obstacles in an adjacent aisle, which is also cluttered but less crowded. Despite potentially having higher resistance due to more obstacles, some shoppers choose this less crowded aisle. As a result, the total number of shoppers passing through the store increases with the addition of this parallel path.

Similarly, adding a load in parallel, even if it has high resistance, reduces the total or effective resistance of a circuit. This is because it provides an additional path without constraining the existing one. While the formula for calculating resistances in parallel is more complex, it has proven to be useful.

$$\frac{1}{R_{total}} = \frac{1}{R_1} + \frac{1}{R_2} + \frac{1}{R_3} + ...$$

Let's consider the following problem.

How much current should pass through switch K when it is closed?

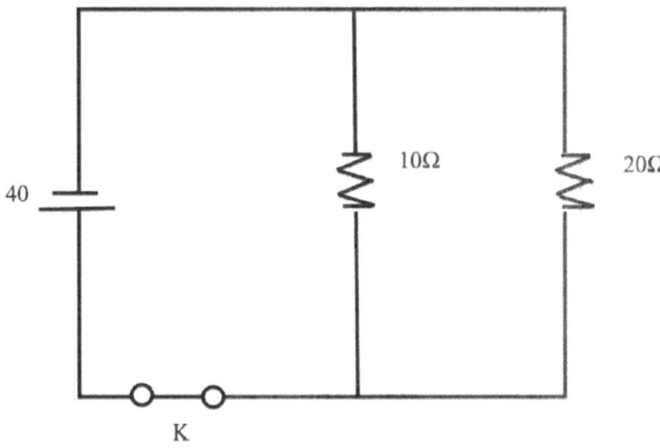

(A) 6 A
(B) 4 A
(C) 2 A
(D) 0 A

Looking at Series-Parallel Circuits

One of the most common arrangements is the series-parallel circuit. In a series-parallel circuit, certain components like an on/off switch are wired in series with several loads that are connected in parallel. The following is an illustration of a series-parallel circuit.

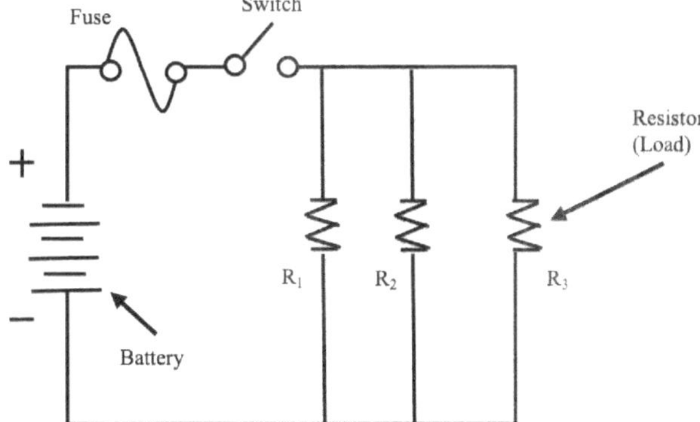

Most residential wiring circuits employ a series-parallel configuration. In this setup, wall outlets are connected in parallel, but all receive power from a circuit breaker wired in series. Turning off the circuit breaker shuts off power to all outlets. However, when the circuit breaker is on, voltage is supplied to all outlets regardless of whether they have loads plugged into them or not.

Calculating the total effective resistance across a series-parallel circuit involves a systematic approach. By applying the relevant formulas for resistors grouped in series or parallel, the circuit is progressively simplified until a single value for resistance is obtained. You can combine Ohm's Law with the other laws about voltage and current to determine how much the voltage drops or how much current goes through a load.

Let's look at an example question.

In the following diagram, all resistors have a 4 Ωresistance. What is the circuit's effective resistance?

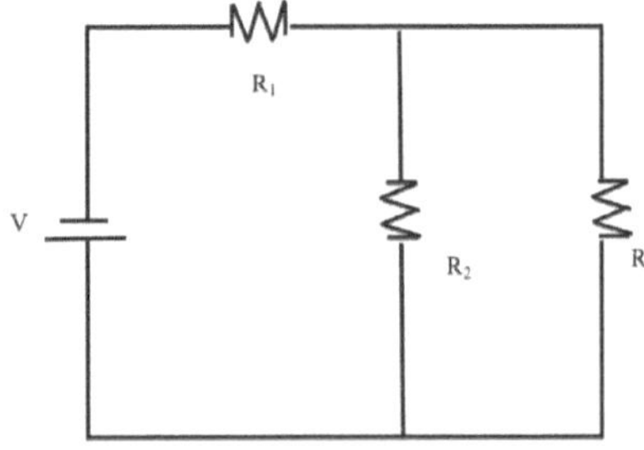

(A) 2Ω
(B) 4Ω
(C) 6Ω
(D) 10Ω

Use the resistance formula to calculate the answer.

$$\frac{1}{R_{23}} = \frac{1}{4} + \frac{1}{4} = \frac{2}{4} = \frac{1}{2} = 2$$

$R_{total} = 4 + 2 = 6Ω$

The answer is C.

Looking at Electrical Power

Electrical power is the measurement of the rate energy is sent to an electric circuit. It is quantified in watts. You can compute it by multiplying the voltage by the current, shown in the formula:

P = IV

The energy carried by each electron in a circuit or wire is determined by the applied voltage. Electrons traveling through higher voltages possess more energy. Conversely, a higher current indicates a greater number of electrons passing through per unit of time. Hence, an equivalent energy delivery rate can be achieved either by fewer electrons per unit of time (lower current) with higher energy per electron (higher voltage), or by more electrons per unit of time (higher current) with lower energy per electron (lower voltage).

This principle applies not only to energy delivery rates (such as in long-distance power lines), but also to energy production rates (as seen in generating plants), and energy consumption rates (as observed in appliances or loads). However, the term "consumption" can be somewhat misleading, as energy is not depleted but rather transformed. For example, a 40 W light bulb doesn't actually consume 40 joules a second. What it does is converts electrical energy into light and heat energy every second, producing two products, not just one.

Let's do an example problem.

If an electric generator powers a cabin at the standard of 120V, what is the power use if the effective resistance is 30Ω?

(A) 100 W
(B) 180 W
(C) 280 W
(D) 480 W

$$I = \frac{120V}{30} = 4A$$
$$P = 4A * 120V = 480W$$

The answer is D, 480 W.

Understanding the metric System and the Standard Electrical Units

According to Ohm's Law and some of the electrical formulas, it's crucial to ensure that the quantities used are in the appropriate units: amperes, ohms, volts, and watts. If any of these quantities are given in different units, they should be converted before performing calculations.

Metric prefixes can be added to any of these base units. If you have a smaller plant, it probably quantifies its output in megawatts (MW). If you are working with a cardiac doctor, they probably measure their patient's heartrate in tiny electrical signals known as milliamps (mA).

You may want to spend some extra time reviewing the Physical Science section in Chapter 9 to make sure you understand the prefixes and the metric system.

UNDERSTANDING THE STRUCTURE OF ELECTRICAL AND ELECTRONIC SYSTEMS

So far, you've acquired fundamental operational and mathematical knowledge about electric circuits. In the upcoming section, you'll further enhance these skills. The discussion will encompass various types of current, safety precautions, and components that impart special characteristics to circuits, enabling them to execute specific functions.

Understanding AC and DC

There are two primary types of current

- Direct current (DC) flows consistently in one direction along a conductor, typically provided by batteries. A lot of your mobile devices use DC power.
- Alternating current (AC) reverses direction multiple times per second as it travels through a conductor. Household electrical outlets supply AC power, offering an efficient means to distribute electricity from power stations to homes and businesses.

In North America, AC power is standardized at 60 cycles per second, known as hertz (Hz). One cycle encompasses the voltage rising from zero to a peak level in one direction, then returning to zero before reaching a peak level in the opposite direction, and finally returning to zero again. The frequency, measured in hertz, indicates how many complete cycles occur within one second. The following illustration is of a 60 Hz AC signal.

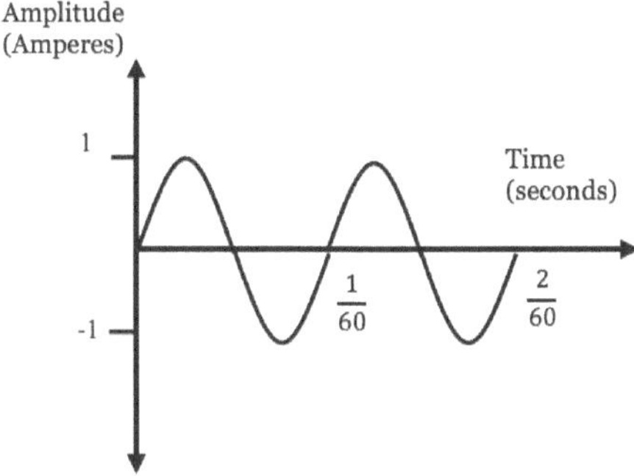

The following is the frequency equation:

$$f = \frac{1}{T}$$

In the given equation, f represents the frequency in hertz (Hz), and T represents the time taken for one cycle in seconds. This equation allows you to calculate either the frequency or the period when provided with one of these values. Therefore, if the frequency is known to be 60 Hz, it can be inferred that one complete cycle takes $\frac{1}{60}$ of a second, as demonstrated by the formula:

$$T = \frac{1}{f}$$

Given that power stations provide AC and most electronic devices operate on DC, you might wonder where and how this conversion occurs. Typically, electronic devices are equipped with built-in AC to DC conversion systems. Later in this section, we'll delve into an essential electrical component known as the diode, which facilitates this conversion process.

The Importance of Grounding

Ensuring the grounding of electrical devices and residential wiring is paramount for safety. Ground serves as the point of lowest potential in a circuit. Given that the potential difference (voltage) is highest between any circuit point and the ground, any stray electricity tends to follow this path, as resistance is minimal. This is crucial for preventing shocks caused by external factors like lightning or internal circuit failures where conducting wires are compromised.

In residential wiring, grounding is established as a standard connection across the entire wiring system, serving as a protective measure against electrical shocks. All wiring grounds are linked to an earth ground, often accomplished through a copper rod driven into the ground or buried conduit. This ensures that electrical current is directed away from panels and equipment in the event of an internal short circuit. The following are both symbols used to mean you need to ground something.

Examining Important Components

The characteristics of an electrical circuit can vary based on the components incorporated within it. This segment will introduce and explore several key components that significantly influence circuit behavior.

Resistors are a critical component in electrical mechanisms. Within an electrical circuit, different components may necessitate varying voltage levels. Additionally, there are instances where regulating current or voltage proves beneficial for controlling specific functions, such as adjusting the volume in a stereo system. To achieve this control, a component called a resistor is employed. As implied by its name, a resistor introduces a defined level of resistance, resulting in a voltage drop when current flows through it.

As previously discussed, resistance plays a crucial role in determining current, as per Ohm's law. Consequently, an increase in resistance (R) leads to a reduction in the current. The following are the symbols used to represent resistors.

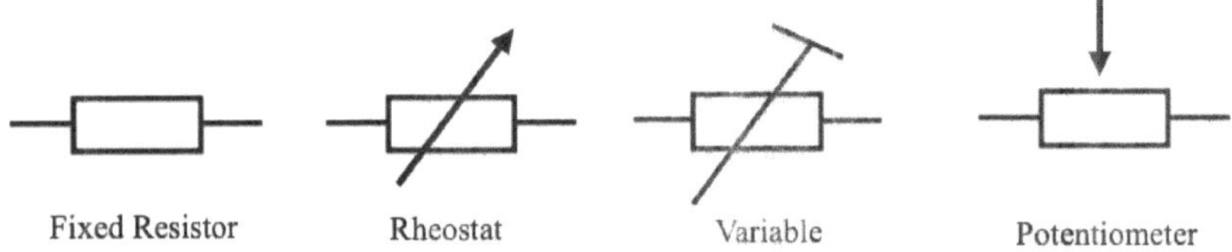

Two primary types of resistors are commonly employed in modern electrical circuits: fixed resistors and variable resistors. Fixed resistors, as their name suggests, maintain a constant resistance value. On the other hand, variable resistors, also known as rheostats or potentiometers, offer the flexibility of adjustable resistance. Potentiometers are particularly useful for modifying the voltage drop across a component within a circuit. In electronic applications, they find application in volume adjustments for devices like TVs, radios, and stereos. Meanwhile, rheostats are valuable for regulating the current flow within a circuit and can serve as light dimmers or speed controllers for small, motorized devices.

Fixed resistors typically employ a color band system to denote their resistance in ohms. Each color band corresponds to a specific numeric value, with the first two bands representing significant digits and the third band indicating the multiplier or the number of zeroes following the digits.

The following table is of the color bands used for fixed resistors.

Color	1st Digit 1st Stripe	2nd Digit 2nd Stripe	Multiplier 3rd Stripe
Black	0	0	x 1
Brown	1	1	x 10
Red	2	2	x 100
Orange	3	3	x 1,000

Yellow	4	4	x 10,000
Green	5	5	x 100,000
Blue	6	6	x 1,000,000
Violet	7	7	-
Grey	8	8	-
White	9	9	-

Considering a resistor with color bands red, violet, and yellow, we can determine its resistance value using the chart. The red band matches 2, the yellow band is four zeros, and the violet band matches 7. If you add up these values, the resistance is 270,000 Ω.

To determine the value of resistance in a simple series circuit with unknown resistance, you can conduct an experimental approach in one of two ways. Firstly, you can measure the resistance directly using an ohmmeter. To do this right, either detach the device or disconnect the power source from the circuit. Ohmmeters have their own power source. Then, attach the ohmmeter leads across both ends of the device to measure the resistance.

If you don't have an ohmmeter, you can connect an ammeter in series with the resistor and a voltmeter in parallel with the resistor. With this setup, you can measure the current (I) and voltage (V). Resistance equals the voltage divided by current (R = V / I).

Understanding Fuses and Circuit Breakers

An electric ground is a crucial circuit safety feature. It serves to divert excess electricity away from the device, thus reducing the risk of electric shock. Fuses and circuit breakers are also vital safety components within circuits. In cases where the current in a circuit becomes too high, wires can overheat, potentially leading to fires.

A short circuit occurs when a conductor bypasses a load within the circuit, resulting in an increase in current flow. This can happen if, for example, the insulation on wires becomes damaged, allowing them to come into direct contact. In a series circuit, a short circuit typically only removes the bypassed load. However, in parallel circuits or if all loads in a series circuit are bypassed, the overall resistance of the circuit decreases significantly, leading to a dangerous surge in current. To mitigate such risks, circuits are often safeguarded with fuses or circuit breakers.

Fuses consist of thin wires that melt when the current exceeds a specified level, effectively cutting off the electricity flow. While this prevents damage to the device, a blown fuse needs to be replaced before the circuit can function again. Fuses with varying current ratings are available to suit different device specifications. The following is what the fuse symbol looks like on schematics and other depictions of circuits and fuses.

Circuit breakers are similar to fuses, and they are reusable. However, they respond more slowly to current surges compared to fuses and typically incur higher installation costs. Despite variations in design, all circuit breakers share a common principle of operation.

One type of circuit breaker employs a bimetallic strip that bends away from its contact within the circuit upon detecting excessive current flow. This action interrupts the circuit, preventing further current passage. Another type of circuit breaker utilizes an electromagnet to achieve a similar effect. When the current surpasses a predetermined threshold, the ferrous material becomes sufficiently magnetized, leading to the opening of the circuit and halting further current flow. Later in this chapter, we will delve into the differences between magnetism and electricity. The following are how circuit breakers are shown on schematics and other images with circuit breakers.

Looking at Capacitors

Capacitors, also referred to as condensers, function as electrical storage devices. They consist of two metal conducting plates separated by a thin insulating material, termed a dielectric. In some cases, air can also act as a dielectric.

A capacitor can retain an electrical charge due to the creation of an electron surplus on its negative plate and an electron deficit on its positive plate by the DC source. This results in an electrostatic attraction between the opposite charges, preserving the charge within the capacitor even after the voltage source is disconnected. When a conductor is connected across the capacitor, it discharges itself by providing a pathway for electrons to move from the negative to the positive plate. While a capacitor permits AC to pass through, it blocks DC, making DC suitable for "charging up" a capacitor.

The capacitive reactance is the resistance provided to the current flow by the capacitor, measured in ohms. It is inversely proportional to the AC signal's frequency. In simpler terms, higher frequencies encounter less resistance to the passage of AC through the capacitor. The following is the symbol generally used for general capacitor.

Capacitance is the capacity of a capacitor as storage (C), and it is measured in farads (F). One farad represents the ability to store one coulomb of electrons under the influence of one volt of electrical potential.

Let's do an example problem.

In the following diagram, the fuse is rated 0.1 A. What happens to the lamp on the circuit when the switch is closed?

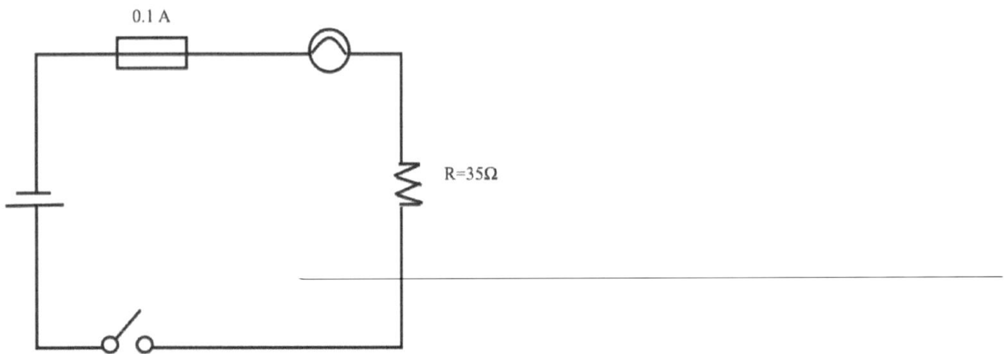

(A) The lamp stays on
(B) The fuse blows, shutting down the light
(C) The fust is unchanged, but the light turns off.
(D) The lamp will keep blinking.

The circuit is larger than the fuse, so the correct answer is B.

Understanding Semiconductors

Semiconductors are elements characterized as having a valence shell with four electrons. Their intermediate conductivity stems from this configuration, which neither allows them to be good conductors nor good insulators. Some of the most common examples of semiconductors are silicon and germanium.

In their pure state, silicon and germanium possess limited utility. However, through a process called doping, impurities are introduced into their crystalline structures, unlocking a plethora of potential applications. Pure silicon boasts a stable crystalline structure where each atom's four valence electrons form strong bonds with surrounding atoms, hindering electron mobility and thus conductivity. Doping disrupts this stability by introducing impurities like phosphorus, arsenic, or antimony, each of which possesses five valence electrons. These additional electrons create "holes" or vacancies in the crystal lattice, enabling the movement of charge carriers. Consequently, doped silicon transforms into an N-type semiconductor, still electrically neutral but capable of conducting electricity due to the presence of mobile charge carriers.

When silicon undergoes doping with elements possessing three valence electrons, such as boron or indium, vacancies, or "holes," emerge in the crystalline lattice where an electron would typically reside. Consequently, regions with an overall positive charge form periodically throughout the structure. As free electrons migrate into these vacancies, they create additional holes in their previous positions. This doping process transforms silicon into a P-type semiconductor material.

Understanding Diodes

When P-type and N-type materials are combined, they form a diode, which acts as an electrical one-way valve at the junction. This junction allows current to flow freely in one direction but blocks it in the opposite direction.

The orientation of the battery in a circuit determines whether the diode is forward-biased or reverse-biased. In the forward-biased condition, the anode (the P-type material) connects with the positive end of the battery. The cathode (the N-type material) should connect to the negative side of the battery.

In a forward-biased condition, the excess electrons at the battery's negative terminal push in the N-type material's free electrons in the direction of the junction. The electrons cross the junction and occupy the "holes" in the P-type material, permitting current to flow.

In a reverse-biased condition where the diode's connection to the battery is reversed, the free electrons in the N-type material move away from the junction, drawn towards the battery's positive terminal. As a result, no electron transfer occurs at the junction, and current flow ceases. The diode is now in a reverse-biased state.

Diodes find widespread application in rectification, which converts AC into DC. As diodes permit current to move in a single direction, they allow either the lower or upper part of the AC waves to pass through. This process is called a half-wave rectification. If you want a more efficient rectification, arrange the diodes into a diamond shape with four corners, creating a full-wave rectifier. A lot of DC power supplies use this configuration. Additionally, diodes play a crucial role in rectification within automotive alternators, where the generated AC is converted into DC to power the vehicle's electrical system.

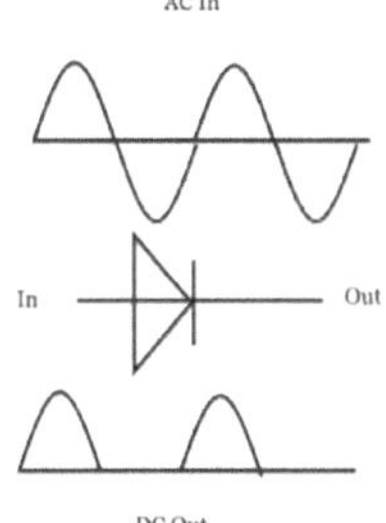

Transistors

The transistor stands as a revolutionary component in the realm of electronic devices, including computers, calculators, and radios. Its versatility enables it to function as an electrical switch, amplifier, or current regulator. Being devoid of moving parts and crafted from solid silicon, it earns the moniker of a solid-state device, ensuring remarkable reliability. Moreover, its compact design allows for the integration of thousands of transistors into a single silicon chip, forming what's known as an integrated circuit—housing all essential circuit components within a single unit. Transistors serve as the cornerstone of the logic operations that power a computer's information processing.

Transistors come in two principal types: NPN and PNP. A thin P-type layer sandwiched between two N-type layers is what makes up an NPN transistor. A PNP transistor, on the other hand, consists of two layers of P-type and one layer of N-type material.

Internally, wires link the three semiconductor material layers, each designated with a specific name. The central layer is consistently labeled as the base, while the outer layers are termed the collector and the emitter. Notably, the transistor symbol includes an arrow identifying the emitter, with the arrow's direction indicating the type of transistor. An easy mnemonic is that the arrow points towards the N-type material.

The transistor operates on the principle of using a small current to regulate a larger current. If you have an NPN transistor, you need to add a positive voltage to the base to activate the transistor, and this lets a substantial current to flow from the collector to the emitter. Once the positive voltage is removed from the base, the transistor deactivates, halting the flow of current.

PNP transistors operate in the reverse manner compared to NPN transistors. To activate a PNP transistor, the base needs a negative voltage to cause the current to reach the collector from the emitter.

Below are additional symbols representing electronic components that you might encounter in the ASVAB. Some of these, such as transformers, inductors, motors, and generators, will be covered in the following section.

Let's look at a sample question.

If a capacitor stores charge, and it is then charged and discharged, what will the output be?

    (A) Direct current
    (B) capacitance
    (C) reactance
    (D) Alternating current

The answer is A.

REVIEWING ELECTRICITY AND MAGNETISM

Electricity and magnetism are intricately connected phenomena. Whenever charges are in motion, they generate magnetic fields, and the electrons moving through a circuit are no exception. This relationship between electricity and magnetism finds application in various devices, from circuit breakers to transformers, motors to generators, and even in everyday items like microphones and doorbells.

A wire carrying an electric current creates its own magnetic field. The strength of this field is determined by how much current passes through the wire. When the wire is wound around a ferrous material and current flows through it, the resulting magnetic field magnetizes the iron core, forming an electromagnet. This induced magnetism is the fundamental principle underlying the operation of many electrical devices mentioned earlier.

Understanding Inductors

By winding a current-carrying wire into a coil, the resulting magnetic field can be intensified. This effect is further amplified when the coil is wound around a ferrous (iron) core, as iron facilitates the movement of magnetic lines of force more effectively than air. Initially, when current flows through the coil, the magnetic field builds up gradually. However, this process encounters resistance when the magnetic field expands, resulting in the voltage coiling in a way that opposes the original current flow. This counter-electromotive force, or counter-emf, slows down the buildup of the magnetic field. On the other hand, interruption to the current causes the collapse of the magnetic field. As a result, the coil's voltage sustains the current flow. This phenomenon, known as self-induction, is a characteristic of electrical components called inductors.

Inductors exhibit resistance to changes in current flow within a circuit. When the current is on the rise, the inductor generates a voltage that opposes this increase. Conversely, if the current decreases, the inductor draws upon the magnetic energy stored in the coil to resist the decline and maintain the current flow. The following is commonly used as the inductor symbol.

The measurement of induction is denoted in Henries, represented by the symbol L. Inductors function in an inverse manner to capacitors: they permit the passage of DC with ease while impeding the flow of AC. This opposition to current flow is termed inductive reactance, measured in ohms, and it escalates with the frequency of the AC signal.

Looking at Transformers

A transformer serves to either step up or step down the voltage within a circuit, leveraging the characteristics of an inductor to achieve this task. As alternating current courses through wires wound around an iron core, it magnetizes the core, creating a fluctuating magnetic field. The dynamic field generates a voltage in a nearby coil of wire. If you change how many turns there are in the primary and secondary coils, as well as their proximity, the transformer changes how much voltage is in the secondary coil relative to the primary coil. The power source connects to the primary coil, and the secondary coil receives the induced electric current. More secondary coils result in higher voltage, and closer proximity enhances the transformer's efficiency in inducing voltage in the secondary coils.

Transformers are essential to supplying electricity. It's more energy-efficient to supply electricity using a low current with a high voltage. However, before reaching households, the voltage must be reduced, or "stepped down," to the standard 120–240 volts required by most household appliances.

Understanding the Basics of Motors and Generators

When a current-carrying wire interacts with a permanent magnetic field, it undergoes a force, which can be utilized to generate mechanical energy. This is how a basic motor operates. The resulting mechanical energy can power various applications, from propelling car tires to spinning fan blades or lifting heavy objects. The force exerted on the wire is directly proportional to the current, the length of the wire, and the strength of the magnetic field. Enhancing any of these factors increases the force exerted on the wire and, consequently, its capacity to perform work. While modern motors often employ electromagnets rather than bare wires, the fundamental principle remains consistent.

Conversely, a generator operates as a motor in reverse. When a wire moves within a permanent magnetic field and forms a complete circuit, it induces an electric current in the wire. A generator makes electrical energy from mechanical energy.

Let's do an example problem.

Which circuit component can change the value of the input voltage when electricity travels from the plant to the residence?

- (A) Generator
- (B) Motor
- (C) Inductor
- (D) Transformer

The answer is D, transformer.

PRACTICING WHAT YOU'VE LEARNED

Now that you've got the basics, you can see test your newly acquired information and put it to the test.

Here are two practice sets with 15 questions. This is similar to how the CAT is setup, although obviously your questions aren't going to change the better you do.

The last section of this chapter details the right answer, and how you could have reached that answer.

Practice 1

1. What component stores electrical energy and blocks DC while allowing AC to pass?

(A) Capacitor
(B) Resistor
(C) Inductor
(D) Diode

2. Which electrical component can be used to limit voltage or current in a circuit?

(A) Inductor
(B) Transformer
(C) Capacitor
(D) Resistor

3. What type of transistor requires a negative voltage at the base to turn it on?

(A) PNP transistor
(B) NPN transistor
(C) FET transistor
(D) Bipolar transistor

4. What is the unit used to measure induction in an inductor?

(A) Ohms
(B) Farads
(C) Henries
(D) Volts

5. How is a diode forward-biased in a circuit?

(A) By connecting the P-type material to the battery's positive terminal
(B) By connecting the N-type material to the battery's negative terminal
(C) By connecting the P-type material to the battery's negative terminal
(D) By connecting the N-type material to the battery's positive terminal

6. Which electrical safety component melts when current exceeds a prescribed amount?

(A) Capacitor
(B) Resistor
(C) Transformer
(D) Fuse

7. What type of current moves in a single direction in a conductor?

(A) Direct current (DC)
(B) Alternating current (AC)
(C) Variable current (VC)
(D) Bipolar current (BC)

8. How does a transformer change voltage in a circuit?

   (A) By altering the current flow
   (B) By converting AC to DC
   (C) By using electromagnets to generate voltage
   (D) By inducing voltage in a secondary coil through a changing magnetic field

9. What measurement unit is used for capacitance?

   (A) Henries
   (B) Ohms
   (C) Farads
   (D) Volts

10. What is the function of a bimetallic strip in a circuit breaker?

   (A) Produces an electric current
   (B) Opens the circuit when too much current flows
   (C) Closes the circuit when too much current flows
   (D) Provides resistance to current flow

11. Which material is common inside a transformer core?

   (A) Copper
   (B) Silver
   (C) Aluminum
   (D) Iron

12. What does an inductor do in a circuit?

   (A) Stores electrical energy
   (B) Blocks DC and passes AC
   (C) Resists change in current flow
   (D) Increases or decreases voltage

13. How does a diode function in a circuit?

   (A) It controls the flow of current in one direction
   (B) It stores electrical energy
   (C) It regulates voltage
   (D) It amplifies electrical signals

14. What is the purpose of a potentiometer?

   (A) To regulate current flow
   (B) To store electrical energy
   (C) To change the voltage drop across a component
   (D) To amplify electrical signals

15. What part is stores electrical charge and block DC while allowing AC to pass?

   (A) Transformer
   (B) Resistor
   (C) Capacitor
   (D) Inductor

Practice 2

1. What electrical component lets the current flow in a single direction?

    (A) Capacitor
    (B) Transformer
    (C) Diode
    (D) Inductor

2. Which component is used to increase or decrease voltage in a circuit?

    (A) Capacitor
    (B) Resistor
    (C) Transformer
    (D) Inductor

3. What type of material is commonly used in the core of a transformer?

    (A) Copper
    (B) Silver
    (C) Aluminum
    (D) Iron

4. Which unit is used to measure resistance?

    (A) Ampere
    (B) Volt
    (C) Ohm
    (D) Farad

5. What is a capacitance's unit of measurement?

    (A) Henry
    (B) Ohm
    (C) Farad
    (D) Volt

6. What component resists change in current flow?

    (A) Resistor
    (B) Capacitor
    (C) Diode
    (D) Transformer

7. How does a potentiometer function in a circuit?

    (A) To regulate current flow
    (B) To store electrical energy
    (C) To amplify electrical signals
    (D) To change resistance and voltage drop

8. What part stores electrical charge and release it when needed?

    (A) Resistor
    (B) Diode
    (C) Capacitor
    (D) Inductor

9. Which component makes electrical energy from mechanical energy?

    (A) Capacitor
    (B) Inductor
    (C) Transformer
    (D) Generator

10. What does an ammeter measure?

    (A) Voltage
    (B) Current
    (C) Resistance
    (D) Capacitance

11. Which component melts when current exceeds a prescribed amount, preventing further electricity flow?

    (A) Capacitor
    (B) Resistor
    (C) Fuse
    (D) Transformer

12. What is the purpose of a relay in an electrical circuit?

    (A) To control the flow of current
    (B) To amplify electrical signals
    (C) To store electrical energy
    (D) To switch on/off a circuit using a small current

13. What type of transistor requires a positive voltage at the base to turn it on?

    (A) PNP transistor
    (B) NPN transistor
    (C) FET transistor
    (D) Bipolar transistor

14. Which component opposes the change in current flow and is measured in Henries?

    (A) Capacitor
    (B) Transformer
    (C) Inductor
    (D) Diode

15. How does the distance from primary and secondary coils affect a transformer's efficiency?

    (A) Increased distance improves efficiency
    (B) Decreased distance improves efficiency
    (C) Distance has no effect on efficiency
    (D) Efficiency is unaffected by the distance

CHECKING YOUR ANSWERS

Now it's time to see how well you did.

Practice 1

1. (A) Capacitor
2. (D) Resistor
3. (A) PNP transistor
4. (C) Henries
5. (A) By connecting the P-type material to the battery's positive terminal
6. (D) Fuse
7. (A) Direct current (DC)
8. (D) By inducing voltage in a secondary coil through a changing magnetic field
9. (C) Farads
10. (B) Opens the circuit when too much current flows
11. (D) Iron
12. (C) Resists change in current flow
13. (A) It controls the flow of current in one direction
14. (C) To change the voltage drop across a component
15. (C) Capacitor

Practice 2

1. (C) Diode
2. (C) Transformer
3. (D) Iron
4. (C) Ohm
5. (C) Farad
6. (A) Resistor
7. (D) To change resistance and voltage drop
8. (C) Capacitor
9. (D) Generator
10. (B) Current
11. (C) Fuse
12. (D) To switch on/off a circuit using a small current
13. (B) NPN transistor
14. (C) Inductor
15. (B) Decreased distance improves efficiency

CHAPTER 11

AUTOMOTIVE INFORMATION

This chapter focuses on the basics you need to know to complete the automotive information section (AI).

A BIT MORE ABOUT THIS SECTION OF THE ASVAB

This section doesn't give you much time to finish the questions, so if you are interested in working with automotives, you will want to really learn the subject, spending some extra time studying on line.

- The ASVAB gives you 7 minutes to finish 10 questions about AI if you take the CAT.
- If you take the traditional ASVAB, you will have 25 questions with just 11 minutes to finish. This means finishing more than two questions a minute.

Automotives are complex and literally has a lot of moving parts. The contemporary automobile stands as a testament to technological advancement. Despite its complexity, at its core, it operates on the same fundamental principles as its predecessors. While modern vehicles incorporate cutting-edge materials and sophisticated computer systems, they still rely on basic systems comprised of numerous smaller parts.

Regardless of its size, an automobile is essentially an amalgamation of countless smaller components. These parts, such as bolts, screws, spark plugs, and belts, are the building blocks of the vehicle. They constitute the foundation upon which the vehicle is constructed, unable to be further disassembled. Systems, on the other hand, consist of interconnected parts working harmoniously to fulfill specific functions.

DELVING INTO ENGINE SYSTEMS

Engine systems are responsible for generating the power required to propel the vehicle and operate its various components. This encompasses engine theory, defining the mechanisms and components essential for engine operation. Additionally, powertrain systems include the cooling system, tasked with dissipating excess heat from the engine, and the lubrication system, which ensures smooth engine operation by circulating motor oil to reduce friction.

Understanding the Basic Engine Theory

The engine commonly utilized in automobiles is referred to as an internal combustion engine. Combustion refers to the rapid ignition of an air-fuel mixture. In the context of internal combustion, this means that fuel is burned inside the engine, and the resulting heat is used directly to drive the engine.

Common fuels employed with internal combustion engines include gasoline, diesel fuel, propane, natural gas, and alcohol, with gasoline and diesel fuel being the most prevalent choices.

For combustion to occur, three essential elements must be present: air, fuel, and a heat source capable of igniting the air-fuel mixture. Without any one of these components, combustion ceases, and the engine stops working. Internal combustion engines transform the chemical energy within the fuel and air into heat energy, which is subsequently converted into mechanical energy. For further details on energy types, refer to chapters 9 and 13.

Examining the Components

The following table provides the most common internal combustion engine components.

Part	Function
Engine block	Provides the structural foundation for the engine's cylinders and reciprocating assembly.
Piston	A cylindrical component with a solid crown (top) that moves up and down within the engine's cylinders. It moves forward due to the pressure generated by hot gases produced during combustion.
Cylinder	Creates a chamber for the piston to move within, facilitating its up-and-down motion as the engine completes its operating cycle.
Piston rings	Create a tight seal for the cylinder and the piston, preventing the leakage of combustion gases. Oil rings ensure the oil cannot get into the combustion chamber.
Wrist pin	Links the piston to the connecting rod and serves as a pivot point for the connecting rod at the small side.
Connecting rod	Joins the piston/wrist pin assembly to the engine's crankshaft. The larger end attaches to the crankshaft's connecting rod journal.
Crankshaft	Transforms the linear motion of the piston into rotary motion, which can be utilized to propel a vehicle or drive various accessories.
Cylinder head	Positioned above the piston, it includes associated ports, the combustion chamber, and the intake and exhaust valves.
Combustion chamber	Situated within the cylinder head directly above the piston, it is where the combustion of the air-fuel mixture occurs.
Intake valve	Allos the air-fuel mixture to move into the combustion chamber. When shut, it seals the combustion chamber from the intake port.
Exhaust valve	Enables the release of gases from the combustion chamber. When closed, it seals the combustion chamber from the exhaust port.
Camshaft	Oversees the opening and closing actions of the exhaust and intake valves. It functions at a speed that is half that of the crankshaft.

The way the linear motion translates into a rotary motion through the combination of the piston-connecting rod to the crankshaft, is similar to a bike rider's leg moving to pedal. Just as the rider's upper leg moves up and down (piston), her lower leg (connecting rod), the bicycle pedals, and the sprocket (crankshaft) transform that straight-line motion into rotary motion, propelling the bicycle's wheels forward.

The four-stroke cycle is used by most internal combustion engines. This means that the piston needs four strokes to get through a single cycle. A piston stroke begins from the top of the cylinder, then extends downwards to the bottom of its travel, and finally returns to TDC.

The intake stroke starts the cycle. The intake valve starts opening at TDC. As the piston descends in the cylinder, it creates a low-pressure area, inducing air intake into the combustion chamber. Concurrently, fuel is injected into the air stream within the intake system. This process ensures the cylinder fills with a cleaner mix of air and fuel. Upon reaching BDC, the intake valve is almost closed, ending of the intake stroke and preparing for the compression stroke.

In the compression stroke, the piston moves up in the cylinder while the engine's valves are shut. With the combustion chamber sealed off, the piston's upward movement compresses the air-fuel mixture, gradually reducing its volume. This compression raises the temperature of the mixture, making it more combustible and enhancing the power of the subsequent stroke, known as the power stroke.

The spark plug start a fire in the air-fuel mixture before the piston reaches TDC, initiating a flame that traverses the combustion chamber. This flame further elevates the temperature of the gases, leading to rapid expansion. This expansion exerts pressure on the piston, propelling it past TDC and initiating the power stroke, which is the cycle's third stroke.

Throughout the power stroke, both valves are closed, ensuring that the generated pressure by the combustion is effectively harnessed to drive the piston toward the BDC. This design prevents the loss of pressure through the valves and maximizes the conversion of energy into mechanical motion.

As the piston approaches BDC, the combustion process should ideally be concluded. Just before reaching BDC, the exhaust valve initiates opening, starting the exhaust stroke, which is the last one. At this stage, the gases within the combustion chamber are fully expended and necessitate expulsion from the engine to prepare for the next cycle.

As the piston commences its upward trajectory, it drives the spent exhaust gases past the open exhaust valve and into the engine's exhaust system. From there, the exhaust gases are eventually directed out into the open atmosphere. As the piston progresses toward top dead center (TDC), it completes a full cycle of events, marking the conclusion of one cycle and the commencement of the next, as the intake stroke resumes.

Examining the Cylinder Arrangement

Automotive engines come in various configurations, primarily categorized by their cylinder arrangement, which denotes the position of cylinders relative to each other. While four, six, and eight cylinders are the most common, engines can also feature three, five, ten, twelve, or even sixteen cylinders.

The inline design is the simplest arrangement, featuring cylinders aligned in a row. This configuration is prevalent in four- and six-cylinder engines, particularly in front-wheel drive cars with transverse-mounted engines. Inline engines are commonly found in small- to medium-sized vehicles.

Engineers also use the horizontally opposed, also known as the flat arrangement. In this setup, all cylinders lie flat in a horizontal plane, with half facing away from the other half and the crankshaft situated between them. Some refer to this as a "boxer" engine since the pistons move in a punching-like motion. A significant advantage of this design is it has a lower center of gravity, which means it is offer better handling and is more stable.

The V-type engine is the most popular design for six- and eight-cylinder engines. These engines feature a crankshaft connected to pistons that are configured in a V-shape. The two rows of cylinders are positioned 60 to 90 degrees apart, each with its own cylinder head, totaling two heads for the engine. This layout offers inherent balance among the engine's moving components, resulting in smooth operation. Additionally, compared to inline engines, V-type engines are more compact in height and length, enhancing aerodynamic flow outside the vehicle. This compact design allows for a lower front end of the vehicle. When an eight-cylinder engine is required, the V-type configuration provides the necessary power while maintaining a compact size.

Examining the Camshaft and It's Location

Another classification used for automotive engines is based on the location of the camshaft. The camshaft regulates the timing of the engine's valves, opening and closing them, and it is powered by the crankshaft through a timing chain or belt. In modern engines, the camshaft may be located in the engine block, while the intake and exhaust valves are typically positioned in the cylinder head. Engines constructed in this manner are referred to as overhead valve or OHV engines because the valves are situated above the piston and combustion chamber. In OHV engines, each valve is operated by a lever called a rocker arm, which is actuated by the camshaft through a pushrod. The OHV setup is compatible with both V and flat engine configurations.

Alternatively, engine designers may place the camshaft above the valves, allowing them to eliminate the pushrods required for the OHV setup. This streamlines and reduces the weight of the valve operation mechanism. A lighter valve train enables higher engine speeds. This design is known as overhead cam or OHC. When a single overhead camshaft is used to control both the intake and exhaust valves of a cylinder, it's called single overhead cam or SOHC. In V-type engines with two cylinder heads, there are typically two camshafts—one positioned above each cylinder head.

To achieve even higher engine speeds, a double overhead cam or DOHC arrangement is employed. This configuration incorporates two camshafts within each cylinder head. One camshaft is responsible for operating the exhaust valves, while the other controls all the intake valves in that head. With this setup, rocker arms, commonly found in SOHC engines, can be eliminated. Instead, the camshafts directly operate the valves through a follower.

Looking at Multiple-Valve Cylinder Heads

Engines are usually classified by how many cylinders are in them. The most common setup is a two-valve cylinder head, featuring one intake valve and one exhaust valve per cylinder. However, for enhanced airflow and engine performance, manufacturers have developed multiple-valve cylinder heads. While some engines utilize three valves per cylinder, the prevalent trend now favors four-valve cylinder heads. With two exhaust valves and two intake per cylinder, the four-valve design offers several advantages. It allows for increased engine operating speeds and facilitates more complete combustion of the air-fuel mixture, thus improving engine power and efficiency.

Reviewing Firing order

The sequence in which cylinders ignite is referred to as the firing order. For a four-cylinder engine, a typical firing order is 1-3-4-2. This signifies that after the first cylinder ignites, the crankshaft completes half a revolution before the third cylinder fires, and so forth. During the initial crankshaft rotation, cylinders 1 and 3 ignite, followed by cylinders 4 and 2 on the subsequent rotation. In a V-8 engine with a firing sequence of 1-8-4-3-6-5-7-2, cylinders 1, 8, 4, and 3 ignite during the first revolution of the crankshaft, followed by cylinders 6, 5, 7, and 2 igniting on the second revolution. Some engines conveniently have their firing order engraved on the intake manifold, simplifying identification. Otherwise, consulting the engine's service information becomes necessary to determine the firing order.

Understanding the Diesel Engine

Diesel engines, also known as compression ignition engines, offer a distinct variation of the internal combustion engine. Unlike gasoline engines, they eschew spark-ignition systems, relying instead on higher compression ratios to trigger combustion. The compression ratio in diesel engines typically ranges from 16:1 to 22:1, generating ample heat of compression to ignite the air-fuel mixture.

Moreover, diesel engines necessitate direct fuel injection into the combustion chamber or a designated pre-combustion chamber. This injection method is crucial because it relies on the intense heat produced by the compressed air within the chamber to ignite the fuel. In contrast, gasoline engines, even those with direct injection systems, rely on spark ignition mechanisms to initiate combustion.

Understanding Engine Operating Conditions

Efficient engine operation relies on the precise blending of air and fuel. If the mixture contains too much fuel and insufficient air, or vice versa, combustion efficiency suffers. Maintaining the ideal balance, known as the stoichiometric ratio, falls within the purview of the engine's fuel system.

The air-fuel ratio quantifies the mass of air relative to the mass of fuel in the mixture. For engines that run on gasoline, the s ratio is 14.7:1, indicating that 14.7 pounds of air combine with 1 pound of fuel for optimal combustion. Fuel particles are atomized and introduced into the incoming air stream by the fuel system.

A lean mixture is the combination of insufficient fuel and excess air. Lean mixtures burn slowly due to increased spacing between fuel molecules, requiring more time for combustion. Additionally, lean mixtures generate higher temperatures, posing a risk of engine damage. A typical lean air-fuel ratio might be 17:1.

A rich air-fuel mixture contains insufficient air and an excess of fuel. Rich mixtures ignite rapidly due to the close proximity of fuel particles, resulting in cooler combustion. While a rich air-fuel mixture may lead to spark plug fouling and the production of black exhaust smoke, it poses less risk to the engine compared to lean mixtures. A typical rich air-fuel mixture might have a ratio of 10:1.

Another critical aspect of engine efficiency is ignition timing, which refers to the moment in the combustion cycle when a spark is produced at the spark plug. This timing is typically described in relation to how the engine's crankshaft is positioned. For instance, if you set the ignition timing at five degrees BTDC, the crankshaft is at five degrees and sparks the ignition before it reaches TDC.

Generally, the spark occurs before the piston reaches TDC during the compression stroke. This timing allows sufficient time for the flame to propagate across the combustion chamber and ignite the air-fuel mixture. By the time the mixture is fully burned, the piston is descending, ready to execute an effective power stroke. Regardless of engine speed, the flame propagation time remains relatively constant. Therefore, at higher speeds, ignition timing must be advanced to ensure optimal downward force on the piston. For instance, an advanced timing setting might be 25 degrees BTDC, initiating the spark when the crankshaft is 25 degrees of rotation before top dead center.

Conversely, retarding the timing involves adjusting the spark to occur later in the combustion cycle. While specific engine conditions may require retarding the timing as a standard operation, excessively retarded ignition timing can negatively impact engine performance.

Combustion in a gasoline engine involves the rapid and complete burning of a compressed air-fuel mixture, and it begins when the ignition system creates a spark. This process entails a flame originating at the spark plug and swiftly traversing the combustion chamber, elevating gas temperatures and building pressure in a uniform and regulated manner.

In typical combustion scenarios, there is no explosion, and ignition is consistently triggered by an electric arc at the spark plug. However, if ignition occurs through means other than the electric arc at the spark plug, it denotes an abnormal occurrence termed pre-ignition. Pre-ignition transpires when the air-fuel mixture ignites prematurely, typically before the scheduled spark plug activation, often due to a heated area within the combustion chamber, like a glowing spark plug electrode or a heated carbon fragment.

Gasoline engine combustion involves the rapid and complete burning of a compressed mixture of air and fuel, which is initiated by a spark from the engine's ignition system. This process starts at the spark plug and quickly spreads across the combustion chamber, heating the gases and generating pressure in a controlled manner.

In normal combustion, ignition occurs through an electric arc at the spark plug, leading to a smooth and controlled combustion process. However, in cases of abnormal combustion, such as pre-ignition, ignition happens prematurely, often due to hot spots within the combustion chamber, like glowing spark plug electrodes or carbon deposits.

Understanding Cooling Systems

There are two primary cooling systems employed in modern automotive engines. The first is air-cooling, which involves circulating air over external cooling fins to dissipate excess heat. The second is water-cooling, where a liquid coolant absorbs heat from the engine and then transfers it to a radiator for dissipation. To maintain optimal emissions and efficiency, it's crucial for cylinder temperatures to reach and maintain operating levels swiftly across various ambient conditions. Hence, most contemporary automotive engines utilize liquid cooling due to its versatility in coolant circuit design, superior heat transfer capabilities, and high heat capacity.

The coolant is the most crucial element within the cooling system. Typically, engine coolant consists of a balanced blend of antifreeze and water in a 50/50 ratio. Frozen coolant can result in severe engine damage, potentially causing a cracked block or cylinder head. Therefore, it's imperative for the coolant to have freeze protection. Ethylene glycol is the most widely used kind of antifreeze. When mixed in a 50/50 ratio with water, ethylene glycol ensures that the coolant won't freeze until it reaches -34°F. Additionally, this mixture elevates the boiling point of the coolant, enhancing its effectiveness in transferring heat, particularly in hot weather conditions.

The following table details the water-cooling system components.

Part	Purpose
Water Pump	Tasked with circulating coolant throughout the cooling system to regulate and manage heat.
Water Jacket	The cylinder head and engine block have hollow areas for coolant flows to absorb heat.
Thermostat	Regulates engine temperature by permitting coolant flow into the radiator whenever the temperature goes over a set level.

Bypass Tube	Enables coolant to return to the water pump after the thermostat closes.
Radiator Hoses	Flexible conduits that facilitate the movement of hot coolant between the engine and the radiator.
Radiator	Facilitates the transfer of heat from the coolant to the air.
Radiator Cap	Maintains pressure within the system and allows coolant exchange between the radiator and the coolant reservoir.
Coolant Recovery Bottle	Serves as a reservoir for coolant, facilitating its movement in and out of the cooling system as engine temperature fluctuates.

Coolant is propelled through the engine via a water pump, which is driven by either a belt or a timing chain. Utilizing power from the engine crankshaft, the water pump directs coolant into the engine block. From there, coolant travels upward to the cylinder head before going through the bypass tube and into the water pump's inlet. This circulation process persists as long as the thermostat remains closed, indicating that the engine is below its operating temperature.

Once the thermostat opens, allowing the engine to reach its operating temperature, hot coolant flows past the thermostat and enters the upper radiator hose. Subsequently, the coolant enters the radiator, where its boiling point is raised through two primary methods. Firstly, coolants composed of a 50/50 mixture of ethylene glycol and water are utilized. Secondly, the pressure within the cooling system is increased, a task managed by the radiator cap. Typically, radiator caps maintain a pressure range of 9 to 16 pounds per square inch (psi). With each 1 psi of pressure added, the coolant's boiling point goes up by approximately 3°F. Consequently, a radiator cap rated at 15 psi elevates the boiling point of the coolant to a range between 212°F and 260°F.

Radiator caps have two distinct valves: a vacuum valve and a pressure valve. As the engine heats up, the coolant within the cooling system expands, leading to increased pressure. When the pressure exceeds the radiator cap's rating, the pressure valve inside the cap will open, allowing coolant to flow from the radiator into the coolant recovery bottle. This process continues until the cooling system pressure returns to the cap's specified rating, causing the coolant level in the recovery bottle to rise accordingly, indicating that the engine has reached its operating temperature.

Upon engine shutdown and subsequent cooling, the coolant contracts, creating a low-pressure environment within the cooling system. To prevent any negative consequences, such as radiator hose collapse, during engine startup, the vacuum valve in the radiator cap activates to equalize the pressure inside the system.

Cooling systems demand regular maintenance to uphold their optimal performance. Essential to this upkeep is the coolant itself, which should be replaced every two to three years to sustain its effectiveness. Maintaining a 50/50 mix of antifreeze and water ensures adequate freeze protection and corrosion resistance, safeguarding the engine against potential damage.

Furthermore, it's vital to inspect the condition of the belt driving the water pump and check for any leaks in the system. Regular examination of hoses and the radiator for visible signs of wear or damage is also recommended. Typically, leaks manifest externally and can be detected as drips from hose connections or faulty components within the cooling system.

Understanding the Lubrication System

The lubrication system is indispensable for engine operation, as it prevents the internal components from generating excessive friction that could lead to engine seizure. Its primary functions include the following:

- Lubrication: The system applies an oil film between moving components to minimize friction and ensure smooth engine operation.
- Cooling: It directs motor oil to areas of high temperature, like the piston undersides, absorbing heat and transferring it to the oil pan or an oil cooler if present.
- Sealing: Motor oil acts as a barrier between the piston, piston rings, and cylinder walls, enhancing compression and sealing combustion gases in the combustion chamber for improved efficiency.
- Cleaning: Special additives in the oil trap contaminants, suspending them in the oil so they can be captured and removed by the engine's oil filter.
- Noise Reduction: Motor oil helps dampen engine noise, contributing to a quieter engine operation.

Engine oil is an important part of the lubrication system as it serves as the primary lubricant for the engine's moving parts. The selection and use of high-quality engine oil are crucial for maintaining optimal engine performance, especially under demanding operating conditions. Engine oil typically consists of two main components additive package and a base oil.

Viscosity is a key characteristic of engine oil. Viscosity is the resistance to flow and is represented by a numerical rating. The Society of Automotive Engineers (SAE) establishes standards for engine oil viscosity. For instance, a 5 SAE viscosity rating would mean that the oil has a relatively low viscosity, so it has a ow resistance to flow. Oil with an SAE 50 oil would have high viscosity.

The American Petroleum Institute (API) quality rating system is another crucial aspect of engine oil classification. This system establishes standards for engine oil quality. For gasoline engines, the API rating typically begins with an "S" prefix, followed by a letter indicating the specific quality standard met by the engine oil.

The earliest gasoline engine oils were rated SA, but as engine oil requirements evolved, higher-quality ratings were introduced. Current gasoline engine oil ratings include SN, SM, SL, and SJ. Utilizing an SA motor oil in today's engines would not provide adequate protection, given the increased demands on modern engines.

Diesel engine oils, on the other hand, are identified with a "C" prefix for their quality ratings. Examples of diesel engine oil ratings include CJ-4, CI-4 plus, CI-4, and CH-4. Engine oils carrying both "S" and "C" ratings are suitable for use in both gasoline and diesel engines, such as engine oil with an SJ/CD rating.

The following are the primary lubrication system components and their functions.

- Oil pan: This component serves as the reservoir for engine oil, typically located at the bottom of the engine.
- Oil pickup tube and screen: Submerged in engine oil, this apparatus filters out large solids and guides oil into the oil pump.
- Oil pump: Positioned to pump oil through the engine oil galleries, it is usually driven by the engine's crankshaft.
- Pressure relief valve: Its function is to avoid excessive pressure buildup within the lubrication system.
- Oil filter: Before oil is distributed to various engine parts, it is filtered by this component, which removes contaminants.
- Oil galleries: These are passages or drillings within the engine assembly that facilitate the transport of oil to critical components.

Two primary designs govern lubrication systems: dry sump and wet sump. While the dry sump system is intricate and mainly reserved for high-performance racing vehicles, the wet sump configuration dominates in commercial and automotive engines. In a wet sump system, the engine's oil reservoir resides at the engine's bottom within the oil pan, where gravity channels all used oil. Here, it undergoes cooling. Subsequently, the oil pump moves oil from the oil pan to the pickup tube and screen into its inlet. Once inside, the oil is pressurized and directed to the engine's oil filter, where particulates are removed. The filtered oil then proceeds to the main oil galleries.

During cold starts, oil exhibits increased resistance to flow, demanding more energy for pumping and elevating engine oil pressure. Should the oil pressure exceed the relief valve's preset value, the valve opens, enabling some oil to return to the oil pan before filtration. This safeguard prevents oil pressure from reaching levels that could harm vital components like the oil filter, bearings, and engine seals.

The engine's reciprocating assembly, comprising pistons, connecting rods, and crankshaft, is the most crucial area in terms of lubrication needs. These components rely on oil to reduce friction and dissipate heat. To fulfill this requirement, the engine features large oil galleries, with passages drilled directly through the crankshaft to supply pressurized oil to the connecting rod bearings. Any excess oil from these bearings is dispersed onto the cylinder walls, aiding in piston lubrication and cooling.

Additionally, separate oil galleries branch from the main oil gallery to supply lubrication to the engine's valve train. This includes components such as the camshaft, lifters, push rods, rocker arms, and valve stems. Additionally, auxiliary components (like the timing chain and camshaft drive) and gear drives for components (like the ignition distributor and oil pump) may also benefit from lubrication provided by this oil flow.

Once the oil has traversed the engine and lubricates the components, it returns to the oil pan at the bottom of the engine. Positioned low in the engine compartment, the oil pan benefits from the cooler ambient air, aiding in dissipating heat from the oil. For applications demanding heavy-duty operation or operating in hot weather conditions, a liquid-to-air heat exchanger, akin to the radiator in the cooling system, may be employed to further cool the engine oil. This cooling process readies the oil for recirculation by the oil pump, thereby sustaining the engine's lubrication system.

Maintaining the lubrication system primarily involves regular oil and filter changes, using high-quality products. This simple yet crucial maintenance practice is the most cost-effective way to reduce engine wear and ensure optimal performance.

On the ASVAB, you might encounter questions testing your knowledge of vehicle systems, including engine operations such as engine theory, cooling systems, and lubrication systems. Let's consider a sample question from this domain.

Let's try a practice question.

When you add engine coolant, what is the typical mixture?

    (A) 20/80 water and soap solution
    (B) 50/50 salt and water
    (C) 50/50 water and ethylene glycol
    (D) 60/40 antifreeze and saltwater

The correct answer is C.

DELVING INTO COMBUSTION SYSTEMS

Combustion systems involve the coordination of several components to facilitate efficient fuel combustion within the engine. This includes the fuel system, which regulates the air-fuel mixture for optimal combustion, the ignition system, responsible for generating and timing the spark that ignites the fuel, and the exhaust system, which removes waste gases from the engine and expels them into the atmosphere.

Understanding the Fuel System

The fuel system plays a crucial role in ensuring the engine operates efficiently by maintaining the correct air-fuel mixture. A properly adjusted mixture is essential for optimal drivability, emission control, fuel economy, and engine longevity.

Traditionally, mechanical methods were used to determine the air-fuel mixture, typically through carburetors, which were reliable but lacked the precision demanded by modern fuel systems. However, advancements in emission control regulations have rendered carburetors obsolete. Electronic fuel injection, controlled by an onboard computer, has replaced carburetors. This sophisticated system incorporates feedback mechanisms, allowing it to swiftly adapt to changing engine conditions with precision and accuracy.

The following are the electronic fuel inject system's primary components and their purposes.

- Electric fuel pump: Positioned inside the fuel tank, this component pressurizes fuel and delivers it to the fuel injectors.
- Fuel filter: Serves to remove impurities before the fuel gets to the fuel rail.
- Fuel rail: Acts as a manifold, supplying pressurized fuel to the inlet of each engine fuel injector.
- Fuel pressure regulator: Maintains optimal pressure within the fuel rail by adjusting to intake manifold vacuum levels. Excess fuel is diverted to the fuel return line and directed back to the fuel tank.
- Fuel injector: Receives electrical signals from the powertrain control module (PCM) and atomizes fuel into the intake air stream. The injector's placement dictates the specific type of fuel injection system employed by the engine.
- Powertrain control module (PCM): Functions as the central computer of the vehicle, governing all engine and transmission-related operations.
- Intake manifold: Distributes air to the intake ports located on the cylinder heads.
- Intake air filter: Removes airborne contaminants that could potentially harm internal engine components. All incoming air moves through the air filter then moves into the engine.
- Throttle body/throttle plate: Linked to the throttle pedal, this unit regulates engine speed and torque output.

In the early stages of fuel injection system development, one or two injectors were typically installed within a throttle body, effectively replacing the carburetor. This setup, known as throttle body injection (TBI), offered reliability but fell short in meeting stringent emission control standards due to limited fuel control capabilities. Today, most engines opt for either multiport fuel injection, with an injector for each cylinder, or direct injection, where fuel is sprayed directly into the combustion chamber.

In multiport fuel injection systems, injectors are positioned within the intake manifold, directing their spray toward the intake valves. This setup ensures superior air-fuel mixing and prevents fuel droplets from settling on the intake manifold runners as air flows toward the cylinder head. In contrast, direct injection employs a high-pressure injector to supply the fuel to the combustion chamber. Here, only air enters the chamber over the intake valves. Direct injection allows precise timing of fuel delivery during the compression stroke and serves as a coolant for the hot compressed air, enabling engines to operate with higher compression ratios for enhanced power and efficiency.

Explaining the Fuel Injection Types

Maintenance for the fuel injection system is minimal. Both the air and fuel filters need to be replaced regularly, and you should consult the manufacturer's guidelines to follow their guidelines for replacement. Additionally, manufacturers may recommend occasional cleaning of the injection system. However, overall, fuel injection systems are designed to deliver dependable performance with little maintenance required.

The ignition system is among the most vital components of a vehicle. It's responsible for initiating combustion, enabling the engine to run. This system must produce high-voltage sparks at precisely the right moment for the engine to operate smoothly and effectively.

The ignition system has two main subsystems:

- The primary ignition system, which works at a low voltage
- The secondary ignition system, which works at a high voltage.

The following are the primary ignition system's primary components, with the ignition system being electronic.

- Battery: Powers the ignition system to start the engine.
- Ignition switch: Controls the power flow to the ignition system, allowing the engine to be turned on and off.
- Primary coil winding: Consists of several hundred turns of heavy wire in the ignition coil, operating at low voltage.
- Ignition module: A transistorized switch that regulates the flow of current in the primary coil winding.
- Reluctor and pickup coil: Generates a signal to activate the ignition module. The reluctor, mounted on the distributor shaft, rotates with the distributor, producing a signal in the stationary pickup coil.
- Distributor: Synchronized with the engine's camshaft, the distributor ensures accurate timing of the spark and directs it to the correct cylinder. The distributor at a rate that is half as fast as the engine.

The following list provides more details:

- Secondary coil winding: Comprising numerous turns of fine wire wrapped around the primary coil winding, it produces high voltage within the ignition coil.
- Coil wire: Conveys high voltage to the distributor cap from the secondary coil winding.
- Distributor cap and rotor: Serve as the conduit for high voltage from the coil wire to each cylinder in the firing sequence. The distributor cap and rotor act as a switching mechanism, allowing one ignition coil to supply all engine cylinders.
- Spark plug wires: Transmit high voltage to each spark plug from the distributor cap.
- Spark plugs: Threaded into the cylinder head and extending into the combustion chamber, they produce the spark needed to initiate combustion. the components for secondary ignition systems.

When the driver turns the ignition switch, current goes from the battery, through the ignition switch, and into the primary coil winding. This current then travels through the ignition module before returning to the battery through the vehicle ground circuit. Within the ignition system, the low-voltage (primary) system governs the high-voltage (secondary) system. The events in the secondary system are contingent upon those in the primary system.

The functioning of the ignition system relies on electromagnetic induction. A magnetic field introduces voltage when it moves across a stationary wire. Additionally, electric current passing through a wire creates a magnetic field. Thus, electric current can create a magnetic field, and vice versa.

With the collapse of the magnetic field in the primary coil winding, a high voltage is induced in the secondary winding. This secondary system voltage typically ranges from 20,000 to 40,000 volts, with modern systems capable of even higher voltages. This significant voltage increase in the ignition coil is achieved due to the secondary winding's large number of wire turns compared to the primary winding. The primary winding's magnetic field interacts with a many secondary winding turns, and creates in a step-up effect.

The high-voltage current in the secondary winding needs to be directed to the correct cylinder in the firing order. This current goes to the distributor cap through the coil wire. From there, it travels to the rotor, which is mounted on top of the distributor shaft and rotates with it. As the engine rotates, the rotor guides the current to the appropriate cylinder and sends it to the spark plug. The spark plug's electrical discharge starts combustion.

As engineers sought to enhance engine performance by streamlining ignition systems, they developed the distributorless ignition system (DIS) as a solution. DIS systems replace the distributor with individual ignition coils, each operating two spark plugs. For example, a V-8 engine would require four separate coils. This design enables each coil to accumulate electric charge more effectively, resulting in a stronger spark upon discharge. However, DIS systems still rely on spark plug wires, which degrade over time and necessitate periodic replacement for optimal system performance.

In the latest ignition system designs, spark plug wires have been entirely eliminated in favor of coil-on-plug ignitions. Here, individual ignition coils are directly mounted onto the spark plugs, with their discharge controlled by the vehicle's computer system. While the integration of computer control has increased system complexity, the removal of moving parts and maintenance-intensive components has significantly improved ignition system reliability. This advancement ensures precise control of spark timing, crucial for maximizing engine efficiency, minimizing emissions, and enhancing overall system dependability.

Understanding the Exhaust Systems

The exhaust system plays a crucial role in eliminating engine waste gases. Its primary functions include facilitating the smooth flow of these gases, dampening the noise produced by the exhaust, and ensuring that both the gases and heat are directed away from the vehicle cabin and its occupants.

The following are the exhaust system's primary components and their purposes.

- Exhaust manifolds: These components are directly connected to the cylinder head's exhaust ports. They primarily manage exhaust heat and noise. Typically crafted from cast iron due to their ability to withstand high temperatures.
- Catalytic converter: This device converts harmful elements in engine exhaust into less harmful compounds like carbon dioxide and water.
- Muffler: Utilizing an expansion chamber and sound-absorbing materials, the muffler reduces the loudness of exhaust noises.
- Tailpipe: Serving as the outlet for exhaust gases into the open atmosphere, the tailpipe is usually positioned at the rear of the vehicle.

As gases exit the engine's exhaust ports, the exhaust manifold gathers them. In a V-type engine, there are typically two exhaust manifolds, one for each cylinder head. These manifolds channel the gases into steel exhaust pipes, which interconnect the key components of the exhaust system.

The exhaust gases then move into the catalytic converter. Within this component, high temperatures facilitate the conversion of toxic exhaust elements into less harmful gases. Some vehicles may employ multiple catalytic converters to adhere to emission control standards.

Upon exiting the catalytic converter, the exhaust gases travel into the muffler. Equipped with expansion chambers, the muffler absorbs and dampens the loud noises produced during the engine's combustion process. Typically positioned towards the rear of the vehicle, the muffler is situated after the catalytic converter and before the tailpipe.

Even after undergoing treatment in the catalytic converter, engine exhaust remains highly toxic. Among the hazardous gases emitted, carbon monoxide stands out as particularly deadly due to its odorless and colorless nature. It's imperative that these gases are directed away from the vehicle's occupants to prevent exposure. Typically, exhaust systems are designed to route these gases to the rear of the vehicle, where they disperse into the open air through the tailpipe.

To optimize engine efficiency and power output, the exhaust system can be fine-tuned to complement the engine's design. Often referred to as header pipes, these "tuned" exhaust manifolds aim to facilitate smoother flow of exhaust gases, allowing the engine to breathe more efficiently. However, while performance gains are achieved, durability may be compromised as header pipes are generally less robust than traditional cast-iron exhaust manifolds.

The exhaust system requires regular inspection for any physical damage, ensuring there are no restrictions or leaks. Additionally, the mounting brackets and clamps should be securely tightened, and any damage should be promptly addressed.

The longevity of the catalytic converter largely depends on the engine's overall health. A well-maintained engine will promote the catalytic converter's durability. Conversely, engine issues such as oil consumption, or problems with the ignition or fuel system, can lead to damage to the catalytic converter.

Let's consider a sample question from the combustion systems section, which encompasses the fuel system, ignition system, and exhaust system.

Let's take a look at an example question.

What does the multiport fuel injection prevent?

- (A) Fuel droplets fall from the exhaust valves
- (B) Fuel droplets fall from the intake manifold runners
- (C) Engine fires
- (D) Engine overheats

The answer is B.

DELVING INTO ELECTRICAL AND CONTROL SYSTEMS

Electrical and control systems play an increasingly vital role in modern vehicles. This encompasses the electrical system, encompassing functions such as starting, charging, lighting, and accessory operations. Moreover, it includes the computer system, which serves as the central control unit managing various vehicle functions, including engine operation, drivetrain control, braking, and suspension.

The importance of the vehicle's electrical system grows every time there is a new model. Many systems that were once mechanically driven are now being redesigned to operate electrically.

The following are the electrical system's major subsystems and their purposes.

- Battery: Stores a chemical form of electrical energy and provides DC for engine starting and accessory operation.
- Starting System: Responsible for cranking the engine to get it started, with the battery and starter motor as its major components.
- Charging System: Supplies electrical current to charge the battery and powers the vehicle's operation, with the alternator being the key component.
- Lighting System: Includes likes like the headlights, marker lights, brake lights, and tail lights.
- Accessories: Comprise the rear-window defogger, windshield wipers, stereo, blower motors, and all other electrically powered accessories.

Battery

The battery is the cornerstone of the entire electrical system. It provides the electrical current necessary to start the engine, supplies current to the electrical system when the load exceeds the alternator's output, and acts as an electrical "shock absorber," preventing voltage spikes during periods of excessive current flow.

An automobile battery consists of lead plates submerged in an electrolyte composed of sulfuric acid and water, hence it is known as a lead-acid battery. Upon discharge, the sulfuric acid in the electrolyte undergoes a chemical reaction, converting into water, and simultaneously, the lead plates undergo a transformation, becoming lead sulfate. Charging the battery reverses this process, restoring the original chemical composition of the lead plates and electrolyte. Working with an automotive battery requires caution due to the substantial electrical energy it contains and the highly corrosive nature of sulfuric acid.

Starting System

To start the engine, a starter motor powered by electricity is employed. The starter solenoid receives an electrical current when the ignition switch is turned on. Next, using the engine's ring gear on the flywheel, the solenoid contacts the starter drive gear. Once the drive is engaged, the solenoid connects the battery to the starter motor, causing it to turn the engine at a sufficient speed to initiate combustion.

Charging System

The charging system provides electrical current to the running engine to recharge the battery and power the vehicle's electrical components. The primary component of the charging system is the alternator. The alternator, driven by a belt connected to the engine's crankshaft, makes electrical energy from mechanical energy. It produces AC, then converts AC to DC for use in the vehicle's electrical system.

A voltage regulator controls the alternator's output. Normal system voltage during engine operation is around 14.5 volts. When accessories such as headlights and the heater motor are turned on, the electrical load increases, causing a drop in system voltage. The voltage regulator detects this drop and responds by increasing the alternator's output to maintain a consistent voltage close to 14.5 volts. As long as the alternator's output matches the electrical load, the system voltage will remain stable.

Lighting System

A vehicle's lighting system includes numerous lights for different purposes. Headlights light up the road, taillights indicate the car's rear to those behind the car, and interior lights let you see the dashboard and instrument panel as necessary. These lights are controlled by switches operated by the driver, which turn the electrical current to the lighting circuits on and off.

The lighting circuits are protected by fuses or circuit breakers. If a circuit draws more current than it is designed to handle, the fuse will "blow" or the circuit breaker will trip, cutting off the current flow. This prevents the wires in the circuit from overheating and can help prevent electrical fires.

Computer System and Accessories

Computer control systems in vehicles function similarly to the human nervous system. Sensors collect data and send it to the computer, which processes this information and then signals actuators to manage various vehicle functions. This is akin to how your ears and eyes return signals to your brain, which then that data to other parts of the body, especially the muscles to move. The following are the major components and their purposes:

- Sensors: These generate signals based on rotational speed, temperature, pressure, and relative position. They are essentially the ears and eyes of the computer system. For instance, an oxygen sensor transmits information to the computer regarding the oxygen levels in the exhaust gases. The computer then analyzes this data to ascertain whether the engine's air-fuel mixture is rich or lean.
- Computer: Often referred to as the powertrain control module (PCM) or engine control unit (ECU), the computer analyzes information received from sensors using predefined strategies (software) and generates commands to regulate vehicle operations. It functions as the system's main control unit.
- Actuators: These receive output signals from the computer and control various vehicle functions. For instance, a fuel injector acts as an actuator by responding to computer signals to inject a specific amount of fuel into the intake air stream.

Automotive technicians communicate with the ECU using a scan tool, which allows them to get information from the computer concerning the operation of the various vehicle systems. The scan tool is connected to the vehicle's computer system through a diagnostic data link that is typically located near the driver's seat.

Let's try an example question.

What kind of current does the alternator produce?

- (A) Alternating
- (B) Direct
- (C) Slow
- (D) Fast

It's in the name, so the answer is A, alternating.

DELVING INTO CHASSIS SYSTEMS

Chassis systems are integral for vehicle performance and safety. These systems include the drivetrain, which transmits power from the engine to the drive wheels, as well as the suspension and steering systems, responsible for maintaining ride quality and controlling vehicle handling. The chassis systems include your brakes.

The Drivetrain System

The engine power the vehicle, but this energy must be efficiently handled to ensure smooth and quick acceleration. This task falls to the vehicle's drivetrain.

A key component of the drivetrain is the transmission, which adjusts engine speed to achieve the desired vehicle speed. Transmissions are either automatic or manual. Drivers who prefer more control over their vehicle's operation often choose a manual transmission, because drivers have control by manually shift gears. In contrast, a car with an automatic transmission is easier to operate because it shifts gears automatically without a driver-operated clutch.

While rear-wheel drive was once the most common design, front-wheel drive has become increasingly popular, especially in small- to medium-sized cars.

While frOnt-whell drive may vary, the following are the usual components found in this part of the Chassis system.

- Clutch: Facilitates the transfer of torque to the transmission from the engine. By engaging and disengaging, the clutch enables smooth starts and gear shifts.
- Transaxle: Contains multiple gears that are selected based on the vehicle's speed and desired acceleration. It combines the functions of the transmission and the drive axle, typically found as separate components in rear-wheel drive vehicles.
- Half shaft: A short shaft that transfers power to the drive wheels from the transaxle. Predominantly used in front-wheel drive vehicles, half shafts allow the wheels to move vertically and turn while transmitting power. Each vehicle has two half shafts, one for each drive wheel.

- Constant-velocity (CV) joints: Positioned at both ends of a half shaft, CV joints can efficiently transmit power at steep angles. The inboard CV joint is located at the transaxle, while the outboard CV joint is situated behind the vehicle wheel.

The following are other components of the drivetrain and their purpose.

- Transmission: Found in rear-wheel and four-wheel drive vehicles, the transmission adjusts engine speed to match the desired vehicle speed.
- Drive shaft: A longer variant of the half shaft, the drive shaft transmits torque to the drive axle from the transmission. It permits limited vertical movement of the vehicle's wheels.
- Universal joints: Situated at each end of the drive shaft, universal joints enable the shaft to function at an angle with the component it drives. Also referred to as U-joints.
- Drive axle: Transfers engine power at a 90-degree angle and distributes it between the two drive wheels. The wheels connect to both sides of the drive axle, which includes a differential allowing the left and right wheels to rotate at varying speeds when the vehicle turns.
- Transfer case: It's location varies based on the car: for four-wheel drive vehicles, it is located between the drive axles and the transmission, and it distributes engine power between the rear and front drive axles.

Transmissions are designed with multiple forward ratios, also known as gears, which determine the speed at which the vehicle travels. For example, a four-speed transmission includes four forward gear ratios and one reverse ratio. When starting from a stop, the driver typically engages low (first) gear to increase engine torque sufficiently for the vehicle to begin moving. As the vehicle gains momentum, the driver shifts to higher gears, such as second and third, to continue accelerating.

High (fourth) gear is typically used for highway cruising, maintaining moderate to low speeds for the most efficient fuel use and to minimize noise. When quick acceleration is needed, downshifting to a lower gear increases engine speed and power output. Selecting the appropriate gear is crucial when operating a vehicle with a manual transmission. Modern transmissions may offer up to eight forward speeds to further optimize performance and fuel efficiency.

Automatic transmissions, while more intricate than manual counterparts, offer greater convenience due to their lack of a driver-operated clutch. Unlike manuals, automatics utilize a torque converter to transmit engine torque to the transmission, employing fluid for power transfer and enabling controlled slippage when the vehicle is stationary.

Gear selection in automatic transmissions is handled internally, with hydraulic pressure activating the appropriate planetary gear sets. The process is electronically controlled by the vehicle's powertrain control module in most modern automatic transmissions, eliminating the need for manual gear shifting by the driver.

Another contemporary automatic transmission is the continuously variable transmission or CVT. Unlike conventional transmissions that need specific gear settings to establish transmission ratios, a CVT employs an input and an output cone, which are connected by a belt or chain. The transmission ratio is based on belt's or chain's position on the cones. Unlike traditional transmissions, which are restricted to fixed gears, a CVT allows for continuous variation of the ratio within the range defined by the diameters of the two cones.

Suspension and Steering System

The latest advancements in suspension and steering systems integrate computerized control to enhance vehicle performance. Despite the heightened complexity introduced by computerization, the fundamental principles governing these systems remain unchanged. The following are the system's major components.

- Springs: Support the vehicle's chassis and enable vertical movement.
- Shock absorbers: Dampen the energy released from the wheel's vertical motion.
- Control arms, also known as A-arms: only present in long-short arm suspension systems, this part stabilizes the vertical position of the steering knuckle as the wheel travels up and down.
- Steering knuckle: uses the ball -joints to link the lower and upper control arms. The wheel hub mounts on the spindle.
- Ball joints: Enable the steering knuckle to move vertically and turn simultaneously through ball-and-socket assemblies.
- Steering linkage: Establishes the steering knuckle and steering wheel connection.
- Wheel hub: Serves as the place where the vehicle's tire assembly is mounted.
- Tire: Contacts the road surface and contributes to vehicle stability and handling by providing a footprint.

To ensure a smooth ride, the vehicle's wheels need to move up and down while maintaining the stability of the chassis. This task falls to the vehicle's springs, which facilitate wheel movement without transmitting road shock. However, as the springs absorb energy during this process, they must release it effectively to prevent the vehicle from bouncing excessively after encountering bumps.

Located between a control arm and the chassis, the shock absorber plays a crucial role in dissipating or absorbing the energy in the spring as it compresses (jounces) or extends (rebounds). Without properly functioning shock absorbers, the vehicle may continue bouncing when you go over a bump. Employing a piston and hydraulic oil, the shock absorber absorbs surplus mechanical energy from the suspension system and converts it into heat, thereby ensuring a smoother ride.

The upper and lower control arms serve as supports for the steering knuckle, with the control arms attached to the chassis through control arm bushings. Positioned between the chassis and either the upper or lower control arm, the long-short arm suspension system's spring supports the vehicle's weight and absorb road shocks. Whichever control arm bears the spring becomes the load-bearing arm, supporting the vehicle's weight as it travels to the steering knuckle and tire assembly.

The upper and lower ball joints connect the steering knuckle to the control arms, functioning like human hip joints where a ball and stud rotate within a socket to enable extensive movement. These ball joints enable the steering knuckle to move freely up and down with the control arms and also facilitate its left and right turning motion via the steering linkage.

The steering linkage establishes the connection between the steering wheel and the steering knuckle, with two primary designs commonly in use: linkage steering and rack and pinion. In long-short arm suspensions, linkage steering connects the steering column to the pitman arm, center link, and idler arm, with tie rods completing the connection to the steering knuckles. On the other hand, smaller vehicles often utilize the rack-and-pinion steering system due to its compact design compared to the linkage steering system.

Tires

Tires play a pivotal role in the suspension and steering system, bearing the weight of the vehicle and establishing the final connection with the road surface. Among tire designs, the radial tire stands out for its stable footprint and minimal rolling resistance.

The tire construction initiates with the beads, circular segments crafted from robust materials like steel wire, encased within rubber. These beads serve as the anchoring points for the tire on the rim. Body plies constitute the tire's main structure, extending from bead to bead. All other tire components, such as the liner (internal sealed surface), sidewalls, and tread, are affixed to these body plies.

Belts, situated between the plies and the tread, play a crucial role in stabilizing the tire's footprint on the road. A secure footprint not only enhances traction across diverse road surfaces but also elevates the vehicle's handling and braking performance.

The key maintenance for suspension and steering systems involves lubricating ball joints, tie rod ends, and other components. This task is typically performed using a hand-operated grease gun, ensuring careful metering of the grease to prevent over-lubrication.

Proper tire maintenance primarily revolves around maintaining correct inflation levels. Inadequate tire pressure poses safety risks and elevates fuel consumption by increasing the rolling resistance. An underinflated tire risks damage and potential blowouts, while overinflation accelerates tire wear. Automotive manufacturers provide recommended inflation pressures for both front and rear tires, facilitating optimal tire care.

Brake Systems

Among all the systems within a car, arguably the most crucial is the brake system. While being unable to move forward poses its challenges, the inability to stop presents an even greater concern.

Brake systems have the following primary components.

- The brake pedal connects the driver's foot to the master cylinder, allowing precise control of braking.
- Positioned in the engine compartment near the driver, the master cylinder generates hydraulic pressure to activate the brake systems at the wheels.
- A fluid reservoir atop the master cylinder supplies fluid to the brake circuits.
- Brake lines, whether they are steel lines running along the chassis or flexible hoses connecting steel lines to brake assemblies, move hydraulic pressure to the brake systems from the master cylinder.

Braking mechanisms harness the kinetic energy of a vehicle's movement, transforming it into heat energy via friction. There are two primary types of brake setups:

- Drum brakes: Utilizing expanding shoes, they engage with a rotating drum to create friction. Disc brakes: Brake pads situated on either side of a spinning disc are brought together, effectively slowing down the vehicle.

These brake systems operate hydraulically. A pumping piston, housed within the master cylinder, responds to the brake pedal's action, exerting pressure on the brake fluid within the system. Because fluids are non-compressible, the brake fluid travels through the brake lines, activating the pistons in the brake assemblies to engage the brakes. The intensity of the braking force corresponds directly to the pressure applied to the brake pedal by the driver, determining the level of fluid pressure and resulting braking power.

In drum brake systems, the components responsible for generating friction are the brake shoes. Each drum brake assembly comprises two brake shoes. When hydraulic pressure is exerted on the wheel cylinder, positioned between the brake shoes, its pistons expand and press against them. Consequently, the brake shoes extend outward, touching the brake drum's inner surface. This interaction creates friction, effectively decelerating the vehicle.

Disc brakes outperform drum brakes primarily because of a superior heat dissipation capabilities. Disc brake pistons are notably larger, enabling them to exert greater force on the brake pads. Additionally, disc brake rotors are designed for enhanced cooling, benefiting from better exposure to the cooler air beneath the vehicle and often featuring air passages to facilitate heat dispersion.

The piston responsible for activating disc brakes resides within a brake caliper. This caliper possesses lateral mobility, allowing it to "float" during brake application and release. This design enables a single piston to engage the pads on both sides of the rotor, akin to the operation of a C-clamp. Similar to tightening a screw on one side of a C-clamp, the entire assembly constricts, firmly gripping the brake rotor.

Braking force is often enhanced with the aid of a brake booster, located between the master cylinder and the brake pedal. By harnessing engine intake manifold vacuum, the brake booster intensifies the force exerted on the master cylinder, enabling higher hydraulic pressures within the brake system with the same input force from the driver's pedal. Certain brake booster setups employ fluid pressure from the power steering system to enhance braking effectiveness.

Similar to other vehicle systems, computerized control is increasingly integrated into braking systems. Modern cars often feature antilock brakes (ABS), designed to prevent wheel lock during intense braking scenarios. Utilizing speed sensors affixed to each wheel, the ABS system relays the relative speeds of all wheels to a dedicated computer (the ABS computer). Should the ABS computer detect a significant variance in wheel speed beyond a predetermined threshold, it activates pumps and valves within the ABS system to adjust brake pressure for the affected wheel or wheels. This technological enhancement grants drivers greater control in slippery conditions, promoting more predictable and safer stopping maneuvers.

Let's do a sample problem.

You will almost certainly control a modern automatic transmission by:

- (A) The engine
- (B) The PCM
- (C) The driver
- (D) Oil pressure

The answer is the PCM, or B.

PRACTICING WHAT YOU'VE LEARNED

Now that you've got the basics, you can see test your newly acquired information and put it to the test.

Here are two practice sets with 10 questions. This is similar to how the CAT is setup, although obviously your questions aren't going to change the better you do.

The last section of this chapter details the right answer, and how you could have reached that answer.

Practice 1

1. What is the purpose of engine oil?

    (A) To cool the engine
    (B) To clean the windshield
    (C) To lubricate engine parts
    (D) To increase fuel efficiency

2. What does ABS stand for?

    (A) Advanced Brake System
    (B) Automatic Braking System
    (C) Antilock Brake System
    (D) Accelerated Braking Solution

3. What does RPM stand for?

    (A) Revolutions Per Mile
    (B) Rounds Per Minute
    (C) Revolutions Per Meter
    (D) Revolutions Per Minute

4. What is the purpose of a catalytic converter?

    (A) To increase engine power
    (B) To reduce emissions
    (C) To cool down the exhaust
    (D) To improve fuel economy

5. What does OBD-II stand for?

    (A) Onboard Diagnostic System II
    (B) Overdrive Boost Device II
    (C) Oxygen Bypass Detector II
    (D) Oil Burnout Detector II

6. What does PSI refer to in tire pressure?

    (A) Pounds per Square Inch
    (B) Pressure Sensing Indicator
    (C) Perimeter Support Index
    (D) Pressure Stability Indicator

7. What is the function of a fuel injector?

    (A) To measure fuel level
    (B) To ignite the fuel
    (C) To regulate tire pressure
    (D) To deliver fuel to the engine

8. What is the purpose of a radiator?

    (A) To filter the air
    (B) To cool the engine
    (C) To cool the cabin
    (D) To power the headlights

9. What does EV stand for?

    (A) Engine Vehicle
    (B) Electric Vehicle
    (C) Eco-friendly Vehicle
    (D) Efficient Vehicle

10. What does AWD stand for?

    (A) All Weather Drive
    (B) Automatic Wheel Drive
    (C) All-Wheel Drive
    (D) Accelerated Wheel Drive

Practice 2

1. What is the function of the brake master cylinder?

    (A) To activate the horn
    (B) To control the air conditioning
    (C) To engage the brakes
    (D) To regulate tire pressure

2. What does CVT stand for?

    (A) Continuously Variable Transmission
    (B) Controlled Vehicle Technology
    (C) Continuous Vibration Transfer
    (D) Constant Velocity Transmission

3. What is the role of a serpentine belt?

    (A) To play music
    (B) To drive the engine components
    (C) To heat the cabin
    (D) To inflate the tires

4. What is the purpose of a throttle body?

    (A) To control air intake
    (B) To adjust tire pressure
    (C) To regulate fuel flow
    (D) To measure engine temperature

5. What is the function of a shock absorber?

    (A) To increase tire grip
    (B) To reduce engine noise
    (C) To control suspension movement
    (D) To regulate fuel injection

6. What is the primary purpose of a radiator cap?

    (A) To prevent coolant leaks
    (B) To regulate tire pressure
    (C) To maintain radiator pressure
    (D) To control engine temperature

7. What does HVAC stand for?

    (A) Heating, Ventilation, and Air Conditioning
    (B) High Velocity Automotive Cooling
    (C) Hot Vehicle Air Control
    (D) Humidity Ventilation and Air Circulation

8. What is the function of an alternator?

    (A) To charge the battery and power electrical systems
    (B) To cool the engine
    (C) To adjust tire pressure
    (D) To ignite the fuel

9. What does RPM gauge measure?

    (A) Radiator Pressure Measurement
    (B) Relative Power Modulation
    (C) Rotations Per Mile
    (D) Revolutions Per Minute

10. What does 4WD stand for?

    (A) 4-Wheel Drive
    (B) 4-Wheel Differential
    (C) 4-Wheel Driveability
    (D) 4-Wheel Drift

CHECKING YOUR ANSWERS

Now it's time to see how well you did.

Practice 1

1. (C) To lubricate engine parts

2. (C) Antilock Brake System

3. (D) Revolutions Per Minute

4. (B) To reduce emissions

5. (A) Onboard Diagnostic System II

6. (A) Pounds per Square Inch

7. (D) To deliver fuel to the engine

8. (B) To cool the engine

9. (B) Electric Vehicle

10. (C) All-Wheel Drive

Practice 2

1. (C) To engage the brakes

2. (A) Continuously Variable Transmission

3. (B) To drive the engine components

4. (A) To control air intake

5. (C) To control suspension movement

6. (C) To maintain radiator pressure

7. (A) Heating, Ventilation, and Air Conditioning

8. (A) To charge the battery and power electrical systems

9. (D) Revolutions Per Minute

10. (A) 4-Wheel Drive

CHAPTER 12

SHOP INFORMATION

This chapter focuses on the basics you need to know to complete the shop information section.

A BIT MORE ABOUT THIS SECTION OF THE ASVAB

To function efficiently in any industrial or technical setting, it's crucial for a technician to accurately identify tools and utilize them safely. Hand tools serve as the cornerstone of industry; without them, productivity would be severely hindered. Even the most intricate machinery would struggle to operate smoothly without access to hand tools for maintenance and repair tasks.

You won't have long for this section, but you won't have to answer too many questions either. If you know this topic well, you will probably be able to get through it fairly quickly.

- The ASVAB gives you 6 minutes to finish 10 questions about AI if you take the CAT.
- If you take the traditional ASVAB, you will have 25 questions with just 11 minutes to finish. This means finishing more than two questions a minute.

KNOWING THE DIFFERENT MEASURING TOOLS

An essential competency for any technician involves making precise measurements.

Steel rules or tape measures are suitable for measuring distances, including to a fraction of an inch. However, they may not suffice for tasks demanding higher precision due to the challenge of reading smaller increments on the scale. Steel rules also prove valuable for laying out straight lines.

For precision down to thousandths of an inch, technicians rely on micrometers. Among the most prevalent is the outside micrometer, designed for measuring the thickness of flat items or the outer diameter of cylindrical objects. Structurally akin to a C-clamp, the micrometer's spindle adjusts the distance between itself and the anvil by rotation. Measurements are taken between the spindle and anvil's opposing faces.

The thimble of the micrometer is directly connected to the spindle, while the sleeve remains stationary, bearing markings to indicate the spindle's movement relative to the anvil. As the thimble extends outward, it reveals the graduations on the sleeve, with the micrometer reading being determined by the thimble's position in relation to the sleeve.

Each complete rotation of the thimble corresponds to precisely 1 inch of spindle movement, equivalent to " or 0.025". Graduations on the sleeve also occur every 0.025" to track thimble rotations. After four rotations, the spindle advances of an inch or 0.100", marked by a prominent "1" on the sleeve, followed by "2" at 0.200", and so forth.

The thimble's exterior features 25 evenly spaced graduations. Given that one full turn of the thimble moves the spindle 0.025", each graduation on the thimble represents 0.001". Reading a micrometer involves adding the sleeve measurement to the aligned number on the thimble relative to the longitudinal line on the sleeve.

A caliper, a dual-legged tool, serves to gauge the distance between two points on an object, and it can also transfer measurements between objects. Varieties include outside calipers, for measuring external dimensions, and inside calipers, for assessing internal dimensions.

A spirit level, also referred to as a bubble level or simply a level, is utilized to ascertain the horizontal (level) or vertical (plumb) orientation of a surface. In its tubular form, a spirit level consists of a fluid-filled tube housing a bubble that aligns at the center when the surface is level. Alternatively, a bullseye spirit level features a fluid-filled circle with a slightly convex face, accommodating a centered bubble. While a tubular spirit level levels solely in the direction of the tube, a bullseye spirit level can ensure surface alignment across a plane.

A steel square is used to measure or lay out angles. A steel square is also known as a carpenter's square or framing square, because carpenters commonly use it to frame stairs and rafters. Its two arms meet at a 90-degree angle. The short arm is referred to as the tongue, and the long arm as the blade (or body).

Let's look at an example question.

You are measuring a wood plank within $\frac{1}{40}$ inch. Which of the following answers is the best tool to do this?

    (A) Measuring tape
    (B) Micrometer
    (C) Spirit level
    (D) Steel rule

The answer is the micrometer, or B.

KNOWING THE DIFFERENT STRIKING TOOLS

The hammer serves the purpose of striking objects to either remove or install them. Virtually every technician utilizes a hammer in some capacity, whether it's for driving nails or loosening components of an assembly. Hammers come in various designs and sizes tailored to their specific applications.

Among the most commonly used by metalworkers and mechanics is the ball-peen hammer, featuring a standard striking face and a rounded end suitable for shaping metal and crafting gaskets. Additionally, mechanics often employ rubber mallets to safeguard parts from damage during striking. Unlike hammers designed for maximum impact, rubber mallets are crafted to install or remove delicate components such as hubcaps without causing surface harm. Carpenters achieve a similar effect using wooden mallets.

Carpenters frequently opt for a claw hammer, renowned for its dual functionality. Featuring a hammer head with two ends. One of the ends drives nails, while the other extracts them. There are various sizes of claw hammers, and they are determined by the head's weight. A standard claw hammer has a 13 oz head. A rough-framing hammer used for framing wooden structures, range in size from 16 to 20 oz.

For heavy-duty tasks such as driving fence posts or demolishing drywall, a sledgehammer is indispensable. Characterized by a long handle and a large steel head, the sledgehammer typically necessitates the use of both hands to wield effectively.

A nail, a pin-shaped metal fastener, features a sharp point on one end and typically a flat head on the other, although headless nails also exist. The section of the nail between the head and the point is referred to as the shank. You drive nails into wood to hold things together.

Rivets serve as metal fasteners for assembling parts. Essentially, a rivet has a pin with a head at one side. To install a rivet, a hole, matching the diameter of the rivet, is drilled through the two pieces to be joined. Once the pieces are securely clamped together, the rivet's head is positioned on a solid surface while the opposite end is shaped into a head using either a hammer or a specialized riveting tool. This process results in a semi-permanent assembly, as the rivet must be drilled out to be removed.

Hammers are frequently paired with chisels, punches, or drifts for various tasks. A chisel typically features a long, sharp edge and is employed for cutting purposes. Punches, on the other hand, are slender tools utilized for driving small fasteners and marking layouts. Drifts serve to strike objects where it's crucial to avoid direct contact between the hammer and the workpiece.

The cold chisel ranks among the most common chisels, featuring a straight, sharp edge ideal for cutting off bolt heads or separating assembly pieces. As cold chisels dull over time, they require periodic sharpening on a bench grinder.

Punches come in assorted designs, with the pin punch and center punch being the most prevalent. Pin punches are cylindrical and straight, available in various sizes ranging from small, such as " in diameter, to larger sizes up to ". Primarily used for driving pins out of holes, pin punches also follow the pin's path through the hole as it forces the pin out.

Center punches are essential for creating small indentations that act as starting points for drilling tasks. By establishing a small indentation with a center punch, the drill bit can maintain its position long enough to initiate a hole. Without this initial marking, attempting to drill metal may result in the drill bit veering off course, failing to hit the intended target.

When using a hammer to manipulate parts within an assembly, there's a risk of damaging the parts if they're directly struck by the hammer. The hardened forged steel head of a ball-peen hammer can easily harm softer materials. To prevent such damage, a drift is employed. Placing the drift against the object and then striking the drift with a hammer ensures the part being manipulated remains unharmed. Drifts are typically crafted from softer metals like mild steel, brass, or even aluminum.

Let's do an example problem.

What kind of hammer would you use to add a hubcap?

 (A) sledge hammer
 (B) Claw hammer
 (C) Pneumatic hammer
 (D) Rubber mallet

The answer is D, rubber mallet.

KNOWING THE DIFFERENT TURNING TOOLS

Screwdrivers, available in diverse sizes and styles, are among the most frequently used tools for installing and removing fasteners. Among the oldest designs is the flat-tip screwdriver, featuring a flat blade designed to turn screws with a single slot across their top.

However, flat-tip screwdrivers have become less popular with the advent of newer designs like Phillips, Robertson, and Torx screwdrivers. A Phillips screwdriver is recognized by its cross-shaped tip, whereas Robertson screwdrivers have a square tip, and Torx screwdrivers are distinguished by their unique tip. These modern designs offer enhanced grip on the fastener, facilitating easier removal and installation. With improved contact between the screwdriver and screw, it's also possible to fasten screws more securely.

Screwdrivers are exclusively used with screws, threaded fasteners that vary in size and shape depending on the project. Screws tighten when turned clockwise. For more information on fastener classification, refer to the "Nuts and Bolts" section later in this chapter. Washers, disk-shaped rings, may accompany screws to shield the work surface and distribute the force applied during screw tightening.

In the realm of wrenches, length directly affects leverage and twisting force (torque), allowing for greater effectiveness in tightening or loosening fasteners. Wrenches generally fall into two categories: open-end and box-end. The open-end wrench prioritizes speed with its open design, facilitating easy sliding onto and off of fasteners like cap screws. For loosening tightly secured fasteners, the box-end wrench is recommended. Its enclosed design wraps completely around the bolt head, ensuring greater surface contact and more even force distribution.

Combination wrenches have a popular design because they feature a box end and an open end on opposite sides. Despite accommodating the same size fastener, technicians can first loosen bolts using the box end and then expedite bolt removal using the open end.

In situations where a specific wrench size isn't available, an adjustable can serve as a versatile alternative.

An alternative to using wrenches for loosening fasteners is employing sockets. Similar to wrenches, sockets are available in two main designs. The six-point design, with its greater wall thickness, is typically favored by mechanics, particularly in smaller socket drive sizes, due to its strength. However, the twelve-point design proves beneficial for specific applications, offering easier alignment with bolt heads in confined spaces.

To choose the right size for a task, one simply needs to measure the distance between two parallel sides of the bolt head. For example, if the bolt head measures $\frac{9}{16}$" across two parallel sides, then a $\frac{9}{16}$" socket is needed to loosen it.

Sockets are available in various drive sizes, determined by the size of the opening that attaches to the drive tool. For instance, if the square end of a socket measures $\frac{3}{8}$ " across, it corresponds to a $\frac{3}{8}$" drive socket. Common drive sizes include $\frac{1}{4}$", $\frac{3}{8}$", $\frac{1}{2}$", and $\frac{3}{4}$". Larger drive sizes cater to the hefty fasteners found in heavy industries.

Sockets offer exceptional versatility as they can be utilized with various drive tools. The ratchet is the most commonly used drive tool for sockets. It rotates the fastener in one direction as the handle is moved back and forth along a narrow arc. Ratchets are reversible, allowing them to be configured to either tighten or loosen fasteners. They prove particularly advantageous in confined spaces, especially when paired with an extension bar, as there's no need to replace the tool with each turn—simply maneuver the handle back and forth.

Furthermore, sockets can be employed with pneumatic (compressed air) power tools, such as an air impact wrench. An air impact wrench swiftly removes fasteners by applying substantial torque (twisting force) and employing a hammering action that loosens fasteners through vibration. It's crucial to note that only impact sockets should be utilized with an air impact wrench.

On the ASVAB, fasteners like screws and bolts are among the commonly tested types of tools, making it worthwhile to familiarize oneself with this broad array of hardware.

Wrenches are essential for working with various threaded fasteners, like nuts and bolts. Bolts feature external threads, whereas nuts possess internal threads. Bolts usually have a hexagonal or square head, necessitating a wrench to hold them steady while tightening the nut with another wrench to secure the assembly. Threaded bolts can only be inserted into nuts or holes with matching threads.

Though the distinction between screws and bolts isn't always clear-cut, it's often based on the head shape and the tool used for installation or removal. Bolts typically have a hexagonal or square head and are paired with a wrench and nut, while screws commonly boast a rounded head with a corresponding indentation for a screwdriver to install or remove them. Both fasteners feature threading along part or all of their shaft.

Nuts are threaded onto bolts to secure assemblies together. They typically feature either a square or hexagonal head and can be secured in place using various methods.

Wing nuts facilitate manual disassembly of components due to the two "wings" attached to the nut, allowing for easy tightening and loosening without the need for hand tools.

Castellated nuts employ a cotter pin to secure them in place. The cotter pin passes through a hole in the bolt or stud threaded by the nut, engaging with the cutouts in the nut to prevent loosening.

Lock nuts feature a nylon insert within their threads, providing sufficient interference to prevent loosening and effectively locking them in place.

The type of thread applied to a fastener varies depending on its diameter and the desired strength of the final product. Thread pitch, which is determined using a thread pitch gauge, identifies the spacing between threads.

Fasteners measured in fractional inches utilize threads distinguished by how many threads there are per inch. There are two primary thread categories: Unified National Coarse (UNC) and Unified National Fine (UNF). UNC or coarse threads feature a relatively low number of threads per inch, while UNF or fine threads boast a higher number of threads per inch.

For instance, a bolt with a diameter of 1/4 inch could have two potential thread pitches. If it were a UNC bolt, it would have 16 threads per inch, whereas if it were a UNF bolt, it would feature 24 threads per inch.

Two additional critical measurements of a fastener are its diameter and length. The diameter refers to the distance across the unthreaded portion of the bolt, providing insight into the size of the hole intended for the fastener's installation. Meanwhile, the length of the bolt signifies the distance from the underside of the bolt head to the end of the bolt, excluding the bolt head itself from the measurement of the bolt's length.

Let's try a practice question.

You are working on something with temporary bolts, and you want to be able to easily remove them without a wrench. What is the best nut for this project?

(A) Castellated nut
(B) Lock nut
(C) Standard nut
(D) Wing nut

The answer is wing nut, or D.

KNOWING THE DIFFERENT FASTENING TOOLS

Retaining rings, also known as snap rings, serve to inhibit the axial movement of cylindrical parts within bores or on shafts. External snap rings are fitted into grooves on shafts, while internal snap rings are positioned within grooves inside a bore. Snap rings are installed and removed using snap-ring pliers.

Soldering is a method used to join metals by bonding a metal alloy to their surfaces. It's a low-temperature process that can be carried out using simple tools and affordable materials.

Solder typically have a lead and tin alloy, with varying proportions of the metals to achieve specific properties, such as the desired melting point. Higher lead percentages yield lower melting points.

The key aspect of soldering lies in cleaning the surfaces to be joined. Any oxides or contaminants on the surfaces can hinder a solid connection. The optimal preparation method involves using a flux, which cleans the surfaces through chemical action. For electrical connections, a rosin flux is recommended, with solder containing the flux within its core referred to as rosin-core solder.

Various tools can generate the required heat to melt solder, but the soldering iron is the most commonly used. Typically electrically powered, soldering irons can draw anywhere from 25 to 100 watts of power. Lower-power irons are suitable for soldering electrical connections, while higher-powered ones are preferred for sheet metal work.

Another tool frequently utilized for soldering electrical connections is the soldering gun. Equipped with a two-step trigger, the soldering gun enables technicians to swiftly select either a low or high heat setting. Soldering guns have rapid warm-up cycle. Additionally, soldering guns often feature a built-in light to illuminate the work area.

When soldering, commence by mechanically cleansing the materials intended for joining. This may entail employing a stainless steel brush or sandpaper. Gently heat the surfaces while applying flux, ensuring thorough cleaning by the flux. Once the flux has adequately cleaned the material, introduce solder. Tin the surfaces by evenly spreading the solder thinly over them, ensuring thorough heating of any areas to be tinned.

Prior to soldering an electrical connection, it's advisable to establish a robust mechanical connection first. This entails stripping insulation from wires and securely twisting them together. Remember, the entire connection must be heated before applying solder. Failure to heat the entire connection may result in the creation of a "cold" solder joint with potentially unpredictable electrical properties.

For achieving high-strength joints when joining metals, welding is the preferred method. Unlike soldering, welding involves melting the base metal of the objects to be joined, requiring very high temperatures.

Two primary welding processes are oxyacetylene welding and electric-arc welding. Oxyacetylene welding employs a torch fueled by oxygen and acetylene. Combusting these gases generates an exceptionally hot flame capable of melting steel and other ferrous materials. Filler rod is melted along with the base metal to form a completed weld.

Electric-arc welding encompasses numerous specialized processes, with stick welding being the simplest and most recognizable type. Stick welding requires two cables and an electric-arc welding machine. One of the cables connects to the workpiece with a ground clamp, and the second attaches to an electrode holder, known as a stinger, which the welder uses to hold the welding rod.

When the welding rod contacts the piece, it creates an electric arc. This arc generates an immense amount of heat and serves three primary purposes. Firstly, it melts the base metal of the material being welded. Secondly, it melts the electrode, making this molten metal a filler in the weld. Lastly, the flux coating on the electrode's exterior burns, creating gases that form a protective shield. This gaseous shield prevents air from interacting with the hot metal, thus preserving the integrity of the weld.

As the welder progresses with the weld, the electrode gradually burns away and diminishes in length. Consequently, the welder must continuously advance the stinger closer to the workpiece until the electrode is fully consumed. Upon depletion of the electrode, the welder terminates the arc, and a fresh electrode is inserted into the stinger to resume welding.

Should additional heat be required for the weld, the welder can augment the electric current supplied by the welding machine. Generally, larger workpieces or electrodes necessitate a greater amount of current to accomplish the weld effectively.

During this process, the electric arc emits a high-intensity light, including ultraviolet light, which can cause "sunburn" on exposed skin and damage the retinas of unprotected eyes. As a safety precaution, welders must cover all exposed skin with protective clothing, preferably made of leather, and wear face shields (helmets) equipped with light filters. These filters are typically very dark, allowing the welder to observe the welding process but blocking out other light sources.

Modern welding helmets feature electronic filters that detect ultraviolet light from the electric arc and rapidly switch from clear to shaded, ensuring continuous protection. This enables welders to keep their helmets down while aligning the next weld, eliminating the need for frequent helmet adjustments.

Another electric-arc welding method gaining popularity is MIG (metal inert gas) welding, also known as wire-feed welding. In MIG welding, a wire electrode is automatically fed from a spool. The process utilizes a bottled inert gas, such as argon, to shield the weld and requires a relatively advanced welding machine.

Let's try an example question.

The flux used in a metal joining processes is used to do what?

    (A) To bond two meal surfaces
    (B) To transfer electrical voltage
    (C) To clean metal surfaces that will be joined
    (D) To transfer electrical currents

The answer is C, to clean the surfaces.

KNOWING THE DIFFERENT GRIPPING TOOLS

In repair operations, there's often a need to grip, twist, bend, or turn objects, some of which may be irregularly shaped or challenging to hold. Pliers are the go-to tool for handling such objects effectively.

Among the most common types of pliers is the combination slip-joint variety. These pliers feature an adjustable joint between the two handles, offering two different gripping positions. This lets them accommodate objects of various sizes. Additionally, some designs may include a built-in wire cutter, enhancing their utility even further.

When using large diameter objects that require gripping or twisting, technicians often turn to adjustable joint pliers to tackle the task. These pliers boast multiple "arc-joints," allowing for adjustment across a wide range of sizes. With their lengthy handles, they provide excellent leverage, resulting in maximum gripping power. They're commonly referred to as water pump pliers or Channellock® pliers.

Meanwhile, lineman pliers are specifically designed for cutting and bending heavy gauge wire. Although they lack size adjustability, they're engineered for optimal leverage at the jaws, making the cutting process more manageable. While electricians primarily use this type of pliers, they find utility across various trades as well.

Diagonal cutters are specialized pliers designed solely for cutting tasks. Their jaws are angled diagonally, making it easier to cut wires straight across. These cutters are typically employed for cutting wire and small cables.

When it comes to gripping small objects in confined spaces, needle nose pliers are the tool of choice. Featuring long, pointed jaws, they offer maximum reach and precision. Needle nose pliers excel at intricate tasks such as soldering circuit boards and handling small components, often incorporating a wire cutter at the base of the jaws.

Locking pliers, commonly known as Vise-Grip® pliers, are indispensable tools used by technicians across various trades. They are adjustable and available in a wide array of jaw designs, capable of securely locking in place to hold or clamp objects. Many locking pliers also come equipped with built-in wire cutters, enhancing their versatility for different applications.

A clamp is a device utilized to exert pressure, preventing movement between various components of a project. For instance, it proves invaluable in securing pieces together during the drying process of glue. Comprising a metal frame, often C-shaped, with a flat-edged screw perpendicular to the frame's bottom, a clamp tightens as the screw is rotated.

Vises share similarities with clamps as they are both intended to secure wood or other materials. However, a notable distinction lies in the fact that vises are mounted onto the workstation and are employed to immobilize a material while sawing, sanding, drilling, or performing other work on it. Usually, one side (jaw) of the vise remains stationary, while the other can be adjusted by rotating a handle.

Let's try a practice question.

You're tasked with a job that involves gripping various sizes of pipes. Which tool would be the most suitable choice for this task?

    (A) Combination slip-join pliers
    (B) Diagonal cutters
    (C) Needle nose pliers
    (D) Adjustable joint pliers

KNOWING THE DIFFERENT CUTTING TOOLS

Cutting materials such as wood necessitates the use of a saw, a tool commonly employed by carpenters across various trades. The crosscut saw, aptly named for its function, cuts across the wood's grain. Its teeth are distinctive in their knife-like cutting action. In contrast, the rip saw is tailored to cut along the grain of the wood, featuring teeth shaped like chisels. Additionally, the teeth are alternately set on alternating sides to create a wide slot, enabling it to cut a straight line regardless of if the grain is curved or not.

A coping saw is ideal for intricate, curving cuts. Its thin, flexible blade is securely held in a wide frame, allowing for precision. Moreover, the blade can be rotated within the frame, enhancing its versatility for intricate cuts on larger materials.

Similarly thin but reinforced with a rigid steel strip along its top edge, the back saw is designed for finer cuts, boasting 14 to 16 teeth per inch. Often paired with a miter box, the back saw facilitates precise cuts at specific angles.

A staple in a mechanic's toolbox is the hacksaw, employed for cutting various metals like steel, aluminum, or copper. Hacksaw blades are interchangeable, with the selection of the appropriate blade crucial for the material being cut.

Consider the blade's teeth per inch (TPI) when choosing a hacksaw blade; higher TPI is advantageous for cutting thinner materials. Additionally, ensure proper blade orientation during installation—teeth should face away from the handle. This configuration allows the hacksaw to cut on the forward stroke. When retracting the hacksaw, ease off downward pressure to avoid blade breakage.

For tasks requiring quicker cuts or handling robust materials, powered saws are often preferred over manual saws. A widely-used powered saw is the circular saw, characterized by its rapidly rotating circular blade. Two common variants of the circular saw exist. The miter saw, a standalone tool, is ideal for creating precise crosscuts in lumber, particularly when angled cuts are required. Distinguishing it from the table saw, which also features a circular blade, is crucial. Unlike the miter saw, a table saw integrates its blade into the table or bench. Moreover, their operational methods differ. When using a miter saw, secure the material firmly and lower the spinning circular blade to execute the cut. Conversely, with a table saw, carefully maneuver the material toward the rotating blade for cutting.

A frequently encountered power saw is the band saw, named for its blade—a single continuous band of metal—that rotates rapidly between two spinning wheels to create uninterrupted cuts in wood or metal. Band saws are employed in timber processing to produce lumber and are adept at cutting both straight and curved lines in various materials. When utilizing a band saw to cut metal, it's essential to employ a coolant wash to maintain the material's temperature and cleanliness, preventing overheating and debris buildup.

Drilling is the process of creating small holes in wood or metal, while boring is used for larger holes in softer materials. Although the processes are similar, they employ different tools.

Drill bits are utilized for drilling holes, with carpenters typically using bits up to 1 inch in diameter, while mechanics often require bits as large as 1/2 inch. For larger holes, specialized tools are used.

Most drill bits cut while rotating clockwise, known as right-hand drill bits, while left-hand drill bits cut in the opposite direction. Left-hand drill bits are often preferred to extract broken bolts that are in threaded holes.

Hole saws are another tool for boring large holes, but unlike drill bits, they are not adjustable and can only drill holes of a specific size.

Hand-operated tools for driving drill bits and other boring tools have largely been replaced by electric drills in modern practice. Electric drills come with various features such as chuck size, reversibility, and speed control.

The chuck of an electric drill is the part that secures the drill bit. Chuck size is identified by the largest diameter bit it can accommodate, with common sizes including $\frac{1}{4}"$, $\frac{3}{8}"$, and $\frac{1}{2}"$. Traditional drill chucks require a chuck key for tightening and loosening, but keyless chucks, which can be adjusted by hand, are becoming more prevalent.

Reversible drills can operate in both clockwise and counterclockwise directions, offering versatility in drilling applications. Variable speed drills allow users to adjust the drilling speed using the trigger.

As battery technology progresses, cordless drills are becoming increasingly prevalent. Their popularity stems from features like extended battery life and portability. Many cordless drills are equipped with a variety of attachments, known as drivers or bits, allowing them to function as powered screwdrivers and socket wrenches.

When using a drill, ensuring the bit is sharp is paramount for safety. Dull bits necessitate more time and pressure to complete tasks, increasing the risk of bit breakage and surface damage to the material being drilled. Drill bits can be sharpened using a special attachment on a bench grinder.

Let's try a practice question.

You need a rounded shape cut from a square piece of wood. What is the best saw to do this?

- (A) Hack saw
- (B) Rip saw
- (C) Coping saw
- (D) Crosscut saw

The answer is A, a coping saw – the rest work for straight cuts.

KNOWING THE DIFFERENT FINISHING TOOLS

In woodworking, achieving precise fits and smooth surfaces often requires shaving off small amounts of material. A plane is a tool commonly employed for this task.

Among the various types of planes available, the jack plane stands out as a versatile option for general woodworking tasks. Featuring a smooth lower surface and an adjustable-depth blade that extends slightly below it through the mouth, the jack plane operates akin to a disposable razor blade, offering flexibility and efficiency in material removal.

When employing a jack plane, the carpenter grips it firmly with both hands and glides it smoothly across the work surface, shaving off a thin layer of wood. To achieve a deeper cut, the blade can be adjusted to extend further from the plane's lower surface.

Wood chisels offer another method for shaping or refining wooden surfaces. Available in various widths, ranging from $\frac{1}{8}$" to 2", wood chisels are typically operated by hand. However, for making deeper cuts, it's advisable to use a soft-faced mallet to gently tap the chisel through the material.

Files serve the purpose of smoothing, polishing, and shaping various materials. Constructed from hardened steel, they feature diagonal rows of teeth. These teeth configurations can be either a single row of parallel teeth (single-cut) or a pattern of crisscrossing teeth (double-cut). Files can be flat, half-round, round, and triangular shapes.

Typically, files are sold without handles. It's crucial to have a handle ready to attach to a file before using it to prevent the file's end from causing injury to the technician's hand. It's important to note that a file cuts only on the forward stroke, so it should be lifted slightly from the work during the backward stroke.

Rasps are akin to extremely coarse files, ideal for trimming, shaping, and smoothing materials with a rough finish. Unlike files, which excel at finer finishes, rasps are tailored for more aggressive work. Round rasps are handy for refining holes, while flat rasps excel at smoothing flat surfaces. With cutting teeth coarse enough to clear sawdust effortlessly, rasps remain prepared for the next stroke at all times.

Let's try a practice question.

You would use a mallet with what finishing tool?

- (A) File
- (B) Jack plane
- (C) Rasp
- (D) Wood chisel

The answer is the jack plane, or B.

PRACTICING WHAT YOU'VE LEARNED

Now that you've got the basics, you can see test your newly acquired information and put it to the test.

Here are two practice sets with 10 questions. This is similar to how the CAT is setup, although obviously your questions aren't going to change the better you do.

The last section of this chapter details the right answer, and how you could have reached that answer.

Practice 1

1. What is the primary function of a tape measure?

    (A) Marking lines
    (B) Cutting materials
    (C) Holding objects together
    (D) Measuring distances

2. Which tool do you use to drive nails into materials?

    (A) Pliers
    (B) Screwdriver
    (C) Hammer
    (D) Wrench

3. What is the main purpose of a chisel?

    (A) Shaping wood
    (B) Measuring angles
    (C) Cutting metal
    (D) Tightening bolts

4. What tool is commonly used to secure bolts and nuts?

    (A) Hammer
    (B) Wrench
    (C) Pliers
    (D) Screwdriver

5. Which tool is best suited for holding objects firmly in place?

    (A) Clamp
    (B) Screwdriver
    (C) Tape measure
    (D) Chisel

6. What is the primary function of a hacksaw?

    (A) Measuring angles
    (B) Cutting wood
    (C) Cutting metal
    (D) Shaping plastic

7. Which tool would you use to smooth and polish surfaces?

    (A) Hammer
    (B) Saw
    (C) File
    (D) Sandpaper

8. What tool is commonly used for tightening screws?

    (A) Pliers
    (B) Screwdriver
    (C) Hammer
    (D) Wrench

9. Which tool is used for removing nails from materials?

    (A) Pliers
    (B) Screwdriver
    (C) Hammer
    (D) Pry bar

10. What tool is commonly used for marking measurements on materials?

    (A) Pliers
    (B) Hammer
    (C) Screwdriver
    (D) Pencil

Practice 2

1. Which tool is used for shaping metal on a lathe?

    (A) Hammer
    (B) Chisel
    (C) Lathe tool
    (D) Saw

2. What tool would you use for gripping and turning bolts and nuts?

    (A) Pliers
    (B) Screwdriver
    (C) Wrench
    (D) Hammer

3. What is the primary function of a vise grip?

    (A) Cutting
    (B) Gripping
    (C) Turning
    (D) Sanding

4. Which tool is used for drilling holes in materials?

    (A) Chisel
    (B) Saw
    (C) Drill
    (D) Pliers

5. What tool would you use to tighten or loosen nuts and bolts?

    (A) Wrench
    (B) Screwdriver
    (C) Hammer
    (D) Pliers

6. Which tool is used for cutting through materials quickly?

    (A) Pliers
    (B) Saw
    (C) File
    (D) Hammer

7. What tool is commonly used for removing excess material from a surface?

    (A) Chisel
    (B) File
    (C) Saw
    (D) Pliers

8. What tool is commonly used for applying force to small objects or materials?

    (A) Hammer
    (B) Wrench
    (C) Pliers
    (D) Screwdriver

9. Which tool is used for tightening or loosening screws with slotted heads?

    (A) Pliers
    (B) Wrench
    (C) Hammer
    (D) Screwdriver

10. What tool is commonly used for cutting through metal pipes?

    (A) Saw
    (B) Chisel
    (C) Hacksaw
    (D) Screwdriver

CHECKING YOUR ANSWERS

Now let's see how you did.

Practice 1

1. (D) Measuring distances

2. (C) Hammer

3. (A) Shaping wood

4. (B) Wrench

5. (A) Clamp

6. (C) Cutting metal

7. (D) Sandpaper

8. (B) Screwdriver

9. (D) Pry bar

10. (D) Pencil

Practice 2

1. (C) Lathe tool

2. (C) Wrench

3. (B) Gripping

4. (C) Drill

5. (A) Wrench

6. (B) Saw

7. (B) File

8. (A) Hammer

9. (D) Screwdriver

10. (C) Hacksaw

CHAPTER 13

MECHANICAL COMPREHENSION

You need to understand applied physics to do well with mechanical comprehension. If you want to go back and review the physics section of Chapter 9, that could help you better understand this chapter.

A BIT MORE ABOUT THIS SECTION OF THE ASVAB

Applied physics involves the practical application of the laws of physics, which is essential in fields such as engineering, architecture, and heavy equipment operation. This chapter builds on the information from the Physical Science section, to explain the functioning of various common mechanical devices.

You get more time for this section.

- The ASVAB gives you 22 minutes to finish 15 questions about AI if you take the CAT.
- If you take the traditional ASVAB, you will have 25 questions with just 19 minutes to finish. This means you have a less than a minute a question.

UNDERSTANDING THE PHYSICS OF MECHANICAL DEVICES

Mechanics are a part of the Physics branch of science and it deals with the laws of motion, energy, and forces. Key concepts in mechanics include velocity, momentum, and Newton's second law, which relates force, mass, and acceleration. All mechanical devices, including the simple machines discussed in this chapter, operate based on the application of force to move or change the position of a mass.

Looking at Mass and Force

In mechanics, it is crucial to understand that mass determines how much force is needed to achieve a specific acceleration. The greater the mass, the more difficult it is to change the object's motion. This is called inertia, and it is one of the properties mass.

A force is essentially a push or pull. Forces are ubiquitous, ranging from obvious examples like a tractor pulling a plow, a baseball being hit, or someone pushing through a crowd, to more subtle ones like Earth's gravity, which constantly exerts a downward force (weight) keeping objects grounded. All forces are vector quantities, meaning they act in a specific direction. Net force is the total force acting on an object.

Without applied force, objects would not move or, if already in motion, would continue moving indefinitely without stopping. Heavier objects, such as a freight train, require significant force to change their motion. The goal of mechanical technology is to efficiently apply forces to move varying amounts of mass.

A Little More about Newton's Laws

One of the most important historical functions of machines has been to amplify the force exerted by humans or animals, allowing them to move large objects that would otherwise be immovable. Ancient structures like the pyramids and Stonehenge demonstrate a long tradition of using mechanical aids to enhance labor.

According to Newton's third law of motion, there is an equal and opposite reaction to every action. This principle applies to all applications of force, including those involving simple machines. For example, when using a rope and pulley to lift a weight, the tension force is transmitted through the rope via a series of action-reaction force pairs, allowing the weight to be lifted.

Let's do a quick example question.

What are simple machines and mechanical devices?

- (A) They are technologies that work around the laws of physics.
- (B) They work with momentum only.
- (C) They function based on the rules of force and mass.
- (D) They transmit and multipole force to increase available energy.

The answer is C.

LOOKING AT THE COMMON TYPES OF FORCE

As previously discussed, weight and mass are distinct concepts. Mass is an intrinsic property of matter, and weight gravity's exerted force on an object. The downward pull that Earth exerts on objects is a specific example of gravity in action. Weight can be easily measured using a spring scale, like those commonly found in bathrooms. For all objects on Earth, weight acts in the same direction: straight down toward the center of the Earth.

The general formula for Newton's Law of Universal Gravitation, covered in chapter 9: General Science, explains why a person's weight varies on planets of different sizes and masses. For instance, astronauts weighed only one-sixth of their Earth weight when they were on the Moon. However, unless otherwise specified, it is assumed that everyday mechanical work is performed on Earth.

For objects in free fall near the Earth's surface, the acceleration due to gravity (g) is approximately $9.8 \ m/s^2$, ignoring air resistance. This value remains consistent regardless of elevation, whether one is on top of Mount Everest or parachuting from a plane, as g does not change significantly with such variations in height.

You can use Newton's Second Law for this, $F = ma$, to derive the weight, the general symbol for force, F, is replaced with the more specific force of weight, W, and acceleration, a, is replaced with the constant acceleration due to gravity, g. This results in the modified version of the formula:

$W = mg$

- W is the weight in newtons (N)
- m is the mass in kilograms (kg)
- g is the acceleration due to gravity ($9.8 \ m/s^2$ or N/kg).

To determine an object's weight in newtons, simply multiply its mass in kilograms by 9.8.

Understanding Friction

There are several forces that resist movement, known as responsive forces. When an object moves along any surface, it rubs against the surface to create kinetic friction. This reaction always moves opposite to the direction that the object is moving. Because kinetic friction opposes motion, it will eventually slow any moving object to a stop. For instance, ice is polished to minimize kinetic friction, but a hockey puck will still eventually slide to a stop unless constantly pushed, even on the slipperiest rink.

When an object is stationary, a force that attempts to slide it across a surface encounters a different type of friction. To push a stationary box across a concrete floor, one must overcome the static friction that opposes the initiation of movement.

Both types of friction depend on two factors: the coefficient of friction, mu, which represents how much two materials resist sliding against each other; and the normal force, which is the opposite and equal force. The formulas for kinetic and static friction, respectively, are:

$F_f = \mu_k F_N$

$F_f = \mu_s F_N$

- F_f is the force of friction.
- μ_k is kinetic friction.
- μ_s is static friction.

The coefficients of friction are constant for any two surfaces in contact and are dimensionless quantities, meaning they have no units. Consequently, the force unit used to measure F_N (whether newtons, pounds, etc.) will also be used to measure the frictional force, F_f.

Slippery surfaces like ice or wet asphalt have low coefficients of both kinetic and static friction with most objects. Friction is directly proportional to these coefficients. This is why it's crucial for drivers to understand that weather conditions can reduce the effectiveness of braking, as it relies on the friction between the wheels and the road surface.

Looking at a Free Body Acting on a Sliding Object

In general, surfaces exhibit a higher coefficient of static friction compared to kinetic friction. This means more force is needed to move a stationary object than to keep it moving. Once the large static friction force is overcome and the object starts to move, the smaller kinetic friction force takes over, reducing the resistance. This explains why moving heavy objects typically involves an initial big push, followed by a smaller, consistent force to keep the object in motion.

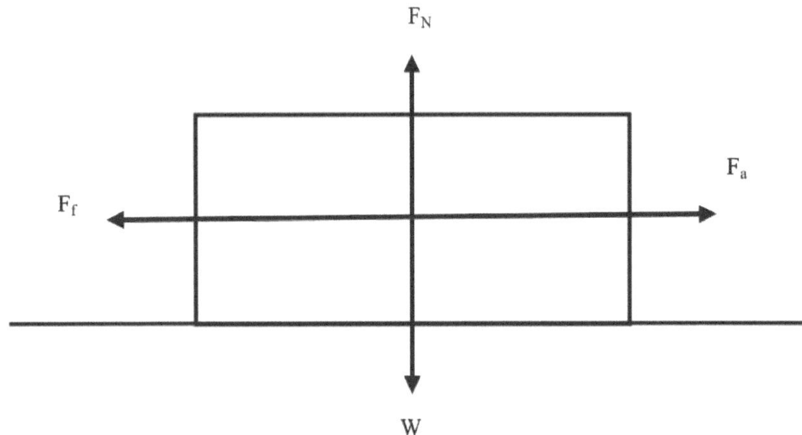

Another notable distinction between static and kinetic friction is that static friction occurs only when an object is being attempted to be moved. Similar to the normal force that counters the weight of objects, static friction responds to an applied force by matching and counteracting it. The formula for static friction actually determines its maximum value, beyond which the force will break and the object will start to move.

Both types of friction forces are directly related to the normal force of the surface. Try this experiment: place your palm flat on a nearby surface like a table and attempt to slide your hand along it. Notice the resistance caused by kinetic friction. Now, apply more pressure by pushing harder against the same surface while trying to slide your hand.

Is it more challenging? If you exert significant pressure against the surface, you might struggle to overcome the force of static friction and slide your hand without reducing the normal force first. When someone leans against a wall, a portion of their weight presses against the wall, which pushes back with an equal and opposite normal force. This normal force creates a corresponding static friction force that keeps you from sliding down walls.

Typically, the normal force arises on flat horizontal surfaces as a reaction to the object pusing down on it. Consequently, the normal force is often equal to the object's weight, causing the friction forces to equal the coefficients of friction multiplied by the object's weight. This assumption holds true except on non-horizontal surfaces, such as slopes, or when additional forces act to push or hold an object up.

Another resisting force is air resistance, also known as drag. Unlike friction, which acts on solid surfaces, drag opposes motion through a fluid like air or water. Two primary factors influence drag: cross-sectional area and speed. High-performance vehicles like cars, boats, and aircraft are streamlined to reduce the surface area encountering air or water resistance. Conversely, parachutes increase surface area to increase drag, allowing skydivers to safely slow down before landing.

Understanding Tension

When using cables to pull an object, the applied force on one end results in an equal pulling force on the attached object from the other end. This transmission of force through rope or similar materials creates internal stretch force known as tension. Tension can be measured in various force units like N or lbs.

Tension is commonly assumed to be evenly distributed across the material. However, in reality, tension is greatest where the force is applied. For instance, consider a 100-pound weight suspended from a 10-pound chain. The bottommost link in the chain bears the weight and thus experiences a tension of 100 pounds.

However, the chain link just above it also has a crucial role. It supports the link holding the 100-pound weight, meaning it must bear a tension equivalent to 100 pounds plus the weight of one chain link. Continuing to the chain's top, where the applied force holds it in place, the link at the top is holds both the 100-pound weight and the weight of the entire chain below. At this point, the tension is 110 pounds.

The key point is that, assuming no vulnerabilities or flaws in the material being used, a cable or rope is most likely to snap near the section where it is being pulled.

Understanding Hydraulic Pressure

Fluid power presents another means of achieving mechanical advantage. Hydraulics utilizes liquids to transmit force.While traditional simple machines serve well for numerous tasks, hydraulics offers increased versatility in redirecting and amplifying forces within intricate systems. To comprehend hydraulics, one must first grasp the notion of pressure. When force is uniformly applied across a defined area, pressure is exerted on that surface. The following formula shows how to determine pressure:

$$P = \frac{F}{A}$$

-   F is force in pounds.
-   A is the area in square inches.
-   P is pressure in pounds per square inch (psi).

For instance, a force of 100 pounds applied over an area of 10 square inches yields a pressure of 10 psi.

Even under immense pressure, liquids experience only minimal changes in volume due to their near-incompressibility. This quality makes liquids highly efficient at transmitting force, as little to no energy is expended in compressing them. Additionally, liquids naturally adopt the shape of their container, filling available space as needed. Pascal's law, formulated by Blaise Pascal in the 1600s, dictates that the pressure within an enclosed fluid remains uniform and acts in all directions.

In practical applications, exerting force on one end of a hydraulic system uniformly increases pressure, thereby creating force in a different direction at another point with an equal area of application. Altering the surface area over which force is exerted can also impact the force's magnitude. In hydraulic systems, a lesser force applied over a larger distance yields a greater force over a shorter distance.

Understanding Torque

Torque is how much force is required to rotate something, but it is not a force on its own.

$$\tau = rF$$

-   τ (the Greek letter tau) is torque.
-   r is the length of the lever.
-   F is the amount of force applied.

Torque is calculated as the result of a perpendicular applied force (F) acting on the lever arm of a rotating object, multiplied by the length of the lever arm (r) measured from the pivot point (fulcrum) to the point of force application. You measure force Newtons (N) or pounds (lb), while the lever arms' length is measured in meters (m) or feet (ft). Therefore, torque can be quantified in standard units such as Newton-meters (N-m) or foot-pounds (ft-lb).

It's important to note that the length measurement for torque is perpendicular to the force applied in rotating the object, not parallel. As a result, torque should not be confused with work, which is measured in N-m (more commonly expressed as joules,

J), representing a force applied through a distance. Additionally, torque should not be equated with energy, which is also measured in joules (J).

Torque is directly linked to the force applied for rotation and is also influenced by the distance from the center of rotation where this force is exerted. Optimal torque is achieved by applying force at the furthest point of a lever arm. For instance, when a mechanic encounters difficulty turning a stubborn bolt despite exerting maximum force, using a longer wrench could offer a solution. While the force applied remains the same, employing a longer lever arm enhances torque, potentially facilitating the loosening of the bolt. Unlike work, which necessitates successful rotation, torque is considered to have been applied regardless of whether the object has moved.

UNDERSTANDING ENERGY, POWER, AND WORK

In a broad context, energy refers to the capacity to induce a change in the state of an object. Mechanical Comprehension, as part of the ASVAB, predominantly focuses on mechanical energy, which encompasses various forms.

Kinetic energy denotes the energy that is coupled with the motion. You might notice a professional pitcher pitches a baseball at 40 m/s has more kinetic energy than pitchers practicing and warming up in the bull pit. Likewise, a car traveling at 10 meters per second carries more kinetic energy than a bicycle at the same speed. Kinetic energy depends on both the mass and velocity of an object. Calculate kinetic energy (KE) with the following formula.

$$KE = \frac{1}{2}mv$$

- KE is kinetic energy. Even though KE is a linear function, it is measured in squared units.
- M is mass measured in kilograms.
- V is speed measures in meters per second.

In a broad context, energy refers to the capacity to induce a change in the state of an object. Mechanical Comprehension, as part of the ASVAB, predominantly focuses on mechanical energy, which encompasses various forms.

Kinetic energy is thought of as the energy that you can see through motion. Use the following formula to calculate the gravitational potential energy.

PE= mgh

- m is mass and is measured in kilograms.
- g is the acceleration caused by gravity ($9.8m/s^2$)
- h is the object's height measured in meters.
- PE is potential energy measured in joules.

Even though an object will take longer to go down a ramp, it still has the same gravitational pull as something dropped without the ramp.

Looking a Work and Power

Work is achieved when force is employed to displace an object. Similar to energy formulas, the formula for calculating work aligns with real-world observations. Greater force exertion to displace an object over a specific distance results in more work done. Likewise, moving an object a greater distance by applying force corresponds to increased work. The following formula is for calculating work.

W = Fd

- W is work measured in joules.
- F is force measured in newtons.
- d is distance measures in meters.

Power equals the rate of work completed, as represented by the following formula.

$$P = \frac{w}{t}$$

LOOKING AT SIMPLE MACHINES

Centuries ago, simple machines emerged to alleviate the effort required for various tasks like lifting heavy objects, assembling parts, or relocating large machines. These tasks often surpassed human capability to apply sufficient force.

Mechanical advantage, defined as "the advantage gained by using a mechanism to transmit force," allows for the amplification of applied force using appropriate equipment. However, while mechanical advantage increases force, it does so by exerting force over a greater distance. This implies that work output cannot surpass work input. Nonetheless, simple machines simplify arduous tasks by dividing them into smaller, manageable parts.

Machine efficiency gauges how much of the input power is converted into movement or force output. A machine achieving 100 percent efficiency would utilize all applied power. However, real-world machines fall short of this ideal due to factors like friction and worn or ill-fitting components.

An inclined plane, often referred to as a ramp, offers a straightforward solution to lifting heavy objects. A task like lifting a 100 kg (approximately 220 pounds) box, which might be challenging or even impossible for a single person, becomes manageable with the assistance of an inclined plane.

The wedge, a modified version of the inclined plane, offers versatility in movement. Unlike the stationary inclined plane, the wedge is engineered for mobility. Its applications range from lifting to splitting and tightening tasks. The mechanical advantage it provides is dependent on the ratio between its length and height.

Increasing the length of the wedge compared to its height, essentially reducing its slope, decreases the vertical lift achieved as the wedge is pushed horizontally. However, this adjustment amplifies the force generated by the wedge, facilitating the lifting of heavy objects. It's important to note that when force is increased by the wedge, the vertical distance it lifts becomes shorter than the horizontal distance it moves.

The wedge finds various applications in tools like knives, chisels, and log splitters.

The lever, another straightforward machine, serves to amplify force or distance and alter direction. With three fundamental types, levers find diverse applications.

The fulcrum, acting as the pivot point, determines the motion of the lever. Force exerted on one end causes movement in the opposite direction. The placement of the fulcrum dictates any mechanical advantage. When positioned closer to the object being lifted, less force is needed, but the effort end must travel a greater distance to accomplish the task.

The following table describes the types of levers.

Class	Description	Image
First Class	A first-class lever enhances force or distance while redirecting it. A child's teeter-totter is a prime example of this lever type.	Object Force Fulcrum
Second Class	A second-class lever amplifies force in the same direction as the force applied to the object. It allows for a smaller force to lift a larger load, albeit over a greater distance. A classic example of this lever type is a wheelbarrow.	Object Force Fulcrum

Third Class	The third-class lever similarly increases the distance traveled by the object in the same direction as the applied force. Unlike the second-class lever, the fulcrum is at one end, and the object is at the other, with the force applied somewhere in between. Common examples include a fishing pole and a catapult.	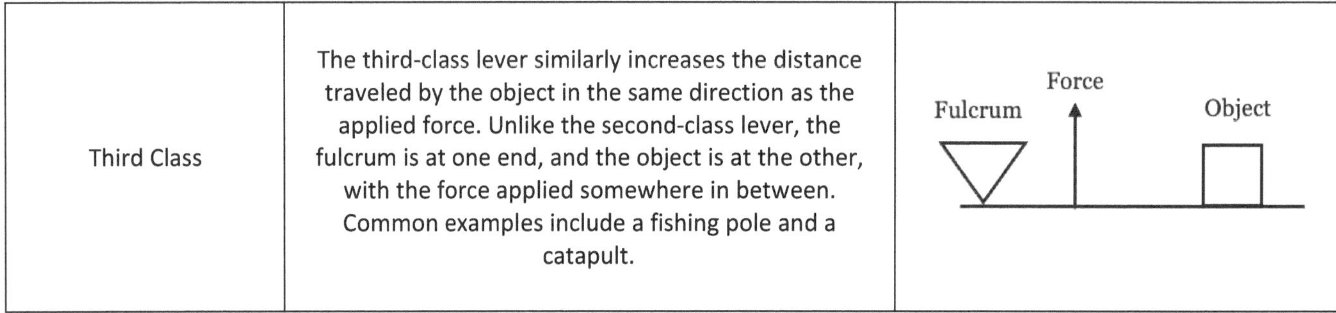

Utilizing force on an object can be facilitated by employing a pulley. This system is designed to alter the direction of force without changing the magnitude. Comprising a wheel over which a rope, belt, or chain is looped, a pulley system proves advantageous in scenarios where pulling downward is easier than pulling upward.

Using a single pulley on its own is relatively rare. Instead, multiple pulleys are often arranged in what is known as a block and tackle configuration to enhance lifting force. In this setup, as depicted here, four pulleys are employed: two are affixed to a fixed object (like a ceiling or building wall), and the other two are attached to the lower block (the load). When the rope is pulled, the lower block moves towards the stationary pulleys. Notably, to lift the load by 1 foot, the rope must be pulled a total of 4 feet. This is because each of the four rope links must shorten by 1 foot to move the lower block by 1 foot. Disregarding friction, this yields a mechanical advantage of 4:1. Hence, if 4 pounds of force are needed to lift a load, only 1 pound needs to be exerted on the rope. It's important to note that the mechanical advantage ratio of a block and tackle system can be determined by counting the number of pulleys.

The wheel and axle mechanism extends beyond a mere freely rotating wheel, like the front wheel of a standard bicycle. Instead, it employs two wheels mounted on a shaft, with one wheel having a larger diameter than the other. This setup enhances mechanical advantage. For instance, a steering wheel in a car exemplifies a wheel and axle.

Typically, the effort is applied to the larger-diameter wheel. Substantial movements of the large wheel yield minor movements of the small wheel. Given the larger wheel's greater circumference, a point on its edge travels farther per revolution compared to a point on the smaller wheel. Consequently, as a large movement translates into a small one, force is amplified. The mechanical advantage is calculated based on the ratio of the wheels' diameters.

Gears serve as versatile tools for transmitting force between two points, offering significant mechanical advantage.

They facilitate the adjustment of torque and rotational speed. The gear ratio, which compares the number of teeth on the large gear to those on the small gear, determines this. In the illustration, with the large gear boasting 18 teeth and the small gear 12, the gear ratio stands at 3:2.

When a large gear is driven by a smaller one, a reduction in speed occurs. Consequently, the large gear rotates more slowly than the small gear, resulting in a slower output speed proportionate to the gear ratio. However, this arrangement entails an associated amplification of torque, with the torque output from the large gear surpassing that applied to the small gear, in accordance with the gear ratio. Therefore, a decrease in speed corresponds to an increase in torque, a relationship often described as inversely proportional by mathematicians.

PRACTICING WHAT YOU'VE LEARNED

Now that you've got the basics, you can see test your newly acquired information and put it to the test.

Here are two practice sets with 15 questions. This is similar to how the CAT is setup, although obviously your questions aren't going to change the better you do.

The last section of this chapter details the right answer, and how you could have reached that answer.

Practice 1

1. A moving car has what type of energy?

(A) Potential energy
(B) Kinetic energy
(C) Thermal energy
(D) Chemical energy

2. Which law of motion states that every action has an equal and opposite reaction?

(A) Newton's First Law
(B) Newton's Second Law
(C) Newton's Third Law
(D) Newton's Law of Gravitation

3. What is the SI unit of force?

(A) Joule
(B) Newton
(C) Watt
(D) Pascal

4. What is the formula for calculating work?

(A) Work = Force × Time
(B) Work = Force × Distance
(C) Work = Mass × Acceleration
(D) Work = Energy ÷ Time

5. Which simple machine is a flat surface tilted at an angle?

(A) Wheel and axle
(B) Pulley
(C) Inclined plane
(D) Lever

6. What is the principle behind the operation of a pulley?

(A) Changes direction of force
(B) Increases force
(C) Increases distance
(D) Reduces friction

7. In a lever, where is the fulcrum located?

(A) At the end where the effort is applied
(B) In the middle
(C) At the end where the load is placed
(D) Anywhere along the lever

8. What type of lever increases force in the same direction as the applied force?

(A) First-class lever
(B) Second-class lever
(C) Third-class lever

(D) Fourth-class lever

9. What property of an object determines its resistance against motion changes?

(A) Mass
(B) Weight
(C) Volume
(D) Density

10. Which force opposes the motion of an object sliding across a surface?

(A) Tension
(B) Friction
(C) Gravity
(D) Buoyancy

11. Torque is measured in what units?

(A) Newton
(B) Watt
(C) Pascal
(D) Newton-meter

12. What is the function of a wedge?

(A) Increases force
(B) Changes direction
(C) Reduces friction
(D) Splits objects

13. What determines the mechanical advantage of a gear system?

(A) Number of teeth on the large gear
(B) Radius of the gear
(C) Material of the gear
(D) Color of the gear

14. What is the principle behind the operation of a wheel and axle?

(A) Changes direction of force
(B) Increases force
(C) Increases distance
(D) Reduces friction

15. What type of energy is stored in an object based on its position?

(A) Kinetic energy
(B) Thermal energy
(C) Potential energy
(D) Chemical energy

Practice 2

1. Which law of motion asserts that, without an external force, an object in motion will usually remain in motion and an object at rest will generally remain at rest?

    (A) Newton's First Law
    (B) Newton's Second Law
    (C) Newton's Third Law
    (D) Pascal's Law

2. What is the SI unit of work?

    (A) Newton
    (B) Joule
    (C) Watt
    (D) Pascal

3. Which simple machine has a rigid bar that pivots moves around a single point?

    (A) Inclined plane
    (B) Wheel and axle
    (C) Pulley
    (D) Lever

4. What determines the mechanical advantage of an inclined plane?

    (A) Length of the plane
    (B) Width of the plane
    (C) Angle of inclination
    (D) Mass of the load

5. What is the function of a screw?

    (A) Increases force
    (B) Changes direction
    (C) Converts rotational motion into linear motion
    (D) Holds objects together

6. What principle describes the relationship between pressure, force, and area in a fluid?

    (A) Newton's Law of Gravitation
    (B) Archimedes' Principle
    (C) Pascal's Law
    (D) Boyle's Law

7. In which type of lever is the effort applied between the load and the fulcrum?

    (A) First-class lever
    (B) Second-class lever
    (C) Third-class lever
    (D) Fourth-class lever

8. What is the formula to determine pressure?

    (A) Pressure = Force × Distance
    (B) Pressure = Force / Area
    (C) Pressure = Mass × Acceleration
    (D) Pressure = Work / Time

9. What is the Imperial system's unit of measurement for torque?

    (A) Foot-pound
    (B) Pound-foot
    (C) Newton-meter
    (D) Kilogram-meter

10. Which type of energy is associated with an object due to its temperature?

    (A) Kinetic energy
    (B) Potential energy
    (C) Thermal energy
    (D) Electrical energy

11. What is the function of a wheel and axle?

    (A) Increases force
    (B) Changes direction
    (C) Increases distance
    (D) Reduces friction

12. What principle states that the buoyant force acting on an object in a fluid is equal to the weight of the fluid displaced by the object?

    (A) Newton's Third Law
    (B) Bernoulli's Principle
    (C) Archimedes' Principle
    (D) Boyle's Law

13. Which type of friction occurs when two surfaces touch while one or both are in motion?

    (A) Static friction
    (B) Rolling friction
    (C) Fluid friction
    (D) Kinetic friction

14. What determines the mechanical advantage of a pulley system?

    (A) Number of ropes
    (B) Diameter of the pulley
    (C) Length of the rope
    (D) Number of pulleys

15. Which law of motion states that the rate of change of momentum of an object is directly proportional to the applied force and takes place in the direction of the force?

    (A) Newton's First Law
    (B) Newton's Second Law
    (C) Newton's Third Law
    (D) Pascal's Law

CHECKING YOUR ANSWERS

Now let's see how you did.

Practice 1

1. (B) Kinetic energy
2. (C) Newton's Third Law
3. (C) Newton
4. (B) Work = Force × Distance
5. (C) Inclined plane
6. (A) Changes direction of force
7. (D) Anywhere along the lever
8. (B) Second-class lever
9. (A) Mass
10. (B) Friction
11. (D) Newton-meter
12. (D) Splits objects
13. (A) Number of teeth on the large gear
14. (B) Increases force
15. (C) Potential energy

Practice 2

1. (A) Newton's First Law
2. (B) Joule
3. (D) Lever
4. (C) Angle of inclination
5. (D) Holds objects together
6. (C) Pascal's Law
7. (C) Third-class lever
8. (B) Pressure = Force / Area
9. (A) Foot-pound
10. (C) Thermal energy
11. (A) Increases force
12. (C) Archimedes' Principle
13. (D) Kinetic friction
14. (D) Number of pulleys
15. (B) Newton's Second Law

CHAPTER 14

ASSEMBLING OBJECTS

This section is entirely unique and there isn't anything else like it on the ASVAB. It is a mix of logical and puzzle solving. The ASVAB evaluates your skills and aptitude to determine which military careers align best with your abilities. For instance, roles such as pilots, mechanics, machine operators, and engineers require strong spatial relationship skills, which involve visualizing how components fit together to form a complete entity. The Assembling Objects (AO) subtest assesses these spatial relationship skills, allowing you to demonstrate your proficiency in assembling items. As always, it's advisable to consult with your recruiter or career counselor to gain insights into which subtests are vital for achieving your career objectives.

A BIT MORE ABOUT THIS SECTION OF THE ASVAB

The skills necessary for the AO test can be acquired through learning. Even individuals who find it challenging to visualize how shapes interconnect can enhance their spatial relationship skills significantly by adopting strategies and engaging in practice. In this chapter, you'll have the chance to develop these skills effectively.

You have more time for this section.

- The ASVAB gives you 15 minutes to finish 25 questions about AI if you take the CAT.
- If you take the traditional ASVAB, you will have 15 questions with just 17 minutes to finish. This means you have a less than a minute a question.

You will have two types of puzzles to solve in this seciton.

- Jigsaw-Puzzle-Type Problems: These assess your skill in determining how an object will appear when its components are assembled into a larger shape.
- Connector-Type Problems: These evaluate your ability to accurately connect two objects using a line at specified points.

In both types of AO items, you'll have five boxes.

- The first box shows drawings of unattached parts.
- The remaining four answer-choice boxes contain drawings of shapes, with only one depicting the shape that would result from correctly assembling all the parts.

WORKING WITH JIGSAW-PUZZLES

The best way to learn about these is to just dive right in. Here's the first example.

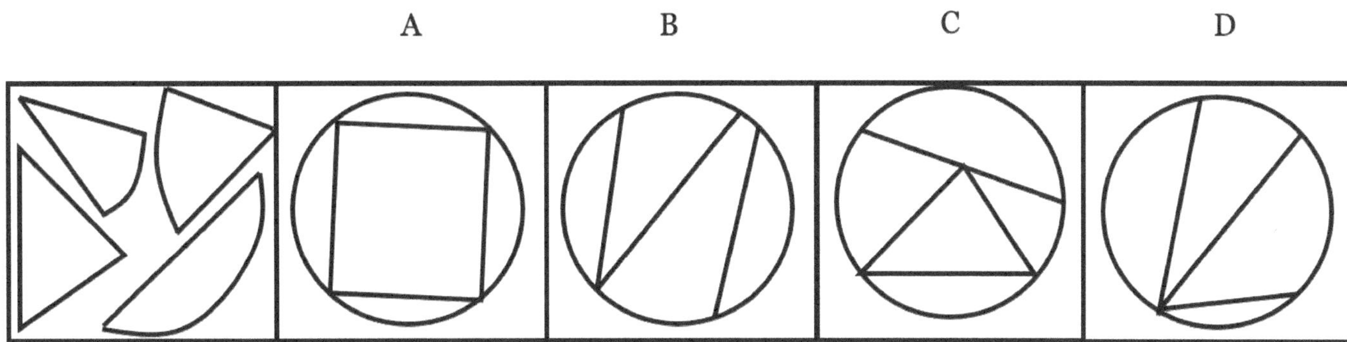

The box on the left contains the parts you need to consider. The boxes with the letters above them are the potential answers. If you were to take all of the pieces and put them together, you would get C. The following is how the parts look if you were to pull C apart, then rotate a few of the components.

You will have to learn how to pull them apart mentally, or you can quickly draw them separated to get a better idea of teach component on your own. At that point, use the process of elimination for the other three images.

Use the following steps to reach your answer.

1. Identify something unique in the shape, then identify it in the pieces.
2. Eliminate all of the options that don't have that shape.
3. Repeat steps 1 and 2 until you've removed all but one option.

Let's take a look at another one.

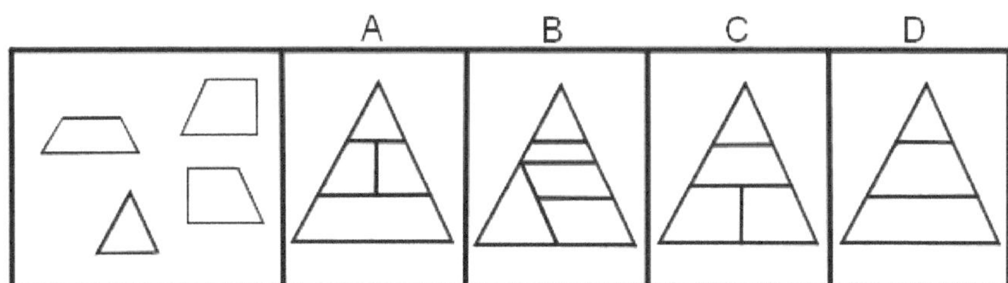

1. The top portion is clearly the triangle, especially since it is the same on all the images.
2. The base must be two pieces because it should be the longest, but the other three pieces are roughly the same size.
3. There are only two options where there are two shapes on the bottom, B and C. You can eliminate B since there aren't two triangle pieces in the beginning. That only leaves C.

Another things you could have done to eliminate B and D is that there are four pieces on the left, but there are only three pieces in D, and five pieces in B.

Here are a few things to keep in mind if you want to be able to quickly eliminate some of the options.

-   If an option has the wrong number of pieces, you can eliminate it.
-   Shapes in the incorrect answers might be inverted left-to-right or top-to-bottom. The correct answer might rearrange or rotate the shapes, but it will not invert any of them.

- Incorrect answers might feature shapes that are either enlarged or shrunken compared to those in the instructions box. The correct answer will maintain the original shapes and sizes.
- Shapes with straight edges in the instructions box might appear with curved edges in some incorrect answers. The correct answer will preserve the straight edges.
- Some incorrect answers may split a shape from the instructions box into two or more smaller shapes. The correct answer will keep the shapes intact as presented in the instructions box.

You can try these with the next couple of puzzles.

1. Which figure best represents how the objects in the first box appear if put together?

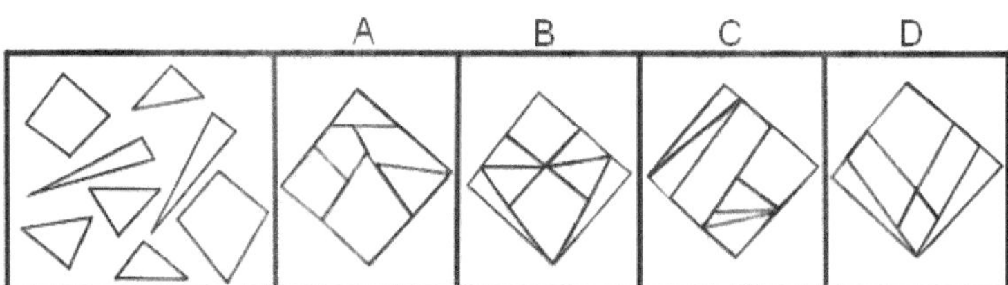

The answer is B. One of the shapes is a diamond, so you can eliminate C. You can eliminate D because it has two diamonds. That just leaves A and B. You can eliminate A because it has the wrong number of shapes.

2. Which figure best represents how the objects in the first box appear if put together?

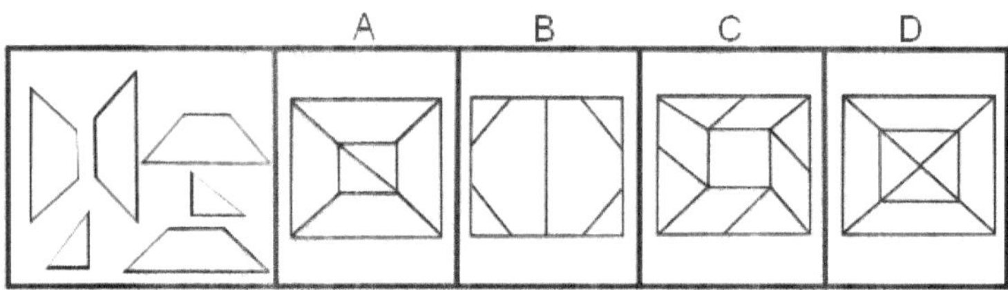

The answer is A. There are only two small triangles, so you can eliminate all by A.

WORKING WITH CONNECTOR-TYPE PROBLEM

Here's the first example of this type of problem. Which images best shows how the objects in the first box will look if you match the letters?

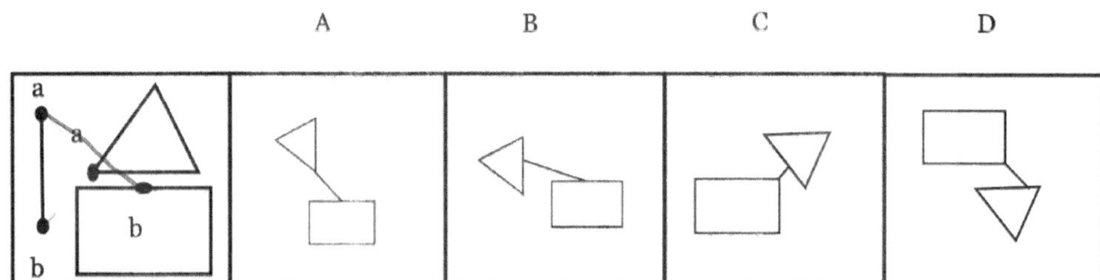

This kind of problems provides two shapes and a line, then asked you to connect them based on the initial shapes with letters representing where to connect. Think of the line like a string that you are trying to use to link the two shapes.

In the image, the answer is A. Notice that the triangle as the point in a corner of the triangle, so you can eliminate B and C, leaving A and D. The problem with D is that the b is in the middle of the line on the rectangle, not the corner. That just leaves A.

Here are a few ways to help you eliminate the wrong answers.

- If a shape's connection point is on the edge in the instructions box, it might be inside the shape in some wrong answers.

- When the connection point is on a corner, it may appear on different corners or along the edge in the incorrect choices.
- Some wrong answers may show shapes flipped horizontally or vertically.
- Shapes may also be resized, either enlarged or reduced, in some incorrect options. The correct answer will preserve the shapes and their original proportions.

You can try these with the next couple of puzzles.

1. Which figure best represents how the objects in the first box appear if put together?

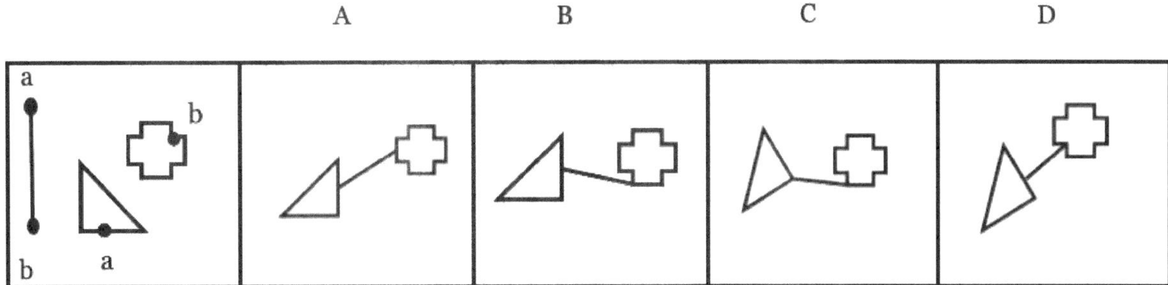

Only D has the line touching the right point on the plus sign, so you can eliminate the other three just based on that one aspect.

2. Which figure best represents how the objects in the first box appear if put together?

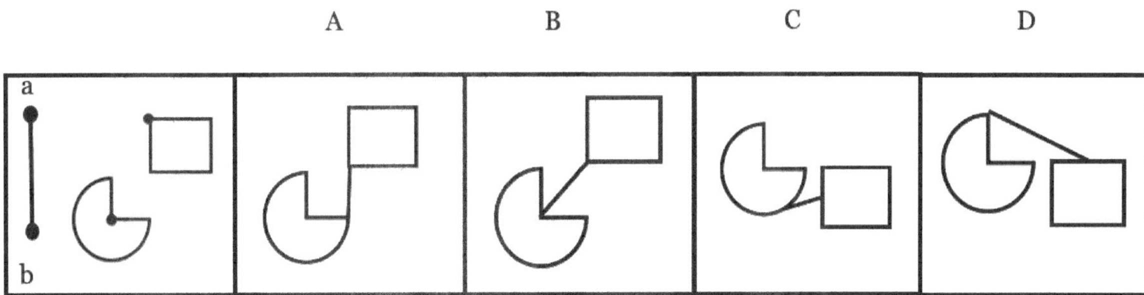

The answer is B.

PRACTICING WHAT YOU'VE LEARNED

Now that you've got the basics, you can see test your newly acquired information and put it to the test.

Here are two practice sets with 10 questions each.

The last section of this chapter details the right answer, and how you could have reached that answer.

Practice 1

1. Which figure best represents how the objects in the first box appear if put together?

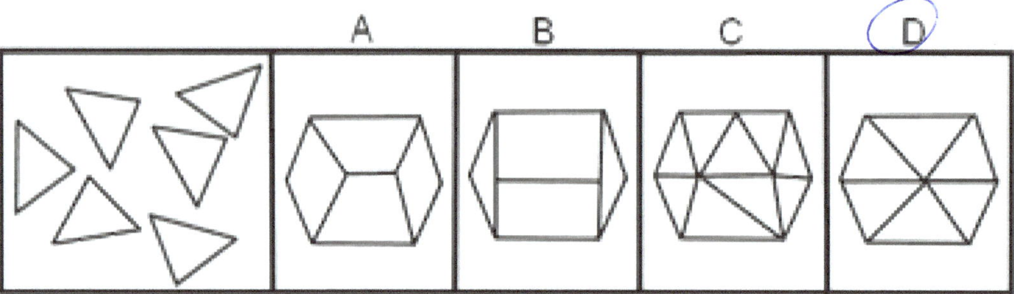

2. Which figure best represents how the objects in the first box appear if put together?

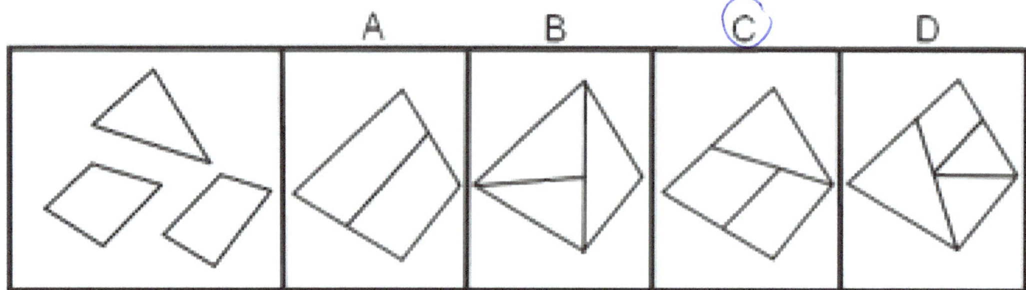

3. Which figure best represents how the objects in the first box appear if put together?

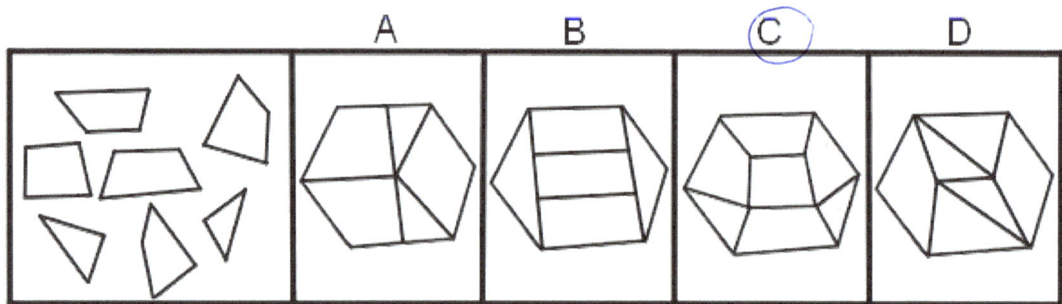

4. Which figure best represents how the objects in the first box appear if put together?

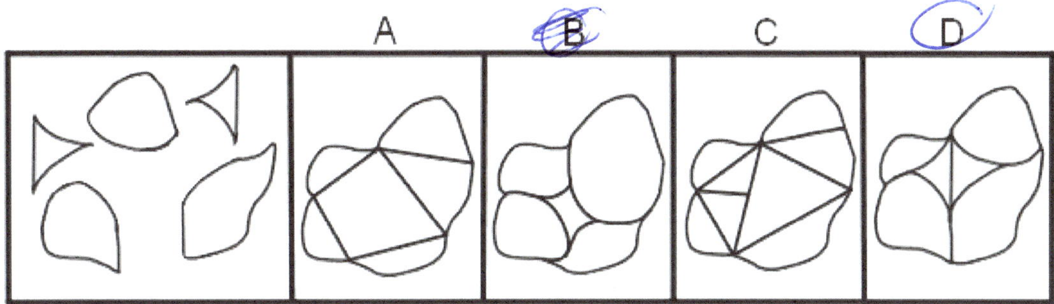

5. Which figure best represents how the objects in the first box appear if put together?

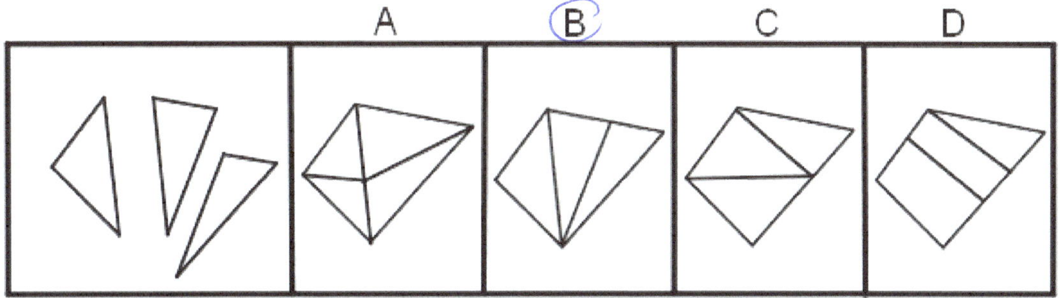

6. Which figure best represents how the objects in the first box appear if put together?

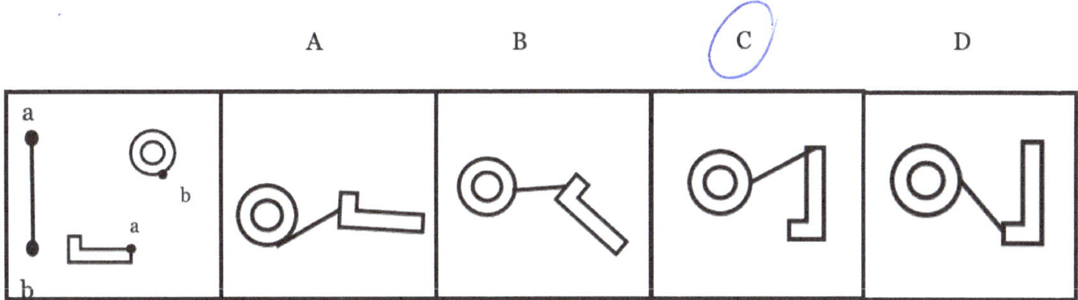

7. Which figure best represents how the objects in the first box appear if put together?

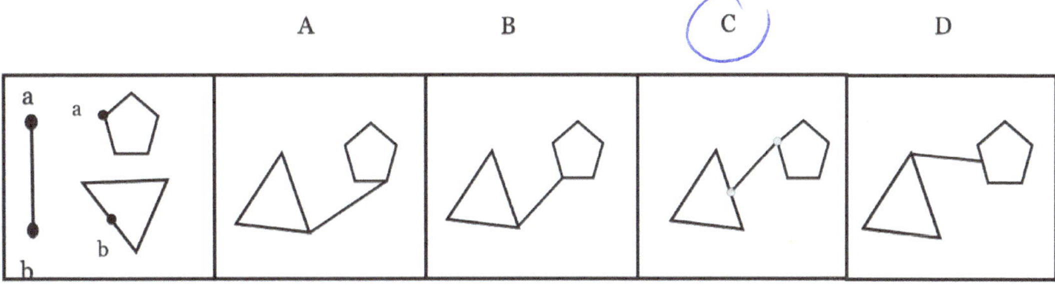

8. Which figure best represents how the objects in the first box appear if put together?

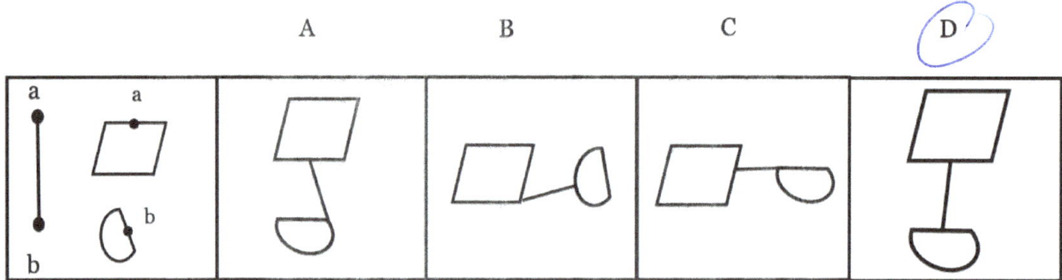

9. Which figure best represents how the objects in the first box appear if put together?

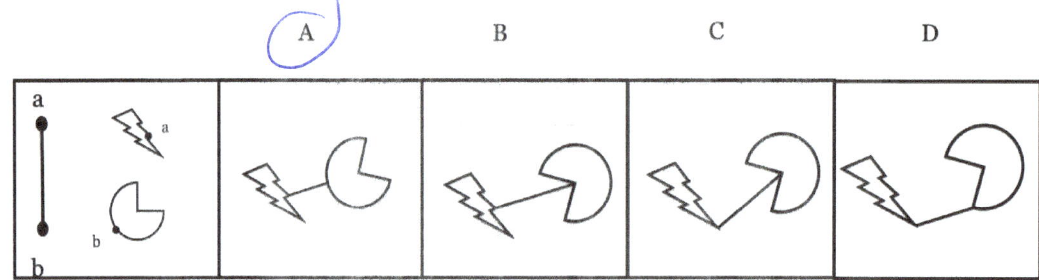

10. Which figure best represents how the objects in the first box appear if put together?

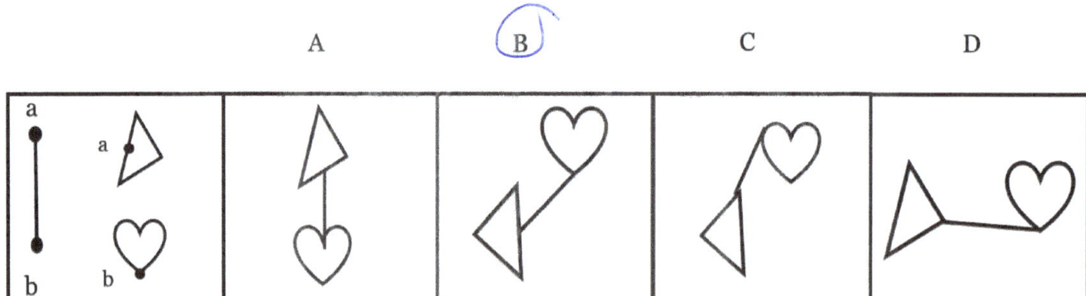

Practice 2

1. Which images best shows how the objects in the first box will look if you match the letters?

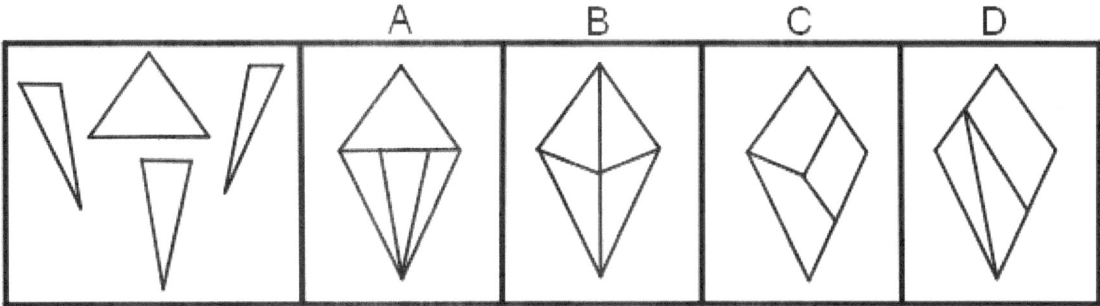

2. Which figure best represents how the objects in the first box appear if put together?

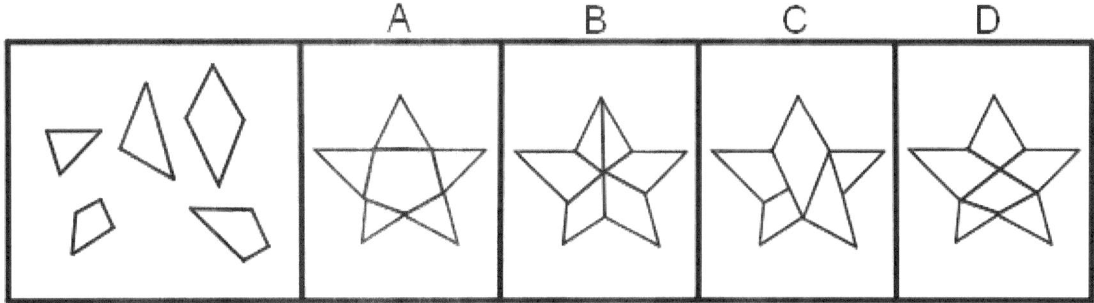

3. Which figure best represents how the objects in the first box appear if put together?

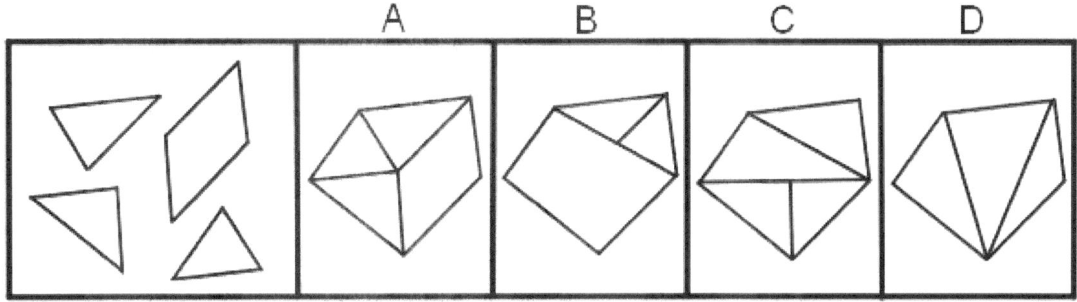

4. Which figure best represents how the objects in the first box appear if put together?

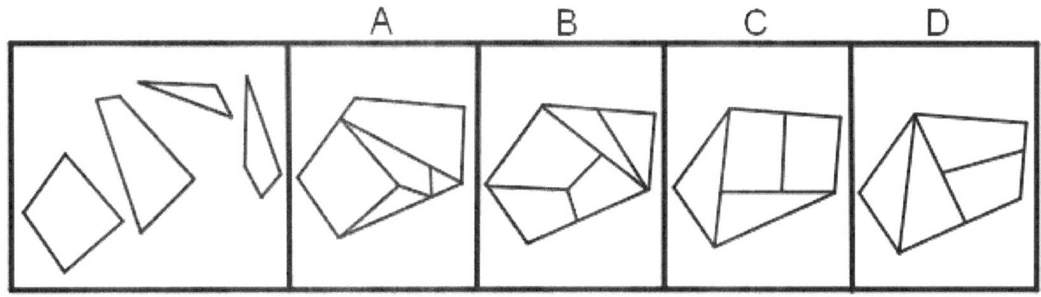

5.   Which figure best represents how the objects in the first box appear if put together?

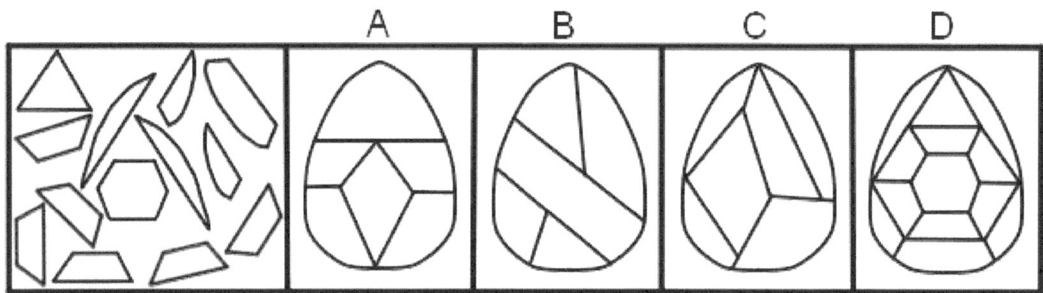

6.   Which figure best represents how the objects in the first box appear if put together?

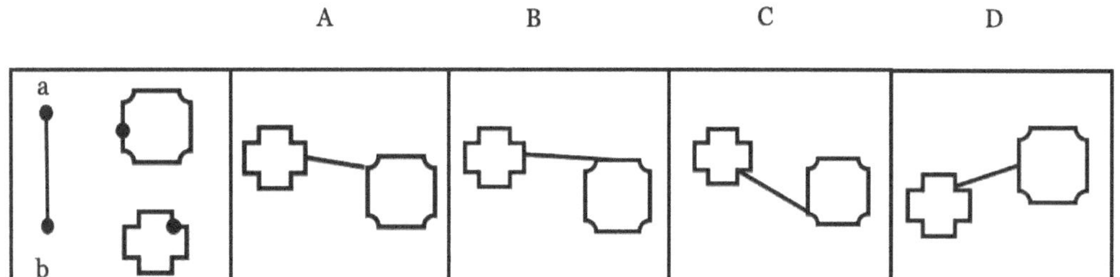

7.   Which figure best represents how the objects in the first box appear if put together?

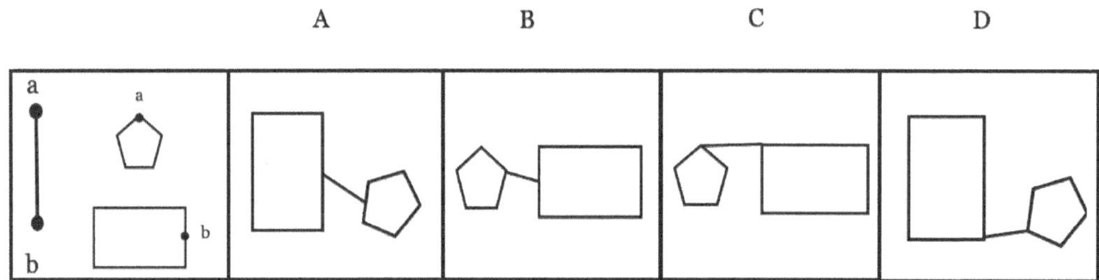

8.   Which figure best represents how the objects in the first box appear if put together?

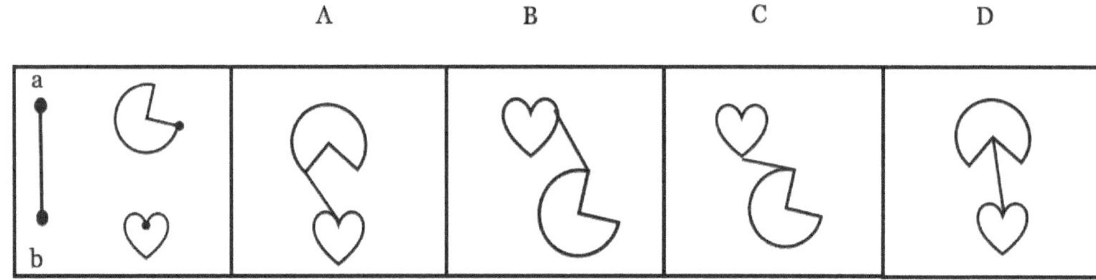

9. Which figure best represents how the objects in the first box appear if put together?

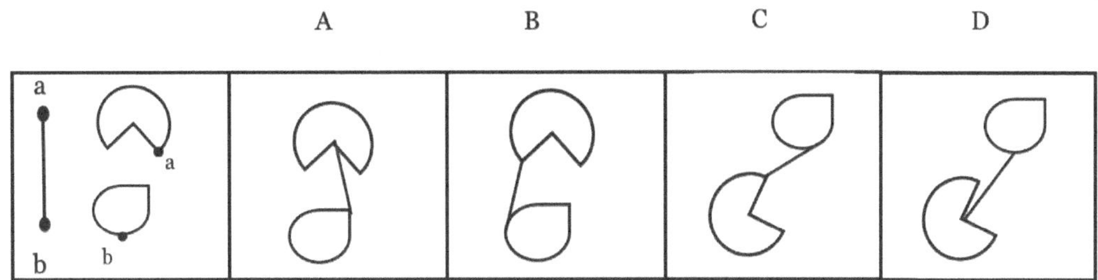

10. Which figure best represents how the objects in the first box appear if put together?

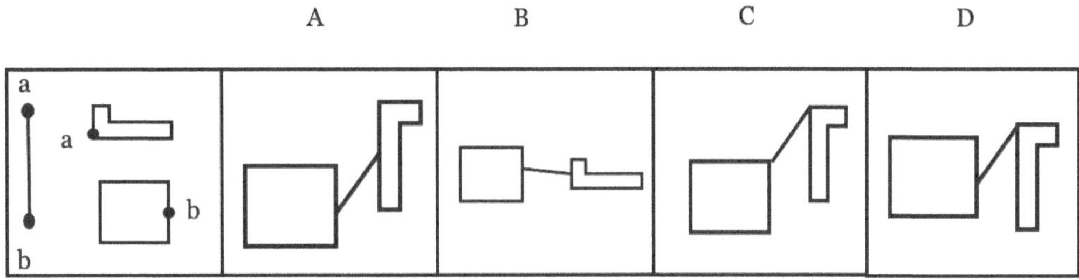

CHECKING YOUR ANSWERS

Now it's time to see how well you did.

Practice 1

1. D
2. C
3. C
4. D
5. B
6. C
7. C
8. D
9. A
10. B

Practice 2

1. A
2. C
3. A
4. A
5. C
6. D
7. B
8. A
9. C
10. D

CHAPTER 15

PRACTICE TESTS

This chapter has four practice tests for all of the sections The o

PRACTICE TEST 1

Here is the first practice test, and each section has between 10 and 15 practice questions.

General Science

1. What is the basic unit of life in all living organisms?

   (A) Atom
   (B) Molecule
   (C) Cell
   (D) Tissue

2. Which process do plants use to convert sunlight into food?

   (A) Respiration
   (B) Photosynthesis
   (Cc) Digestion
   (D) Fermentation

3. DNA stands for:

   (A) Deoxyribonucleic Acid
   (B) Deoxynucleic Acid
   (C) Dioxyribonucleic Acid
   (D) Deoxynucleic Agent

4. Which physiological system in humans is in charge of blood circulation throughout the body?

   (A) Digestive system
   (B) Respiratory system
   (C) Circulatory system
   (D) Nervous system

5. What layer of the Earth lies directly below the crust?

   (A) Core
   (B) Mantle
   (C) Outer core
   (D) Inner core

6. What is the most abundant gas in the Earth's atmosphere?

   (A) Oxygen
   (B) Carbon dioxide
   (C) Nitrogen
   (D) Hydrogen

7. The Milky Way is an example of a:

   (A) Star
   (B) Planet
   (C) Galaxy
   (D) Comet

8. What causes the phases of the Moon?

   (A) Earth's shadow
   (B) Moon's rotation
   (C) Moon's position relative to Earth and Sun
   (D) Earth's rotation

9. What is the force that pulls objects toward the Earth?

   (A) Magnetism
   (B) Gravity
   (C) Friction
   (D) Tension

10. Which of the following is an example of kinetic energy?

   (A) A book on a shelf
   (B) A moving car

(C) A stretched rubber band
(D) A charged battery

11. What is the speed of light in a vacuum?

(A) 3,000 km/s
(B) 30,000 km/s
(C) 300,000 km/s
(D) 3,000,000 km/s

12. Which law states that for every action, there is an equal and opposite reaction?

(A) Newton's First Law
(B) Newton's Second Law
(C) Newton's Third Law
(D) Law of Conservation of Energy

13. What is the chemical symbol for water?

(A) O
(B) H2O
(C) CO2
(D) NaCl

14. Which element is most abundant in the Earth's crust?

(A) Iron
(B) Silicon
(C) Oxygen
(D) Carbon

15. What is the pH level of a neutral substance?

(A) 0
(B) 7
(C) 14
(D) 3

Arithmetic Reasoning

1. Solve for x: $2x - 3 = 7$

(A) $x = 2$
(B) $x = 4$
(C) $x = 5$
(D) $x = 6$

2. What is the value of $|-7|$?

(A) -7
(B) 0
(C) 7
(D) 14

3. What is the result of $\frac{4}{9} \div \frac{2}{3}$?

(A) $\frac{2}{9}$
(B) $\frac{3}{6}$
(C) $\frac{6}{9}$
(D) $\frac{6}{27}$

4. What is the product of the first 5 prime numbers?

(A) 120
(B) 210
(C) 300
(D) 420

5. What is the value of $\sqrt{100}$?

(A) 10
(B) 50
(C) 100
(D) 200

6. If $a = 2$ and $b = 3$, what is the value of $a^3 - b^2$?

(A) 1
(B) 5
(C) 11
(D) 13

7. What is the sum of the first 20 positive even integers?

(A) 200
(B) 210
(C) 220
(D) 230

8. What is the cube root of 64?

(A) 4
(B) 6
(C) 8
(D) 12

9. What is $\frac{3}{4} + \frac{2}{3}$?

(A) $\frac{5}{7}$
(B) $\frac{5}{6}$
(C) $\frac{11}{12}$
(D) $\frac{1}{2}$

10. What is the value of $|-15|$?

(A) -15
(B) 0
(C) 15
(D) 30

11. Solve for x: $2x + 5 = 17$

(A) 6
(B) 7
(C) 8
(D) 9

12. What is the value of $|-12|$?

(A) -12
(B) 0
(C) 12
(D) 24

13. What is the result of $\frac{\frac{3}{4}*5}{6}$?

    (A) $\frac{5}{12}$
    (B) $\frac{8}{8}$
    (C) $\frac{1}{2}$
    (D) $\frac{7}{12}$

14. What is the sum of the first 10 positive even integers?

    (A) 110
    (B) 120
    (C) 130
    (D) 140

15. What is the square root of 169?

    (A) 10
    (B) 12
    (C) 13
    (D) 14

Word Knowledge

1. Adversary most closely means

    (A) Ally
    (B) Opponent
    (C) Supporter
    (D) Friend

2. He approached the challenge with aplomb, showing no signs of hesitation.

    (A) Fear
    (B) Confidence
    (C) Anger
    (D) Doubt

3. The debate was marked by acrimonious exchanges between the candidates.

    (A) Friendly
    (B) Harsh
    (C) Informal
    (D) Confusing

4. Benevolent

    (A) Malevolent
    (B) Cruel
    (C) Kind
    (D) Indifferent

5. Candid most closely means

    (A) Deceptive
    (B) Frank
    (C) Reserved
    (D) Evasive

6. Debilitate most closely means

    (A) Strengthen
    (B) Invigorate
    (C) Weaken
    (D) Heal

7. Elated most closely means

    (A) Depressed
    (B) Overjoyed
    (C) Indifferent
    (D) Worried

8. The efficacy of the new drug was proven in clinical trials.

    (A) Cost
    (B) Effectiveness
    (C) Safety
    (D) Complexity

9. The CEO's capricious decisions often left the employees feeling uncertain.

    (A) Predictable
    (B) Thoughtful
    (C) Unpredictable
    (D) Rash

10. Flourish most closely means

    (A) Wither
    (B) Decline
    (C) Prosper
    (D) Fail

11. Hostile most closely means

    (A) Friendly
    (B) Benevolent
    (C) Aggressive
    (D) Amicable

12. Indifferent most closely means

    (A) Concerned
    (B) Passionate
    (C) Apathetic
    (D) Interested

13. Lucid most closely means

    (A) Confusing
    (B) Clear
    (C) Muddled
    (D) Obscure

14. Meticulous most closely means

    (A) Careless
    (B) Negligent
    (C) Thorough
    (D) Haphazard

15. She was known for her <u>altruistic</u> behavior, always
helping those in need.

    (A) Selfish
    (B) Charitable
    (C) Lazy
    (D) Skeptical

Paragraph Comprehension

1. Sarah was a voracious reader who frequently found herself engrossed in a book. Her favorite genre was fantasy, and she loved immersing herself in magical worlds filled with dragons, wizards, and mythical creatures. One day, while browsing the shelves of her local bookstore, Sarah stumbled upon a dusty old tome hidden behind a row of books. Curious, she picked up the book and began to read, unaware that its pages held the key to unlocking a fantastical adventure of her own.

    What is Sarah's favorite genre of book?

    (A) Mystery
    (B) Romance
    (C) Fantasy
    (D) Science fiction

2. The human brain is a complex organ responsible for controlling bodily functions, processing information, and regulating emotions. It consists of billions of neurons that communicate through electrical impulses and chemical signals. Understanding the structure and function of the brain is essential for advancing neuroscience and developing treatments for neurological disorders.

    What is the main theme of the passage?

    (A) The structure of the human brain
    (B) The importance of mental health
    (C) The functions of the nervous system
    (D) The history of neuroscience

3. Anna, Brian, and Claire are three friends who have different pets. Anna has a cat, Brian has a dog, and Claire has a fish. If the statements are true, which pet does Brian have?

    What pet does Brian have?

    (A) Cat
    (B) Dog
    (C) Fish
    (D) Cannot be determined

4. Renewable energy sources, such as solar, wind, and hydroelectric power, offer sustainable alternatives to fossil fuels. Unlike fossil fuels, which are limited and contribute to climate change, renewable energy sources are plentiful and emit little or no greenhouse gases. The transition to renewable energy is critical for reducing the consequences of climate change and guaranteeing a sustainable future.

    What is the main theme of the passage?

    (A) The benefits of fossil fuels
    (B) The importance of renewable energy
    (C) The history of energy production
    (D) The impact of climate change

5. Mahatma Gandhi was a leader of the Indian independence movement against British rule, advocating for nonviolent civil disobedience. Born in India in 1869, Gandhi's philosophy of Satyagraha, or "truth force," inspired movements for civil rights and freedom across the world. What was Mahatma Gandhi's approach to achieving independence for India?

    What was Mahatma Gandhi's approach to achieving independence for India?

    (A) Violent rebellion against British rule
    (B) Peaceful protests and nonviolent resistance
    (C) Military conquest and warfare
    (D) Economic sanctions and trade restrictions

6. Jane Goodall is a British primatologist and conservationist known for her groundbreaking research on chimpanzees in Tanzania. Born in London in 1934, Goodall's discoveries about chimpanzee behavior challenged prevailing scientific beliefs and revolutionized our understanding of primate intelligence and social interactions. What field of study is Jane Goodall renowned for?

What field of study is Jane Goodall renowned for?

(A) Botany
(B) Anthropology
(C) Primatology
(D) Astronomy

7. Biometric technology uses unique biological characteristics, such as fingerprints and facial features, for authentication and identification purposes. This technology is increasingly being integrated into security systems, mobile devices, and access control mechanisms for enhanced security and convenience. What is the primary advantage of biometric technology?

What is the primary advantage of biometric technology?

(A) It requires complex passwords and security codes
(B) It relies solely on physical tokens for authentication
(C) It is vulnerable to identity theft and fraud
(D) It offers enhanced security and convenience

8. Quantum computing leverages the principles of quantum mechanics to perform complex calculations at speeds exponentially faster than classical computers. While still in its early stages of development, quantum computing has the potential to revolutionize fields such as cryptography, optimization, and drug discovery. What distinguishes quantum computing from classical computing?

What distinguishes quantum computing from classical computing?

(A) Quantum computing operates at exponentially faster speeds
(B) Quantum computing relies on binary logic
(C) Quantum computing is limited by Moore's Law
(D) Quantum computing uses classical bits for data storage

9. Wearable technology refers to electronic devices that can be worn on the body, either as accessories or as clothing. These devices, such as smartwatches and fitness trackers, collect data and provide information or feedback to the user. How does wearable technology enhance user experiences?

How does wearable technology enhance user experiences?

(A) By restricting movement and flexibility
(B) By limiting access to real-time data and information
(C) By providing personalized feedback and insights
(D) By increasing reliance on external devices for daily activities

10. Four students—Jack, Kate, Lily, and Mike—are lined up for a school photo. Jack is standing between Kate and Lily. Mike is standing next to Lily. Whose place is at the far end of the line?

Whose place is at the far end of the line?

(A) Jack
(B) Kate
(C) Lily
(D) Mike

Mathematics Knowledge

1. Solve for x: 5x + 7 = 2x + 16

    (A) 1
    (B) 2
    (C) 3
    (D) 4

2. What is the area of a triangle with a base of 10 cm and a height of 8 cm?

    (A) 20 cm²
    (B) 40 cm²
    (C) 60 cm²
    (D) 80 cm²

3. Simplify: 2(x + 3) - 4(x - 2)

    (A) -2x + 10
    (B) 2x + 10
    (C) -2x - 2
    (D) 2x − 2

4. Find the volume of a cylinder with a radius of 3 cm and a height of 7 cm. (Use π = 3.14)

    (A) 138.26 cm³
    (B) 197.82 cm³
    (C) 205.92 cm³
    (D) 263.76 cm³

5. Solve for y: 4y - 9 = 3y + 5

    (A) 14
    (B) -14
    (C) 4
    (D) -4

6. What is the measure of each interior angle of a regular octagon?

    (A) 120°
    (B) 135°
    (C) 140°
    (D) 150°

7. If f(x) = 2x² - 3x + 1, what is f(2)?

    (A) 1
    (B) 3
    (C) 5
    (D) 7

8. Find the circumference of a circle with a diameter of 12 cm. (Use π = 3.14)

    (A) 18.84 cm
    (B) 24.52 cm
    (C) 37.68 cm
    (D) 75.36 cm

9. Simplify: (3x² - 2x + 4) - (x² + 3x - 2)

    (A) 2x² - 5x + 6
    (B) 4x² + x + 6
    (C) 2x² - 5x + 2
    (D) 4x² + x + 2

10 What is the area of a circle with a radius of 5 cm? (Use π = 3.14)

    (A) 15.7 cm²
    (B) 25.12 cm²
    (C) 78.5 cm²
    (D) 314 cm²

11. Solve for x: $(3x + 1)/2 = 4$

    (A) 1
    (B) 2
    (C) 3
    (D) 4

12. What is the volume of a cone with a radius of 4 cm and a height of 9 cm? (Use π = 3.14)

    (A) 150.72 cm³
    (B) 200.96 cm³
    (C) 301.44 cm³
    (D) 452.16 cm³

13. If the length of a rectangle is three times its width and the perimeter is 48 cm, what is the width of the rectangle?

    (A) 4 cm
    (B) 6 cm
    (C) 8 cm
    (D) 12 cm

14 Simplify: 5(x - 2) - 3(x + 4)

    (A) 2x - 22
    (B) 2x - 14
    (C) 8x - 22
    (D) 8x - 14

15. Find the length of the hypotenuse in a right triangle with legs of 9 cm and 12 cm.

    (A) 13 cm
    (B) 14 cm
    (C) 15 cm
    (D) 16 cm

Electronics Information

1. What is the charge of an electron?

(A) Positive
(B) Negative
(C) Neutral
(D) It varies

2. Where are electrons located in an atom?

(A) In the nucleus
(B) Orbiting the nucleus
(C) In the proton
(D) In the neutron

3. What is the primary role of electrons in electrical conduction?

(A) To produce light
(B) To carry electrical charge
(C) To neutralize protons
(D) To form chemical bonds

4. What is the standard unit of electric current?

(A) Volt
(B) Ohm
(C) Watt
(D) Ampere

5. What is a common source of direct current (DC)?

(A) Power plants
(B) Car battery
(C) Household outlets
(D) Solar panels

6. What is the flow of electrical charge called?

(A) Voltage
(B) Resistance
(C) Current
(D) Capacitance

7. What is the unit of electrical potential difference?

(A) Ohm
(B) Ampere
(C) Watt
(D) Volt

8. Which device measures voltage?

(A) Ammeter
(B) Voltmeter
(C) Ohmmeter
(D) Thermometer

9. What is the voltage across a 10-ohm resistor with a current of 2 amperes?

(A) 5 volts
(B) 10 volts
(C) 20 volts
(D) 200 volts

10. Which component stores electrical energy in an electric field?

(A) Resistor
(B) Capacitor
(C) Inductor
(D) Transistor

11. What is the function of a diode?

(A) To store energy
(B) To allow current to flow in one direction only
(C) To amplify signals
(D) To resist current flow

12. What type of transistor is commonly used in digital circuits?

(A) Bipolar Junction Transistor (BJT)
(B) Field-Effect Transistor (FET)
(C) Zener Diode
(D) Varistor

13. What is the relationship defined by Ohm's Law?

(A) $V = IR$
(B) $P = IV$
(C) $R = V + I$
(D) $V = I + R$

14. What is the main purpose of a resistor in a circuit?

(A) To increase voltage
(B) To store charge
(C) To limit current
(D) To produce light

15. How is electrical power calculated in a circuit?

(A) $P = IV$
(B) $P = IR$
(C) $P = I/R$
(D) $P = V/I$

Auto and Shop

1. What tool is used to fasten pieces of wood together with nails or staples?

    (A) Screwdriver
    (B) Hammer
    (C) Drill press
    (D) Pliers

2. Which device is commonly used to remove excess material from a workpiece?

    (A) Lathe
    (B) Saw
    (C) Planer
    (D) Clamp

3. What is the primary function of a lathe in a workshop?

    (A) Cutting metal
    (B) Sanding wood
    (C) Shaping cylindrical objects
    (D) Welding pieces together

4. Which tool is used to cut curves and intricate shapes in wood or metal?

    (A) Circular saw
    (B) Bandsaw
    (C) Bench grinder
    (D) Belt sander

5. What is the purpose of a bench grinder in a workshop?

    (A) Cutting metal
    (B) Sharpening tools
    (C) Smoothing rough surfaces
    (D) Measuring angles

6. Which type of wrench is commonly used for turning nuts and bolts in tight spaces?

    (A) Socket wrench
    (B) Adjustable wrench
    (C) Pipe wrench
    (D) Box-end wrench

7. What tool is used to hold workpieces securely in place while working on them?

    (A) Vise
    (B) Clamp
    (C) Pliers
    (D) Screwdriver

8. What is the function of a miter saw in woodworking?

    (A) Sanding wood
    (B) Cutting metal
    (C) Cutting precise angles in wood
    (D) Drilling holes

9. Which device is used to measure angles accurately in woodworking?

    (A) Tape measure
    (B) Protractor
    (C) Combination square
    (D) Level

10. What is the primary function of a router table in a workshop?

    (A) Cutting metal
    (B) Sanding wood
    (C) Shaping edges of wood
    (D) Welding pieces together

11. Which tool is used to remove and install screws with a specific drive type?

    (A) Screwdriver
    (B) Pliers
    (C) Drill press
    (D) Allen wrench

12. What is the purpose of a jointer in woodworking?

    (A) Sanding wood
    (B) Planing wood surfaces
    (C) Cutting precise angles
    (D) Joining wood edges

13. Which device is used to create holes in metal or wood with precision?

    (A) Hammer drill
    (B) Circular saw
    (C) Jigsaw
    (D) Drill press

14. What is the function of a welding machine in a workshop?

    (A) Cutting metal
    (B) Joining metal pieces together
    (C) Smoothing metal surfaces
    (D) Bending metal

15. Which tool is commonly used for cutting large sheets of wood or metal into smaller pieces?

    (A) Jigsaw
    (B) Circular saw
    (C) Bandsaw
    (D) Bench grinder

Mechanical Comprehension

1. Which simple machine is a ramp?

   (A) Lever
   (B) Wheel and axle
   (C) Inclined plane
   (D) Pulley

2. What is the mechanical advantage of a lever if the effort arm is 5 meters long and the load arm is 2 meters long?

   (A) 2.5
   (B) 3
   (C) 2
   (D) 1

3. Which type of lever has the fulcrum positioned between the effort and the load?

   (A) First-class lever
   (B) Second-class lever
   (C) Third-class lever
   (D) It depends on the specific setup

4. What is the purpose of a wedge in a mechanical system?

   (A) To amplify force
   (B) To change the direction of force
   (C) To increase the distance over which force is applied
   (D) To split objects apart

5. If a wheel has a radius of 0.5 meters and is turned through an angle of 60 degrees, what is the distance traveled by a point on its circumference?

   (A) $\pi/6$ meters
   (B) $\pi/3$ meters
   (C) $\pi/2$ meters
   (D) π meters

6. What type of force opposes the motion of objects rolling against each other?

   (A) Frictional force
   (B) Tension force
   (C) Gravitational force
   (D) Magnetic force

7. Which of the following is an example of a third-class lever?

   (A) Tweezers
   (B) A nutcracker
   (C) A wheelbarrow
   (D) A crowbar

8. What is the mechanical advantage of an inclined plane if its length is 6 meters and its height is 2 meters?

   (A) 6
   (B) 4
   (C) 3
   (D) 2

9. What is the purpose of a gear in a mechanical system?

   (A) To change the direction of force
   (B) To amplify force
   (C) To transfer rotational motion between shafts
   (D) To reduce friction between moving parts

10. If a screw has 10 threads per centimeter and is turned through 5 complete rotations, how far does it advance?

   (A) 50 centimeters
   (B) 25 centimeters
   (C) 10 centimeters
   (D) 5 centimeters

11. Which simple machine is best suited for lifting heavy objects?

   (A) Lever
   (B) Wheel and axle
   (C) Pulley
   (D) Inclined plane

12. What is the advantage of using a lever?

   (A) It decreases the force needed to lift an object
   (B) It increases the distance over which force is applied
   (C) It changes the direction of force
   (D) It allows for precise control of force

13. What is the mechanical advantage of a wheel and axle system if the radius of the wheel is 4 meters and the radius of the axle is 1 meter?

   (A) 1
   (B) 2
   (C) 3
   (D) 4

14. How does lubricating a machine reduce friction?

   (A) By increasing the coefficient of friction
   (B) By decreasing the coefficient of friction
   (C) By increasing the weight of the moving parts
   (D) By increasing the force of friction

15. If a lever has a mechanical advantage of 2 and the effort force applied is 50 newtons, what is the load force that can be lifted?

   (A) 25 newtons
   (B) 50 newtons
   (C) 100 newtons
   (D) 200 newtons

PRACTICE TEST 2

Here is the first practice test, and each section has between 10 and 15 practice questions.

General Science

1. Which organelle is known as the powerhouse of the cell?

   (A) Nucleus
   (B) Mitochondrion
   (C) Ribosome
   (D) Endoplasmic Reticulum

2. Which process do cells use to divide and reproduce?

   (A) Photosynthesis
   (B) Mitosis
   (C) Osmosis
   (D) Transcription

3. What is the term for the genetic material that carries information in living organisms?

   (A) RNA
   (B) Protein
   (C) DNA
   (D) Lipid

4. Which body system is responsible for breaking down food into nutrients?

   (A) Circulatory system
   (B) Respiratory system
   (C) Nervous system
   (D) Digestive system

5. Which planet is known as the Red Planet?

   (A) Venus
   (B) Mars
   (C) Jupiter
   (D) Saturn

6. What is the main cause of seasons on Earth?

   (A) Earth's distance from the Sun
   (B) Earth's tilt on its axis
   (C) The Moon's gravity
   (D) Solar flares

7. What type of rock is formed from molten lava?

   (A) Sedimentary
   (B) Metamorphic
   (C) Igneous
   (D) Limestone

8. What do we call a system of stars, dust, and gas held together by gravity?

   (A) Nebula
   (B) Comet
   (C) Galaxy
   (D) Asteroid belt

9. What is the unit of force in the International System of Units (SI)?

   (A) Joule
   (B) Newton
   (C) Watt
   (D) Pascal

10. Which of the following describes an object in motion staying in motion unless acted on by an outside force?

    (A) Inertia
    (B) Acceleration
    (C) Momentum
    (D) Friction

11. What phenomenon causes the bending of light as it passes from one medium to another?

    (A) Reflection
    (B) Refraction
    (C) Diffraction
    (D) Dispersion

12. What is the formula to calculate the speed of an object?

    (A) Speed = Distance × Time
    (B) Speed = Distance / Time
    (C) Speed = Time / Distance
    (D) Speed = Distance + Time

13. Which element has the chemical symbol "O"?

    (A) Gold
    (B) Osmium
    (C) Oxygen
    (D) Olivine

14. What type of bond involves the sharing of electron pairs between atoms?

    (A) Ionic bond
    (B) Hydrogen bond
    (C) Covalent bond
    (D) Metallic bond

15. What is the chemical formula for table salt?

    (A) NaCl
    (B) H2O
    (C) CO2
    (D) C6H12O6

Arithmetic reasoning

1. What is $\frac{1}{3}$?

   (A) $\frac{1}{6}$
   (B) $\frac{1}{3}$
   (C) $\frac{2}{3}$
   (D) $\frac{4}{3}$

2. If a = 4 and b = 6, what is $a^2 - b^2$?

   (A) 16
   (B) 18
   (C) 20
   (D) 24

3. What is the prime factorization of 48?

   (A) $2^3 * 3$
   (B) $2^4 * 3$
   (C) $2^4 * 3^2$
   (D) $2^5 * 3$

4. What is 0.75 * 100?

   (A) 65
   (B) 70
   (C) 75
   (D) 80

5. What is the value of 5! ?

   (A) 120
   (B) 240
   (C) 360
   (D) 720

6. Solve for x: 3x - 2 = 13

   (A) 5
   (B) 6
   (C) 7
   (D) 8

7. What is the value of |-20|?

   (A) -20
   (B) 0
   (C) 20
   (D) 40

8. What is $\frac{3}{5} - \frac{1}{5}$?

   (A) $\frac{1}{10}$
   (B) $\frac{1}{5}$
   (C) $\frac{2}{5}$
   (D) $\frac{3}{5}$

9. What is the cube root of 64?

   (A) 2
   (B) 4
   (C) 6
   (D) 8

10. What is $\frac{4}{5}$ of 100?

    (A) 60
    (B) 70
    (C) 80
    (D) 90

11. What is 2 * 3 + 4?

    (A) 8
    (B) 10
    (C) 12
    (D) 14

12 What is the sum of the first 20 positive odd integers?

    (A) 400
    (B) 420
    (C) 440
    (D) 460

13. What is $\frac{7}{8}$ of 64?

    (A) 42
    (B) 48
    (C) 56
    (D) 63

14. What is 3^3?

    (A) 6
    (B) 9
    (C) 18
    (D) 27

15. If x = 9 and y = 2, what is $x^2 + 2xy + y^2$?

    (A) 115
    (B) 135
    (C) 155
    (D) 175

Word Knowledge

1. The judge's <u>impartial</u> ruling ensured that both parties were treated fairly.

    (A) Biased
    (B) Unfair
    (C) Neutral
    (D) Hasty

2. Obsolete most closely means

    (A) Modern
    (B) Current
    (C) Outdated
    (D) Fashionable

3. Pacify most closely means

    (A) Agitate
    (B) Incite
    (C) Soothe
    (D) Irritate

4. Quaint most closely means

    (A) Modern
    (B) Common
    (C) Charming
    (D) Ordinary

5. Reclusive most closely means

    (A) Sociable
    (B) Withdrawn
    (C) Outgoing
    (D) Extroverted

6. Succinct most closely means

    (A) Verbose
    (B) Lengthy
    (C) Concise
    (D) Wordy

7. His <u>insatiable</u> appetite for knowledge led him to read countless books.

    (A) Satisfied
    (B) Unquenchable
    (C) Limited
    (D) Simple

8. The <u>precarious</u> ladder made him nervous as he climbed to the roof.

    (A) Stable
    (B) Wobbly

    (C) Secure
    (D) Short

9. Tenacious most closely means

    (A) Yielding
    (B) Persistent
    (C) Indifferent
    (D) Weak

10. Ubiquitous most closely means

    (A) Rare
    (B) Scarce
    (C) Omnipresent
    (D) Uncommon

11. Vivid most closely means

    (A) Dull
    (B) Vague
    (C) Bright
    (D) Faint

12. Her <u>voracious</u> reading habits meant she finished several books a week.

    (A) Slow
    (B) Limited
    (C) Greedy
    (D) Sporadic

13. Prudent most closely means

    (A) Reckless
    (B) Careful
    (C) Foolish
    (D) Rash

14. Resilient most closely means

    (A) Rigid
    (B) Flexible
    (C) Weak
    (D) Fragile

15. The athlete's _____ behavior on and off the field made him a role model for many.

    (A) Notorious
    (B) Exemplary
    (C) Indifferent
    (D) Apathetic

Paragraph Comprehension

1. Renaissance polymath Leonardo da Vinci was well-known for his proficiency in a wide range of subjects, including anatomy, engineering, sculpting, and painting. Born in Vinci, Italy, in 1452, da Vinci's artistic masterpieces such as the "Mona Lisa" and "The Last Supper" continue to captivate audiences worldwide. What was Leonardo da Vinci primarily known for?

What was Leonardo da Vinci primarily known for?

(A) His contributions to engineering
(B) His expertise in sculpture
(C) His mastery of the violin
(D) His achievements in mathematics

2. The Great Barrier Reef is the world's largest coral reef system, stretching over 2,300 kilometers along the coast of Queensland, Australia. There are more than 1,500 different types of fish and 400 different kinds of coral found there, among other rich marine life. The reef is a UNESCO World Heritage Site and attracts millions of visitors each year for snorkeling, scuba diving, and eco-tours.

What is the main theme of the passage?

(A) The geography of Australia
(B) The importance of marine biodiversity
(C) The history of coral reefs
(D) The tourism industry in Queensland

3. There are four siblings—Amy, Ben, Chloe, and David. Amy is taller than Ben but shorter than Chloe. Ben is taller than David. Who is the shortest?

Who is the shortest sibling?

(A) Amy
(B) Ben
(C) Chloe
(D) David

4. Evolution is the process by which species change over time through genetic variation, natural selection, and adaptation. It is the central concept in biology and explains how life on Earth has diversified and evolved over billions of years. The theory of evolution, first proposed by Charles Darwin in the 19th century, has since been supported by a vast body of scientific evidence.

What is the main theme of the passage?

(A) The history of biology
(B) The theory of natural selection
(C) The process of evolution
(D) The contributions of Charles Darwin

5. Steve Jobs was a co-founder of Apple Inc. and a visionary entrepreneur who revolutionized the technology industry with products like the iPhone, iPad, and Macintosh computer. Born in California in 1955, Jobs' relentless pursuit of innovation and design perfection transformed Apple into one of the world's most valuable companies. What company did Steve Jobs co-found?

What company did Steve Jobs co-found?

(A) Microsoft Corporation
(B) Amazon.com, Inc.
(C) Apple Inc.
(D) Google LLC

6. The wheel, invented by ancient civilizations thousands of years ago, revolutionized transportation and enabled the development of wheeled vehicles. This simple yet ingenious invention laid the foundation for modern transportation systems and facilitated the movement of goods and people. What invention revolutionized transportation and enabled the development of wheeled vehicles?

What invention revolutionized transportation and enabled the development of wheeled vehicles?

(A) The telephone
(B) The printing press

(C) The wheel
(D) The internet

7. Alexander Fleming, a Scottish chemist, developed penicillin in 1928, and its capacity to treat bacterial infections completely changed medicine. The creation of more life-saving antibiotics was made possible by Fleming's unintentional discovery of penicillin's antibiotic qualities. Who discovered penicillin?

Who discovered penicillin?

(A) Alexander Fleming
(B) Louis Pasteur
(C) Marie Curie
(D) Jonas Salk

8. American politician Barack Obama led the country as its 44th president from 2009 to 2017. The Affordable Care Act and the economic stimulus package are only two of the noteworthy legislative accomplishments that defined Obama's term. What distinguishes Barack Obama's presidency?

What distinguishes Barack Obama's presidency?

(A) His focus on space exploration
(B) His achievements in the culinary arts
(C) His contributions to classical music
(D) His legislative accomplishments and leadership style

9. A teacher gives five students—Emma, Finn, Grace, Henry, and Isabella—different colored pencils: blue, green, red, yellow, and pink. Emma's pencil is blue, and Finn's is green. Grace's pencil is not pink. Henry's pencil is not yellow. Whose pencil is pink?

Whose pencil is pink?

(A) Emma
(B) Finn
(C) Grace
(D) Henry

10. The light bulb, invented by Thomas Edison in 1879, revolutionized the way we illuminate our world. Edison's incandescent bulb, with its tungsten filament and glass enclosure, provided a reliable and efficient source of artificial light. What invention is credited to Thomas Edison?

What invention is credited to Thomas Edison?

(A) The telephone
(B) The automobile
(C) The light bulb
(D) The microwave oven

Mathematics Knowledge

1. If y = 2x + 3 and y = 5, what is the value of x?

    (A) 1
    (B) 2
    (C) 3
    (D) 4

2. What is the area of a rectangle with length l = 2x + 3 and width w = x - 1 when x = 2?

    (A) 5
    (B) 10
    (C) 15
    (D) 20

3. Solve for x: 2x + 4 = 3x - 1

    (A) -3
    (B) -5
    (C) 5
    (D) -1

4. The circumference of a circle is given by 2 πr. If the radius r is 7 cm, what is the circumference?

    (A) 22 cm
    (B) 28 cm
    (C) 44 cm
    (D) 49 cm

5. Simplify: 3(x + 4) - 2(2x - 3)

    (A) x + 18
    (B) x + 6
    (C) x + 12
    (D) x + 24

6. Find the area of a triangle with a base of 2x + 1 and a height of x + 3 when x = 2.

    (A) 8.5
    (B) 12
    (C) 15
    (D) 19

7. If the equation of a line is y = 3x + 2, what is the slope of the line?

    (A) 1
    (B) 2
    (C) 3
    (D) 4

8. Find the volume of a rectangular prism with length l = 3x, width w = x + 1, and height h = 2 when x = 2.

    (A) 24
    (B) 30
    (C) 36
    (D) 42

9. Solve for x in the equation $(x - 2)/3 = 4$.

    (A) 10
    (B) 12
    (C) 14
    (D) 16

10 What is the length of the hypotenuse of a right triangle with legs of lengths 6 cm and 8 cm?

    (A) 10 cm
    (B) 12 cm
    (C) 14 cm
    (D) 16 cm

11. If f(x) = x^2 - 4x + 3, what is f(2)?

    (A) -1
    (B) 0
    (C) 1
    (D) 2

12. Find the surface area of a cube with a side length of s = 4 cm.

    (A) 48 cm²
    (B) 64 cm²
    (C) 96 cm²
    (D) 128 cm²

13 Solve for x: 4(x - 3) = 2(x + 2)

    (A) 2
    (B) 3
    (C) 4
    (D) 5

14. Find the perimeter of a rectangle with length l = 5x and width w = 2x + 1 when x = 2.

    (A) 26 cm
    (B) 28 cm
    (C) 30 cm
    (D) 32 cm

15. If y = $2x^2$ - 3x + 1, what is the value of y when x = -1?

    (A) 2
    (B) 3
    (C) 4
    (D) 5

Electronics Information

1. What is the charge of an electron?

   (A) Positive
   (B) Negative
   (C) Neutral
   (D) It varies

2. Where are electrons located in an atom?

   (A) In the nucleus
   (B) Orbiting the nucleus
   (C) In the proton
   (D) In the neutron

3. What is the primary role of electrons in electrical conduction?

   (A) To produce light
   (B) To carry electrical charge
   (C) To neutralize protons
   (D) To form chemical bonds

4. What is the term for the number of electrons in the outermost shell of an atom?

   (A) Atomic number
   (B) Valence electrons
   (C) Isotopes
   (D) Neutrons

5. Which particle has the least mass in an atom?

   (A) Proton
   (B) Neutron
   (C) Electron
   (D) Quark

6. What is the standard unit of electric current?

   (A) Volt
   (B) Ohm
   (C) Watt
   (D) Ampere

7. What is a common source of direct current (DC)?

   (A) Power plants
   (B) Car battery
   (C) Household outlets
   (D) Solar panels

8. What is the flow of electrical charge called?

   (A) Voltage
   (B) Resistance
   (C) Current
   (D) Capacitance

9. Which type of current alternates its direction periodically?

   (A) Direct current (DC)
   (B) Alternating current (AC)
   (C) Static electricity
   (D) None of the above

10. What safety device is used to protect circuits from excessive current?

   (A) Capacitor
   (B) Resistor
   (C) Fuse
   (D) Transformer

11. What is the unit of electrical potential difference?

   (A) Ohm
   (B) Ampere
   (C) Watt
   (D) Volt

12. Which device measures voltage?

   (A) Ammeter
   (B) Voltmeter
   (C) Ohmmeter
   (D) Thermometer

13. What is the voltage across a 10-ohm resistor with a current of 2 amperes?

   (A) 5 volts
   (B) 10 volts
   (C) 20 volts
   (D) 200 volts

14. What is the term for a device that increases or decreases voltage in a circuit?

   (A) Capacitor
   (B) Resistor
   (C) Transformer
   (D) Diode

15. How is voltage related to current and resistance according to Ohm's Law?

   (A) $V = IR$
   (B) $P = IV$
   (C) $R = V/I$
   (D) $I = V/R$

Auto and Shop

1. What is the purpose of a bandsaw in a workshop?

   (A) Cutting metal
   (B) Measuring angles
   (C) Welding pieces together
   (D) Painting surfaces

2. Which tool is used to remove rust and old paint from metal surfaces?

   (A) Drill press
   (B) Bench grinder
   (C) Wire brush
   (D) Sander

3. What is the function of a bench vise in a workshop?

   (A) Measuring angles
   (B) Holding workpieces securely
   (C) Cutting metal sheets
   (D) Sanding surfaces

4. Which type of screwdriver is designed for use with Phillips head screws?

   (A) Flathead screwdriver
   (B) Allen wrench
   (C) Torx screwdriver
   (D) Phillips screwdriver

5. What is the primary purpose of a table saw in woodworking?

   (A) Sanding wood
   (B) Cutting wood into smaller pieces
   (C) Drilling holes in wood
   (D) Painting wooden surfaces

6. Which type of wrench is adjustable and can fit various sizes of nuts and bolts?

   (A) Combination wrench
   (B) Box end wrench
   (C) Adjustable wrench
   (D) Socket wrench

7. What is the purpose of a router in woodworking?

   (A) Cutting metal
   (B) Sanding wood
   (C) Carving decorative patterns
   (D) Drilling holes

8. Which tool is used to smooth rough edges and surfaces of wood?

   (A) Mallet
   (B) Chisel
   (C) Plane
   (D) Clamp

9. What is the function of a belt sander in a workshop?

   (A) Cutting metal
   (B) Sanding large surfaces quickly
   (C) Measuring angles
   (D) Holding workpieces in place

10. Which type of pliers is commonly used for gripping and turning objects?

   (A) Needle-nose pliers
   (B) Slip-joint pliers
   (C) Locking pliers
   (D) Diagonal cutting pliers

11. What is the purpose of a hacksaw in a workshop?

   (A) Cutting wood
   (B) Cutting metal
   (C) Drilling holes
   (D) Measuring angles

12. Which tool is used to accurately measure lengths and angles in woodworking?

   (A) Tape measure
   (B) Protractor
   (C) Combination square
   (D) Compass

13. What is the primary purpose of a welder in a workshop?

   (A) Cutting metal
   (B) Joining metal pieces together
   (C) Smoothing metal surfaces
   (D) Bending metal

14. Which tool is used to drill holes of various sizes in different materials?

   (A) Drill press
   (B) Circular saw
   (C) Jigsaw
   (D) Hammer drill

15. What is the function of a rotary tool (such as a Dremel) in a workshop?

   (A) Painting surfaces
   (B) Sanding small areas
   (C) Cutting large pieces of wood
   (D) Measuring angles

Mechanical Comprehension

1. What basic device is most suitable for splitting wood?

   (A) Lever
   (B) Pulley
   (C) Wedge
   (D) Inclined plane

2. What is the mechanical advantage of a lever if the effort arm is 8 meters long and the load arm is 2 meters long?

   (A) 4
   (B) 6
   (C) 2
   (D) 8

3. Which type of lever has the load positioned between the fulcrum and the effort?

   (A) First-class lever
   (B) Second-class lever
   (C) Third-class lever
   (D) It depends on the specific setup

4. What is the primary function of a pulley?

   (A) To amplify force
   (B) To change the direction of force
   (C) To increase the distance over which force is applied
   (D) To split objects apart

5. If a wheel has a diameter of 1 meter and is turned through an angle of 90 degrees, what is the distance traveled by a point on its circumference?

   (A) $\pi/2$ meters
   (B) π meters
   (C) 2π meters
   (D) 4π meters

6. When two items slide against one another, what type of force prevents this motion?

   (A) Frictional force
   (B) Tension force
   (C) Gravitational force
   (D) Magnetic force

7. Which of the following is an example of a second-class lever?

   (A) Nutcracker
   (B) Fishing rod
   (C) Tweezers
   (D) Bottle opener

8. What is the mechanical advantage of an inclined plane if its length is 10 meters and its height is 2 meters?

   (A) 5
   (B) 6
   (C) 4
   (D) 8

9. What is the purpose of a gear in a mechanical system?

   (A) To change the direction of force
   (B) To amplify force
   (C) To transfer rotational motion between shafts
   (D) To reduce friction between moving parts

10. If a screw has 5 threads per centimeter and is turned through 3 complete rotations, how far does it advance?

   (A) 3 centimeters
   (B) 5 centimeters
   (C) 15 centimeters
   (D) 25 centimeters

11. Which simple machine is best suited for lifting heavy objects?

   (A) Lever
   (B) Wheel and axle
   (C) Pulley
   (D) Inclined plane

12. What is the advantage of using a lever?

   (A) It decreases the force needed to lift an object
   (B) It increases the distance over which force is applied
   (C) It changes the direction of force
   (D) It allows for precise control of force

13. What is the mechanical advantage of a wheel and axle system if the radius of the wheel is 6 meters and the radius of the axle is 2 meters?

   (A) 1
   (B) 2
   (C) 3
   (D) 4

14. How does lubricating a machine reduce friction?

   (A) By increasing the coefficient of friction
   (B) By decreasing the coefficient of friction
   (C) By increasing the weight of the moving parts
   (D) By increasing the force of friction

15. If a lever has a mechanical advantage of 3 and the effort force applied is 60 newtons, what is the load force that can be lifted?

   (A) 20 newtons
   (B) 30 newtons
   (C) 60 newtons
   (D) 180 newtons

PRACTICE TEST 3

Here is the first practice test, and each section has between 10 and 15 practice questions.

General Science

1. What is the primary function of the roots in a plant?

   (A) Photosynthesis
   (B) Absorbing water and nutrients
   (C) Producing seeds
   (D) Releasing oxygen

2. What molecule carries genetic information in living organisms?

   (A) RNA
   (B) DNA
   (C) Protein
   (D) Carbohydrate

3. Which system in the human body is responsible for transporting oxygen and nutrients to cells?

   (A) Respiratory system
   (B) Digestive system
   (C) Circulatory system
   (D) Nervous system

4. What process do animals use to convert glucose into energy?

   (A) Photosynthesis
   (B) Fermentation
   (C) Cellular respiration
   (D) Transpiration

5. What is the name of our galaxy?

   (A) Andromeda
   (B) Milky Way
   (C) Sombrero
   (D) Whirlpool

6. What is the solid, outer layer of the Earth called?

   (A) Mantle
   (B) Core
   (C) Crust
   (D) Lithosphere

7. What causes tides on Earth?

   (A) Wind
   (B) Sun's gravity
   (C) Earth's rotation
   (D) Moon's gravity

8. What is the name of the phase when the Moon is completely illuminated?

   (A) New Moon
   (B) First Quarter
   (C) Full Moon
   (D) Last Quarter

9. What is the force that opposes motion between two surfaces that are in contact?

   (A) Gravity
   (B) Magnetism
   (C) Friction
   (D) Inertia

10. Which of the following is an example of potential energy?

   (A) A moving bicycle
   (B) A flying bird
   (C) A compressed spring
   (D) Flowing water

11. What is the basic unit of electric current in the International System of Units (SI)?

   (A) Volt
   (B) Watt
   (C) Ohm
   (D) Ampere

12. What law states that energy cannot be created or destroyed, only transformed?

   (A) Law of Inertia
   (B) Law of Universal Gravitation
   (C) Law of Conservation of Energy
   (D) Law of Acceleration

13. What is the chemical symbol for the element carbon?

   (A) Ca
   (B) C
   (C) Co
   (D) Cu

14. Which element is a noble gas?

   (A) Hydrogen
   (B) Oxygen
   (C) Nitrogen
   (D) Helium

15. What type of reaction occurs when an acid and a base combine to form water and a salt?

   (A) Decomposition
   (B) Synthesis
   (C) Neutralization
   (D) Combustion

Arithmetic reasoning

1. What is 5 * 8 + 3?

   (A) 35
   (B) 40
   (C) 43
   (D) 45

2. What is the value of | -35|?

   (A) -15
   (B) 0
   (C) 35
   (D) 30

3. What is the result of $\frac{\frac{4}{8}*3}{8}$?

   (A) $\frac{3}{10}$
   (B) $\frac{3}{20}$
   (C) $\frac{6}{10}$
   (D) $\frac{6}{20}$

4. What is the sum of the first 10 positive integers?

   (A) 45
   (B) 55
   (C) 65
   (D) 75

5. What is the square root of 144?

   (A) 10
   (B) 12
   (C) 14
   (D) 16

6. What is $\frac{2}{3} + \frac{5}{6}$?

   (A) $\frac{3}{4}$
   (B) $\frac{7}{9}$
   (C) $\frac{7}{12}$
   (D) $\frac{11}{12}$

7. If a = 7 and b = 4), what is $a^2 - b^2$?

   (A) 41
   (B) 45
   (C) 49
   (D) 53

8. What is the prime factorization of 36?

   (A) $2^3 * 3$
   (B) $2^2 * 3^2$
   (C) $2 * 3^3$
   (D) $2^2 * 3$

9. What is 0.6 * 0.25?

   (A) 0.0125
   (B) 0.15
   (C) 0.25
   (D) 1.5

10. What is the value of 4!?

    (A) 20
    (B) 24
    (C) 120
    (D) 720

11. Solve for x : 2x - 3 = 9

    (A) 3
    (B) 6
    (C) 8
    (D) 12

12. What is the value of |-3|?

    (A) -30
    (B) 0
    (C) 3
    (D) 9

13. What is $\frac{5}{6} - \frac{1}{3}$?

    (A) $\frac{1}{2}$
    (B) $\frac{1}{3}$
    (C) $\frac{1}{6}$
    (D) $\frac{1}{9}$

14. What is the cube root of 64?

    (A) 4
    (B) 6
    (C) 8
    (D) 10

15 What is $\frac{7}{8}$ of 64?

    (A) 42
    (B) 48
    (C) 56
    (D) 63

Word Knowledge

1. Contemplate most closely means

   (A) Ignore
   (B) Consider
   (C) Forget
   (D) Reject

2. Exquisite most closely means

   (A) Ordinary
   (B) Ugly
   (C) Beautiful
   (D) Common

3. Vivacious most closely means

   (A) Dull
   (B) Lively
   (C) Boring
   (D) Dead

4. Intricate most closely means

   (A) Simple
   (B) Complex
   (C) Easy
   (D) Clear

5. Elusive most closely means

   (A) Obvious
   (B) Clear
   (C) Evasive
   (D) Easy

6. Melancholy most closely means

   (A) Happy
   (B) Sad
   (C) Excited
   (D) Energetic

7. Arduous most closely means

   (A) Easy
   (B) Simple
   (C) Difficult
   (D) Quick

8. Benevolent most closely means

   (A) Kind
   (B) Cruel
   (C) Malevolent
   (D) Rude

9. Ephemeral most closely means

   (A) Lasting
   (B) Permanent
   (C) Brief
   (D) Enduring

10. Garrulous most closely means

    (A) Silent
    (B) Talkative
    (C) Quiet
    (D) Reserved

11. Obsolete most closely means

    (A) Current
    (B) Modern
    (C) Outdated
    (D) New

12. Scrutinize most closely means

    (A) Ignore
    (B) Glance
    (C) Examine
    (D) Overlook

13. Tedious most closely means

    (A) Exciting
    (B) Interesting
    (C) Boring
    (D) Engaging

14. Tranquil most closely means

    (A) Noisy
    (B) Calm
    (C) Chaotic
    (D) Stormy

15. Versatile most closely means

    (A) Limited
    (B) Restricted
    (C) Adaptable
    (D) Narrow

Paragraph Comprehension

1. Michael was passionate about history and loved exploring ancient ruins and artifacts. On a trip to Egypt, he marveled at the majestic pyramids of Giza, wondering about the secrets they held. As he wandered through the ancient temples of Luxor, Michael felt a deep connection to the past, imagining what life was like for the people who lived there thousands of years ago.

Where did Michael go on his trip?

    (A) Greece
    (B) Italy
    (C) Egypt
    (D) Turkey

2. The creation of computer systems with the ability to carry out tasks that normally require human intelligence is known as artificial intelligence (AI). Speech recognition, picture processing, and decision-making are some of these activities. Applications ranging from virtual assistants to self-driving cars are incorporating AI technologies.

What is artificial intelligence's main objective?

    (A) To replicate human intelligence exactly
    (B) To perform tasks that humans are incapable of
    (C) To assist humans in performing tasks more efficiently
    (D) To replace humans in all aspects of work and life

3. Five friends—Emma, Finn, Grace, Henry, and Isabella—are seated in a row. Emma is seated next to Finn but not next to Grace. Grace is not at either end of the row. Finn is seated next to Henry. Whose place is in the middle?

Whose place is in the middle?

    (A) Emma
    (B) Finn
    (C) Grace
    (D) Henry

4. With the invention of the telephone in 1876, Alexander Graham Bell transformed communication by enabling people to converse across long distances. Bell's invention laid the foundation for modern telecommunications and paved the way for future innovations in the field. What invention is credited to Alexander Graham Bell?

What invention is credited to Alexander Graham Bell?

    (A) The telephone
    (B) The automobile
    (C) The light bulb
    (D) The airplane

5. Serena Williams is an American professional tennis player with 23 Grand Slam singles titles, making her one of the most successful athletes in tennis history. Off the court, Williams is an advocate for gender equality and has used her platform to address issues of racial injustice. What is Serena Williams known for besides her tennis career?

What is Serena Williams known for besides her tennis career?

    (A) Her contributions to marine biology
    (B) Her advocacy for gender equality and racial justice
    (C) Her achievements in space exploration
    (D) Her innovations in renewable energy

6. Robotics is a branch of technology that involves the design, construction, and operation of robots. These machines can perform tasks autonomously or under human control, ranging from industrial manufacturing to space exploration. What defines robotics as a field of technology?

What defines robotics as a field of technology?

    (A) The design and construction of computer software
    (B) The development of artificial intelligence algorithms
    (C) The creation and operation of autonomous machines

(D) The study of biological systems and organisms

7. The intricate network of cells, tissues, and organs that makes up the human immune system defends the body against diseases like bacteria, viruses, and fungus. It is divided into two primary branches: the adaptive immune system, which mounts a targeted response to certain pathogens, and the innate immune system, which offers prompt defense against infections. Immunizations are vital instruments in boosting immunity and averting infectious illnesses.

What is the main theme of the passage?

(A) The structure of pathogens
(B) The function of vaccines
(C) The history of infectious diseases
(D) The components of the immune system

8. Amelia Earhart was an aviation pioneer and the first female aviator to fly solo across the Atlantic Ocean. Born in Kansas in 1897, Earhart set numerous aviation records and championed women's rights. Her disappearance during an attempted flight around the world in 1937 remains one of the greatest mysteries in aviation history. What was Amelia Earhart best known for?

What was Amelia Earhart best known for?

(A) Her exploration of the ocean depths
(B) Her advocacy for women's rights
(C) Her discoveries in space exploration
(D) Her achievements in aviation

9. Five friends—Alex, Ben, Chris, Dan, and Eric—went to a restaurant together. Alex sat next to Ben and Chris. Dan sat next to Chris and Eric. If the statements are true, who sat between Alex and Dan?

Who sat between Alex and Dan?

(A) Ben
(B) Chris
(C) Eric
(D) Cannot be determined

10. James Watt created the steam engine in the eighteenth century, which revolutionized manufacturing, transportation, and agriculture during the Industrial Revolution. Watt made the design of the steam engine more practical and efficient for a range of uses. Who invented the steam engine?

Who invented the steam engine?

(A) James Watt
(B) Nikola Tesla
(C) Thomas Edison
(D) Henry Ford

Mathematics Knowledge

1. If y = 3x + 2 and y = 11, what is the value of x?

   (A) 1
   (B) 2
   (C) 3
   (D) 4

2. What is the area of a rectangle with a length of 2x + 4 and a width of x - 1 when x = 3?

   (A) 14
   (B) 18
   (C) 20
   (D) 24

3. Solve for x: 5x - 3 = 2x + 9

   (A) 2
   (B) 3
   (C) 4
   (D) 5

4. The circumference of a circle is given by 2πr. If the radius r is 5 cm, what is the circumference?

   (A) 15.7 cm
   (B) 25.6 cm
   (C) 31.4 cm
   (D) 37.6 cm

5. Simplify: 4(x - 2) + 3(2x + 1)

   (A) 10x - 5
   (B) 10x + 2
   (C) 10x - 6
   (D) 10x + 3

6. Find the area of a triangle with a base of x + 2 and a height of 3x - 1 when x = 4.

   (A) 28
   (B) 36
   (C) 42
   (D) 48

7. If the equation of a line is y = -2x + 5, what is the slope of the line?

   (A) -2
   (B) -1
   (C) 1
   (D) 2

8. Find the volume of a rectangular prism with a length of 3x, a width of 2x + 1, and a height of x + 2 when x = 2.

   (A) 60
   (B) 70
   (C) 80
   (D) 90

9. Solve for x in the equation $(x + 1)/2 = 3$.

   (A) 5
   (B) 6
   (C) 7
   (D) 8

10. What is the length of the hypotenuse of a right triangle with legs of lengths 9 cm and 12 cm?

   (A) 12 cm
   (B) 13 cm
   (C) 14 cm
   (D) 15 cm

11. If f(x) = x^2 - 2x + 1, what is f(3)?

   (A) 1
   (B) 2
   (C) 3
   (D) 4

12. Find the surface area of a cube with a side length of s = 6 cm.

   (A) 216 cm²
   (B) 144 cm²
   (C) 108 cm²
   (D) 96 cm²

13. Solve for x: 3(x - 2) = 2(x + 1)

   (A) 2
   (B) 3
   (C) 4
   (D) 5

14. Find the perimeter of a rectangle with a length of 4x + 1 and a width of 2x - 3 when x = 2.

   (A) 20 cm
   (B) 22 cm
   (C) 24 cm
   (D) 26 cm

15. If y = x^2 - 5x + 6, what is the value of y when x = 1?

   (A) 0
   (B) 1
   (C) 2
   (D) 3

Electronics Information

1. What is the fundamental unit of charge carried by an electron?

    (A) 1 Ampere
    (B) 1 Coulomb
    (C) 1 Volt
    (D) 1 Electron

2. What is the approximate mass of an electron in kilograms?

    (A) 1.67×10^{-27} kg
    (B) 9.11×10^{-31} kg
    (C) 6.02×10^{23} kg
    (D) 3.0×10^{8} kg

3. What is the SI unit of electric current?

    (A) Volt
    (B) Watt
    (C) Ampere
    (D) Ohm

4. What is the term for the opposition to the flow of electric current in a circuit?

    (A) Voltage
    (B) Resistance
    (C) Capacitance
    (D) Conductance

5. What is the formula for calculating voltage in a circuit?

    (A) V = IR
    (B) V = I/R
    (C) V = I + R
    (D) V = I * R

6. Which material is typically used as an insulator in electrical wires?

    (A) Copper
    (B) Aluminum
    (C) Rubber
    (D) Silver

7. What is the term for a device that converts AC to DC?

    (A) Capacitor
    (B) Transformer
    (C) Rectifier
    (D) Resistor

8. What is the unit of measurement for electrical power?

    (A) Ohm
    (B) Ampere
    (C) Watt
    (D) Volt

9. Which law states that the total current entering a junction equals the total current leaving the junction in a circuit?

    (A) Kirchhoff's Voltage Law
    (B) Ohm's Law
    (C) Coulomb's Law
    (D) Kirchhoff's Current Law

10. What is the term for a circuit that has two or more paths for current to flow?

    (A) Parallel circuit
    (B) Series circuit
    (C) Closed circuit
    (D) Open circuit

11. What is the relationship between voltage, current, and resistance described by Ohm's Law?

    (A) V = IR
    (B) I = VR
    (C) R = VI
    (D) I = RV

12. Which component is used to store electrical charge in a circuit?

    (A) Resistor
    (B) Capacitor
    (C) Diode
    (D) Transistor

13. What is the term for the movement of electrons through a conductor?

    (A) Voltage
    (B) Current
    (C) Resistance
    (D) Capacitance

14. Which type of current constantly changes direction?

    (A) Direct current (DC)
    (B) Alternating current (AC)
    (C) Variable current (VC)
    (D) Steady current (SC)

15. What is the measure of the potential energy difference between two points in a circuit?

    (A) Current
    (B) Resistance
    (C) Voltage
    (D) Power

Auto and Shop

1. Which type of wrench is commonly used for tightening and loosening nuts and bolts in automotive work?

    (A) Phillips screwdriver
    (B) Adjustable wrench
    (C) Socket wrench
    (D) Pliers

2. What is the purpose of an air compressor in an automotive workshop?

    (A) To inflate tires
    (B) To power pneumatic tools
    (C) To cool the engine
    (D) To charge the battery

3. What tool is used to measure the thickness of brake pads?

    (A) Vernier caliper
    (B) Micrometer
    (C) Feeler gauge
    (D) Depth gauge

4. Which type of jack is commonly used for lifting vehicles during maintenance?

    (A) Hydraulic floor jack
    (B) Bottle jack
    (C) Scissor jack
    (D) Hi-lift jack

5. What is the purpose of a torque wrench in automotive work?

    (A) To measure engine temperature
    (B) To tighten bolts to a specific torque value
    (C) To diagnose engine problems
    (D) To check tire pressure

6. Which of the following is not a component of a typical braking system?

    (A) Brake rotor
    (B) Brake drum
    (C) Brake pedal
    (D) Fuel injector

7. What is the function of a carburetor in an internal combustion engine?

    (A) To filter air
    (B) To regulate fuel flow
    (C) To ignite the air-fuel mixture
    (D) To control exhaust emissions

8. Which type of lubricant is commonly used in automotive engines?

    (A) Hydraulic fluid
    (B) Brake fluid
    (C) Engine oil
    (D) Power steering fluid

9. What tool is used to remove and install spark plugs?

    (A) Spark plug wrench
    (B) Allen wrench
    (C) Box-end wrench
    (D) Torque wrench

10. What is the purpose of a timing belt or timing chain in an engine?

    (A) To regulate fuel flow
    (B) To control engine temperature
    (C) To synchronize the opening and closing of engine valves
    (D) To adjust tire pressure

11. Which type of oil filter is commonly used in automotive engines?

    (A) Paper filter
    (B) Mesh filter
    (C) Foam filter
    (D) Cloth filter

12. What is the function of a radiator in a vehicle's cooling system?

    (A) To regulate engine temperature
    (B) To filter air
    (C) To generate power
    (D) To cool transmission fluid

13. What tool is used to remove and install wheels on a vehicle?

    (A) Lug wrench
    (B) Screwdriver
    (C) Pliers
    (D) Hammer

14. What is the primary purpose of a catalytic converter in a vehicle's exhaust system?

    (A) To reduce harmful emissions
    (B) To increase fuel efficiency
    (C) To regulate engine temperature
    (D) To improve engine performance

15. Which component of the suspension system helps stabilize the vehicle during turns?

    (A) Shock absorber
    (B) Strut
    (C) Sway bar
    (D) Spring

Mechanical Comprehension

1. Which simple machine is best suited for raising a flag?

   (A) Lever
   (B) Pulley
   (C) Wheel and axle
   (D) Inclined plane

2. What is the mechanical advantage of a lever if the effort arm is 6 meters long and the load arm is 3 meters long?

   (A) 1
   (B) 2
   (C) 3
   (D) 4

3. Which type of lever has the load situated between the fulcrum and the effort?

   (A) First-class lever
   (B) Second-class lever
   (C) Third-class lever
   (D) It depends on the specific setup

4. What is the primary function of a wedge?

   (A) To amplify force
   (B) To change the direction of force
   (C) To increase the distance over which force is applied
   (D) To split objects apart

5. If a wheel has a radius of 0.5 meters and is rotated through an angle of 45 degrees, what is the distance traveled by a point on its circumference?

   (A) $\pi/4$ meters
   (B) $\pi/2$ meters
   (C) π meters
   (D) 2π meters

6. What type of force opposes the motion of objects sliding against each other?

   (A) Frictional force
   (B) Tension force
   (C) Gravitational force
   (D) Magnetic force

7. Which of the following is an example of a third-class lever?

   (A) Nutcracker
   (B) Wheelbarrow
   (C) Fishing rod
   (D) Bottle opener

8. What is the mechanical advantage of an inclined plane if its length is 8 meters and its height is 2 meters?

   (A) 2
   (B) 4
   (C) 6
   (D) 8

9. What is the purpose of a gear in a mechanical system?

   (A) To change the direction of force
   (B) To amplify force
   (C) To transfer rotational motion between shafts
   (D) To reduce friction between moving parts

10. If a screw has 8 threads per centimeter and is turned through 4 complete rotations, how far does it advance?

    (A) 16 centimeters
    (B) 32 centimeters
    (C) 64 centimeters
    (D) 128 centimeters

11. Which simple machine is best suited for raising heavy loads vertically?

    (A) Lever
    (B) Wheel and axle
    (C) Pulley
    (D) Inclined plane

12. What is the advantage of using a lever?

    (A) It decreases the force needed to lift an object
    (B) It increases the distance over which force is applied
    (C) It changes the direction of force
    (D) It allows for precise control of force

13. What is the mechanical advantage of a wheel and axle system if the radius of the wheel is 3 meters and the radius of the axle is 1 meter?

    (A) 1
    (B) 2
    (C) 3
    (D) 4

14. How does lubricating a machine reduce friction?

    (A) By increasing the coefficient of friction
    (B) By decreasing the coefficient of friction
    (C) By increasing the weight of the moving parts
    (D) By increasing the force of friction

15. If a lever has a mechanical advantage of 4 and the effort force applied is 80 newtons, what is the load force that can be lifted?

    (A) 20 newtons
    (B) 40 newtons
    (C) 80 newtons
    (D) 320 newtons

PRACTICE TEST 4

Here is the first practice test, and each section has between 10 and 15 practice questions.

General Science

1. What is the process by which green plants use sunshine to produce food from CO2 and water?

(A) Respiration
(B) Transpiration
(C) Photosynthesis
(D) Fermentation

2. Which of the following is responsible for carrying oxygen in the blood?

(A) Hemoglobin
(B) Platelets
(C) Plasma
(D) White blood cells

3. What is the basic unit of inheritance in living organisms?

(A) Chromosome
(B) Gene
(C) Allele
(D) Nucleotide

4. Which of the following is an example of a vertebrate?

(A) Spider
(B) Fish
(C) Butterfly
(D) Snail

5. What is the process by which water changes from a liquid to a gas at the surface of the Earth?

(A) Condensation
(B) Precipitation
(C) Evaporation
(D) Sublimation

6. What type of rock is formed from the cooling and solidification of magma or lava?

(A) Sedimentary
(B) Metamorphic
(C) Igneous
(D) Granite

7. Which planet is known as the "Morning Star" or "Evening Star"?

(A) Venus
(B) Mars
(C) Jupiter
(D) Saturn

8. What is the name of the imaginary line that divides the Earth into the Northern and Southern Hemispheres?

(A) Equator
(B) Prime Meridian
(C) Tropic of Cancer
(D) Tropic of Capricorn

9. What is the SI unit of force?

(A) Newton
(B) Joule
(C) Watt
(D) Pascal

10. Which of Newton's laws asserts that, absent the intervention of an outside force, any object will continue to be at rest or in uniform motion in a straight line?

(A) Newton's First Law
(B) Newton's Second Law
(C) Newton's Third Law
(D) Law of Gravitation

11. What is the SI unit of energy?

(A) Newton
(B) Joule
(C) Watt
(D) Pascal

12. What is the rate at which work is done or energy is transferred or converted?

(A) Force
(B) Power
(C) Momentum
(D) Velocity

13. What is the chemical symbol for gold?

(A) Ag
(B) Au
(C) Pb
(D) Hg

14. Which of the following is an example of a noble gas?

(A) Nitrogen
(B) Oxygen
(C) Helium
(D) Carbon

15. What type of bond is formed by the transfer of electrons between atoms?

(A) Ionic bond
(B) Covalent bond
(C) Metallic bond

(D) Hydrogen bond

Arithmetic Reasoning

1. What is 3 * 4 + 5?

   (A) 11
   (B) 17
   (C) 19
   (D) 23

2. What is the sum of the first 20 positive odd integers?

   (A) 400
   (B) 420
   (C) 440
   (D) 460

3. What is $\frac{3}{4}$ of 96?

   (A) 54
   (B) 64
   (C) 72
   (D) 81

4. What is 2^5?

   (A) 16
   (B) 32
   (C) 64
   (D) 128

5. If x = 8 and y = 3, what is $x^2 - 2xy + y^2$?

   (A) 33
   (B) 35
   (C) 39
   (D) 41

6. What is 7 * 9 + 3?

   (A) 63
   (B) 64
   (C) 65
   (D) 66

7. What is the result of $\frac{\frac{4}{5}*3}{8}$?

   (A) $\frac{3}{10}$
   (B) $\frac{3}{20}$
   (C) $\frac{6}{10}$
   (D) $\frac{6}{20}$

8. What is the sum of the first 10 positive integers?

   (A) 45
   (B) 55
   (C) 65
   (D) 75

9. What is the square root of 144?

   (A) 10
   (B) 12
   (C) 14
   (D) 16

10. What is $\frac{2}{3} + \frac{5}{6}$?

   (A) $\frac{3}{4}$
   (B) $\frac{7}{9}$
   (C) $\frac{7}{12}$
   (D) $\frac{11}{12}$

11. If a = 7 and b = 4, what is $a^2 - b^2$?

   (A) 41
   (B) 45
   (C) 49
   (D) 53

12. What is 0.6 * 0.25 ?

   (A) 0.0125
   (B) 0.15
   (C) 0.25
   (D) 1.5

13. Solve for x: 2x - 3 = 9

   (A) 3
   (B) 6
   (C) 0
   (D) 12

14. What is $\frac{5}{6} - \frac{1}{3}$?

   (A) $\frac{1}{2}$
   (B) $\frac{1}{3}$
   (C) $\frac{1}{6}$
   (D) $\frac{1}{9}$

15 What is the sum of the first 20 positive odd integers?

   (A) 400
   (B) 420
   (C) 440
   (D) 460

Word Knowledge

1. The teacher was impressed by the student's <u>astute</u> observation during the lecture.

    (A) Silly
    (B) Clever
    (C) Lazy
    (D) Boring

2. Wistful most closely means

    (A) Happy
    (B) Joyful
    (C) Nostalgic
    (D) Unconcerned

3. Zealous most closely means

    (A) Apathetic
    (B) Indifferent
    (C) Enthusiastic
    (D) Lethargic

4. The athlete's performance was <u>remarkable</u>, earning praise from all the spectators.

    (A) Ordinary
    (B) Unusual
    (C) Inadequate
    (D) Slow

5. Her <u>benevolent</u>* actions towards the less fortunate earned her the community's respect.

    (A) Cruel
    (B) Generous
    (C) Selfish
    (D) Indifferent

6. Cogent most closely means

    (A0 Weak
    (B) Convincing
    (C) Confusing
    (D) Unclear

7. The scientist made a <u>pivotal</u> discovery that changed the course of the research.

    (A) Minor
    (B) Central
    (C) Useless
    (D) Slow

8. Ephemeral most closely means

    (A) Permanent
    (B) Lengthy
    (C) Temporary
    (D) Unusual

9. The student's <u>meticulous</u> approach to the project ensured that every detail was perfect.

    (A) Careless
    (B) Hasty
    (C) Thorough
    (D) Lazy

10. Placate most closely means

    (A) Irritate
    (B) Soothe
    (C) Ignore
    (D) Challenge

11. His <u>tenacity</u> in pursuing his goals despite numerous obstacles was truly inspiring.

    (A) Timidity
    (B) Persistence
    (C) Indifference
    (D) Uncertainty

12. The novel was set in a <u>dystopian</u> future where society had collapsed.

    (A) Ideal
    (B) Chaotic
    (C) Utopian
    (D) Harmonious

13. Tenacious most closely means

    (A) Weak
    (B) Unyielding
    (C) Flexible
    (D) Doubtful

14. Altruistic most closely means

    (A) Selfish
    (B) Generous
    (C) Greedy
    (D) Cautious

15. Insipid most closely means

    (A) Tasty
    (B) Dull
    (C) Exciting
    (D) Complex

Paragraph Comprehension

1. The old, abandoned mansion at the edge of town had always intrigued Emma. Legend had it that the mansion was haunted by the ghost of its former owner, a wealthy businessman who mysteriously disappeared years ago. Determined to uncover the truth, Emma ventured into the mansion one stormy night, armed with nothing but a flashlight and her courage.

What legend surrounded the old mansion?

   (A) It was cursed by a witch
   (B) It was haunted by a ghost
   (C) It was built by pirates
   (D) It was inhabited by vampires

2. There are five people—Alice, Ben, Chris, Dan, and Emma—each with a different favorite color: blue, green, red, yellow, and purple. Ben's favorite color is blue. Chris's favorite color is not red or purple. Dan's favorite color is not green or yellow. Whose favorite color is red?

Whose favorite color is red?

   (A) Alice
   (B) Ben
   (C) Chris
   (D) Dan

3. Five friends—Ella, Frank, George, Helen, and Isaac—each have a different type of pet: cat, dog, fish, hamster, and rabbit. Helen has a hamster, and Frank has a cat. Ella's pet is not a fish or a rabbit. Whose pet is a dog?

Whose pet is a dog?

   (A) Ella
   (B) Frank
   (C) George
   (D) Isaac

4. Renewable energy sources, such as solar, wind, and hydroelectric power, offer sustainable alternatives to fossil fuels. Unlike fossil fuels, which contribute to climate change and air pollution, renewable energy sources are abundant and produce little to no greenhouse gas emissions. Transitioning to renewable energy is crucial for mitigating the effects of climate change and reducing dependence on finite resources.

What is the main theme of the passage?

   (A) The benefits of fossil fuels
   (B) The importance of environmental conservation
   (C) The challenges of renewable energy
   (D) The advantages of solar power

5. The transistor, invented by John Bardeen, Walter Brattain, and William Shockley in 1947, revolutionized electronics and paved the way for modern computing and telecommunications. Transistors replaced bulky vacuum tubes, making electronic devices smaller, faster, and more reliable. Who invented the transistor?

Who invented the transistor?

   (A) John Bardeen, Walter Brattain, and William Shockley
   (B) Alexander Graham Bell
   (C) Thomas Edison
   (D) Nikola Tesla

6. The compass, invented by ancient Chinese civilization, revolutionized navigation by enabling sailors to determine direction accurately. This magnetic instrument played a crucial role in maritime exploration and trade routes, shaping the course of human history. What invention revolutionized navigation by enabling sailors to determine direction accurately?

What invention revolutionized navigation by enabling sailors to determine direction accurately?

   (A) The telephone
   (B) The printing press
   (C) The compass
   (D) The internet

7. Marie Curie was an innovative chemist and physicist who made significant contributions to the field of radioactivity study. Curie, who was born in Poland in 1867, is the only person to have received the Nobel Prize in both medicine and physics, making her the first woman to do so. What distinguished Marie Curie's contributions to science?

What distinguished Marie Curie's contributions to science?

(A) Her discovery of penicillin
(B) Her research on radioactivity
(C) Her development of the telephone
(D) Her invention of the light bulb

8. The airplane, invented by the Wright brothers, Orville and Wilbur, in 1903, ushered in the age of aviation and transformed travel and transportation. The Wright Flyer, their first powered aircraft, made history with its successful flight at Kitty Hawk, North Carolina. Who are credited with inventing the airplane?

Who are credited with inventing the airplane?

(A) Alexander Graham Bell
(B) The Wright brothers
(C) Thomas Edison
(D) Henry Ford

9. Malala Yousafzai is a Pakistani activist for female education and the youngest Nobel Prize laureate. Despite facing threats from the Taliban, Yousafzai continues to advocate for girls' right to education worldwide. What has Malala Yousafzai been recognized for?

What has Malala Yousafzai been recognized for?

(A) Her achievements in professional sports
(B) Her contributions to culinary arts
(C) Her advocacy for girls' education
(D) Her innovations in technology

10. The 18th century saw the start of the Industrial Revolution, which had a profound impact on human history. It was marked by the shift from rural, agrarian economies to industrialized ones, propelled by advances in manufacturing and technology. Significant social, economic, and environmental changes—such as increasing pollution, urbanization, and the advent of capitalism—were brought about by the Industrial Revolution.

What is the main theme of the passage?

(A) The history of technology
(B) The impact of urbanization
(C) The consequences of the Industrial Revolution
(D) The development of capitalism

Mathematics Knowledge

1. Solve for x: 6x - 4 = 8x + 10

    (A) -5
    (B) 5
    (C) -7
    (D) 7

2. What is the measure of each exterior angle of a regular decagon?

    (A) 30°
    (B) 36°
    (C) 45°
    (D) 60°

3. Simplify: $(2x - 8)/2$

    (A) x^2 - 4
    (B) x^2 - 8
    (C) 2x - 4
    (D) x - 4

4. What is the perimeter of an equilateral triangle with a side length of 7 cm?

    (A) 14 cm
    (B) 21 cm
    (C) 28 cm
    (D) 35 cm

5. Solve for y: 2y + 3 = 4y - 5

    (A) 4
    (B) 5
    (C) -4
    (D) -5

6. Find the area of a trapezoid with bases a = 4x and b = 2x + 2 and height h = 3 when x = 1.

    (A) 10
    (B) 12
    (C) 14
    (D) 16

7. Simplify: 2(3x - 4) + 5(2x + 1)

    (A) 16x + 1
    (B) 16x - 3
    (C) 16x - 2
    (D) 16x + 3

8. Find the circumference of a circle with radius r = 5 cm. (Use π = 3.14)

    (A) 15.7 cm
    (B) 25.12 cm
    (C) 31.4 cm
    (D) 62.8 cm

9. If y = $(2x - 1)/3$ what is the value of y when x = 4?

    (A) 1
    (B) 2
    (C) 3
    (D) 4

10. Find the volume of a sphere with radius r = 3 cm. (Use π = 3.14 and V = $\frac{4}{3}\pi r^3$)

    (A) 84.78 cm³
    (B) 100.48 cm³
    (C) 113.04 cm³
    (D) 126.48 cm³

11. Find the area of a trapezoid with bases 2x + 3 and x + 1 and height x when x = 3.

    (A) 18
    (B) 21
    (C) 24
    (D) 27

12. Simplify: 3(2x + 1) - 4(x - 2)

    (A) 2x + 11
    (B) 2x + 9
    (C) 2x + 8
    (D) 2x + 7

13. Find the circumference of a circle with radius r = 6 cm. (Use π = 3.14)

    (A) 37.68 cm
    (B) 36.12 cm
    (C) 37.08 cm
    (D) 38.16 cm

14. If y = $\frac{3x-1}{2}$, what is the value of y when x = 5?

    (A) 6
    (B) 7
    (C) 8
    (D) 9

15. Find the volume of a sphere with radius r = 4 cm. (Use π = 3.14 and V = $\frac{4}{3}\pi r^3$)

    (A) 200.96 cm³
    (B) 268.08 cm³
    (C) 301.44 cm³
    (D) 321.28 cm³

Electronics Information

1. What subatomic particle carries a negative charge?

   (A) Proton
   (B) Neutron
   (C) Electron
   (D) Photon

2. What is the SI unit of electric charge?

   (A) Joule
   (B) Coulomb
   (C) Ohm
   (D) Watt

3. Which of the following materials is a good conductor of electricity?

   (A) Glass
   (B) Rubber
   (C) Copper
   (D) Plastic

4. What is the term for the force that drives electric current?

   (A) Resistance
   (B) Capacitance
   (C) Voltage
   (D) Inductance

5. What is the unit of electrical resistance?

   (A) Ampere
   (B) Ohm
   (C) Volt
   (D) Watt

6. According to which law, the voltage across two points and the current flowing through a conductor between them are directly proportional?

   (A) Faraday's Law
   (B) Ohm's Law
   (C) Coulomb's Law
   (D) Kirchhoff's Law

7. What is the term for a closed path through which an electric current flows?

   (A) Conductor
   (B) Insulator
   (C) Circuit
   (D) Transformer

8. Which component is used to store electrical energy in a circuit?

   (A) Resistor
   (B) Capacitor
   (C) Diode
   (D) Transistor

9. What is the measure of the rate of flow of electric charge?

   (A) Voltage
   (B) Current
   (C) Resistance
   (D) Power

10. What is the term for the energy per unit charge carried by an electric current?

   (A) Voltage
   (B) Current
   (C) Resistance
   (D) Power

11. Which type of current flows in one direction only?

   (A) Direct current (DC)
   (B) Alternating current (AC)
   (C) Variable current (VC)
   (D) Steady current (SC)

12. What is the unit of electrical power?

   (A) Ampere
   (B) Ohm
   (C) Watt
   (D) Volt

13. Which component only permits one direction of current flow?

   (A) Capacitor
   (B) Resistor
   (C) Diode
   (D) Inductor

14. What is the term for the opposition to the flow of electric current in a circuit?

   (A) Voltage
   (B) Resistance
   (C) Capacitance
   (D) Conductance

15. What is the term for the phenomenon where a current-carrying conductor generates a magnetic field around it?

   (A) Inductance
   (B) Capacitance
   (C) Resistance
   (D) Conductance

Auto and Shop

1. In an automobile, what component ignites the air-fuel mixture in the engine?

    (A) Carburetor
    (B) Spark plug
    (C) Piston
    (D) Alternator

2. What tool is used to measure tire pressure?

    (A) Jack
    (B) Lug wrench
    (C) Tire gauge
    (D) Socket wrench

3. Which type of oil is typically used in a car's engine?

    (A) Transmission oil
    (B) Brake fluid
    (C) Engine oil
    (D) Power steering fluid

4. What does ABS stand for in relation to car brakes?

    (A) Automatic Braking System
    (B) Anti-Brake System
    (C) Advanced Braking System
    (D) Anti-Lock Braking System

5. Which of the following is not a component of a manual transmission system?

    (A) Gear shifter
    (B) Clutch pedal
    (C) Torque converter
    (D) Flywheel

6. What tool is used to remove and install nuts or bolts?

    (A) Screwdriver
    (B) Wrench
    (C) Pliers
    (D) Hammer

7. Which type of headlight bulb tends to last longer?

    (A) Halogen
    (B) LED
    (C) HID (Xenon)
    (D) Incandescent

8. What does the alternator do in a car?

    (A) Supplies power to the spark plugs
    (B) Charges the battery
    (C) Controls the air-fuel mixture
    (D) Converts fuel into mechanical energy

9. Which tool is used to remove and install tires?

    (A) Lug wrench
    (B) Jack
    (C) Tire iron
    (D) Torque wrench

10. What component of a car's braking system presses against the brake rotor to slow or stop the vehicle?

    (A) Brake caliper
    (B) Brake drum
    (C) Brake booster
    (D) Brake pedal

11. What is the purpose of a serpentine belt in a car engine?

    (A) To control the air-fuel mixture
    (B) To transfer power from the engine to various accessories
    (C) To adjust the timing of the engine
    (D) To cool the engine

12. Which type of motor oil is recommended for high-performance engines or extreme conditions?

    (A) Synthetic oil
    (B) Conventional oil
    (C) Semi-synthetic oil
    (D) Mineral oil

13. What is the purpose of a catalytic converter in a car's exhaust system?

    (A) To regulate the air-fuel mixture
    (B) To reduce harmful emissions
    (C) To increase fuel efficiency
    (D) To control engine temperature

14. Which component of the suspension system absorbs shocks and vibrations from the road?

    (A) Shock absorber
    (B) Strut
    (C) Spring
    (D) Sway bar

15. What does the acronym HVAC stand for in automotive terminology?

    (A) High-Voltage Alternating Current
    (B) Hot Vapors Air Conditioning
    (C) Heating, Ventilation, and Air Conditioning
    (D) Hybrid Vehicle Access Control

Mechanical Comprehension

1. Which simple machine is commonly used to raise heavy loads vertically?

(A) Lever
(B) Pulley
(C) Wheel and axle
(D) Inclined plane

2. What is the mechanical advantage of a lever if the effort arm is 10 meters long and the load arm is 2 meters long?

(A) 2
(B) 3
(C) 4
(D) 5

3. Which type of lever has the effort applied between the fulcrum and the load?

(A) First-class lever
(B) Second-class lever
(C) Third-class lever
(D) It depends on the specific setup
4. What is the primary function of a wedge in a mechanical system?

(A) To amplify force

(B) To change the direction of force
(C) To increase the distance over which force is applied
(D) To split objects apart

5. If a wheel has a diameter of 2 meters and is rotated through an angle of 90 degrees, what is the distance traveled by a point on its circumference?

(A) π meters
(B) 2π meters
(C) 4π meters
(D) 8π meters

6. If two objects are sliding against one other, what type of force opposes it?

(A) Frictional force
(B) Tension force
(C) Gravitational force
(D) Magnetic force

7. Which of the following is an example of a third-class lever?

(A) Nutcracker
(B) Wheelbarrow
(C) Fishing rod
(D) Bottle opener

8. What is the mechanical advantage of an inclined plane if its length is 6 meters and its height is 2 meters?

(A) 2
(B) 3
(C) 4
(D) 5

9. What is the purpose of a gear in a mechanical system?

(A) To change the direction of force
(B) To amplify force
(C) To transfer rotational motion between shafts
(D) To reduce friction between moving parts

10. If a screw has 10 threads per centimeter and is turned through 4 complete rotations, how far does it advance?

(A) 4 centimeters
(B) 16 centimeters
(C) 40 centimeters
(D) 100 centimeters

11. Which simple machine is best suited for splitting wood?

(A) Lever
(B) Wedge
(C) Pulley
(D) Inclined plane

12. What is the advantage of using a lever?

(A) It decreases the force needed to lift an object
(B) It increases the distance over which force is applied
(C) It changes the direction of force
(D) It allows for precise control of force

13. In a wheel and axle system, what mechanical benefit does a system with a radius of five meters for the wheel and one meter for the axle offer?

(A) 1
(B) 2
(C) 3
(D) 5

14. How does lubricating a machine reduce friction?

(A) By increasing the coefficient of friction
(B) By decreasing the coefficient of friction
(C) By increasing the weight of the moving parts
(D) By increasing the force of friction

15. If a lever has a mechanical advantage of 6 and the effort force applied is 120 newtons, what is the load force that can be lifted?

(A) 20 newtons
(B) 40 newtons
(C) 60 newtons
(D) 720 newtons

PRACTICE TEST 1 ANSWERS

General Science

1. (C) Cell
2. (B) Photosynthesis
3. (A) Deoxyribonucleic Acid
4. (C) Circulatory system
5. (B) Mantle
6. (C) Nitrogen
7. (C) Galaxy
8. (C) Moon's position relative to Earth and Sun
9. (B) Gravity
10. (B) A moving car
11. (C) 300,000 km/s
12. (C) Newton's Third Law
13. (B) H2O
14. (C) Oxygen
15. (B) 7

Arithmetic reasoning

1. (C) x = 5
2. (C) 7
3. (C) $\frac{6}{9}$
4. (B) 210
5. (A) 10
6. (A) 1
7. (C) 220
8. (A) 4
9. (B) $\frac{5}{6}$
10. (C) 15
11. (C) 8
12. (C) 12
13. (A) $\frac{5}{12}$
14. (B) 120
15. (C) 13

Word Knowledge

1. (B) Opponent
2. (B) Confidence
3. (B) Harsh
4. (C) Kind
5. (B) Frank
6. (C) Weaken
7. (B) Overjoyed
8. (B) Effectiveness
9. (C) Unpredictable
10. (C) Prosper
11. (C) Aggressive
12. (C) Apathetic
13. (B) Clear
14. (C) Thorough
15. (B) Charitable

Paragraph Comprehension

1. (C) Fantasy
2. (A) The structure of the human brain
3. (B) Dog
4.(B) The importance of renewable energy
5. (B) Peaceful protests and nonviolent resistance
6. (C) Primatology
7. (D) It offers enhanced security and convenience
8. (A) Quantum computing operates at exponentially faster speeds
9. (C) By providing personalized feedback and insights
10. (B) Kate

Mathematics Knowledge

1. (C) 3
2. (B) 40 cm²
3. (A) -2x + 10
4. (B) 197.82 cm³
5. (A) 14
6. (B) 135°
7. (C) 5
8. (B) 37.68 cm
9. (C) $2x^2 - 5x + 2$
10. (C) 78.5 cm²
11. (C) 3
12. (A) 150.72 cm³
13. (B) 6 cm
14. (B) 2x - 14
15. (C) 15 cm

Electronics Information

1.(B) Negative
2. (B) Orbiting the nucleus
3. (B) To carry electrical charge
4. (D) Ampere
5. (B) Car battery
6. (C) Current

7. (D) Volt
8. (B) Voltmeter
9. (C) 20 volts
10. (B) Capacitor
11. (B) To allow current to flow in one direction only

12. (B) Field-Effect Transistor (FET)
13. (A) V = IR
14. (C) To limit current
15. (A) P = IV

Auto and Shop

1. (B) Hammer
2. (C) Planer
3. (C) Shaping cylindrical objects
4. (B) Bandsaw
5. (B) Sharpening tools
6. (A) Socket wrench
7. (A) Vise
8. (C) Cutting precise angles in wood

9. (C) Combination square
10. (C) Shaping edges of wood
11. (A) Screwdriver
12. (D) Joining wood edges
13. (D) Drill press
14. (B) Joining metal pieces together
15. (B) Circular saw

Mechanical Comprehension

1. (C) Inclined plane
2. (B) 3
3. (A) First-class lever
4. (D) To split objects apart
5. (B) $\pi/3$ meters
6. (A) Frictional force
7. (A) Tweezers
8. (D) 2

9. (C) To transfer rotational motion between shafts
10. (C) 10 centimeters
11. (C) Pulley
12. (A) It decreases the force needed to lift an object
13. (D) 4
14. (B) By decreasing the coefficient of friction
15. (C) 100 newtons

PRACTICE TEST 2 ANSWERS

General Science

1. (B) Mitochondrion
2. (B) Mitosis
3. (C) DNA
4. (D) Digestive system
5. (B) Mars
6. (B) Earth's tilt on its axis
7. (C) Igneous
8. (C) Galaxy

9. (B) Newton
10. (A) Inertia
11. (B) Refraction
12. (B) Speed = Distance / Time
13. (C) Oxygen
14. (C) Covalent bond
15. (A) NaCl

Arithmetic reasoning

1. (C) $\frac{2}{3}$
2. (D) 20
3. (C) $2^4 * 3^2$
4. (C) 75
5. (A) 120
6. (B) 5
7. (C) 20
8. (C) $\frac{2}{5}$

9. (B) 4
10. (C) 80
11. (B) 10
12. (B) 420
13. (B) 56
14. (D) 27
15. (C) 155

Word Knowledge

1. (C) Neutral
2. (C) Outdated
3. (C) Soothe
4. (C) Charming
5. (B) Withdrawn

6. (C) Concise
7. (B) Unquenchable
8. (B) Wobbly
9. (B) Persistent
10. (C) Omnipresent

11. (C) Bright
12. (C) Greedy
13. (B) Careful

14. (B) Flexible
15. (B) Exemplary

Paragraph Comprehension

1. (A) His contributions to engineering
2. (B) The importance of marine biodiversity
3. (D) David
4. (C) The process of evolution
5. (C) Apple Inc.

6. (C) The wheel
7. (A) Alexander Fleming
8. (D) His legislative accomplishments and leadership style
9. (A) Emma
10. (C) The light bulb

Mathematics Knowledge

1. (A) 1
2. (B) 10
3. (A) -5
4. (C) 44 cm
5. (C) x + 12
6. (B) 12
7. (C) 3
8. (C) 36

9. (C) 14
10. (A) 10 cm
11. (B) 0
12. (C) 96 cm²
13. (A) 2
14. (A) 26 cm
15. (C) 4

Electronics Information

1. (B) Negative
2. (B) Orbiting the nucleus
3. (B) To carry electrical charge
4. (B) Valence electrons
5. (C) Electron
6. (D) Ampere
7. (B) Car battery
8. (C) Current

9. (B) Alternating current (AC)
10. (C) Fuse
11. (D) Volt
12. (B) Voltmeter
13. (C) 20 volts
14. (C) Transformer
15. (A) V = IR

Auto and Shop

1. (A) Cutting metal
2. (C) Wire brush
3. (B) Holding workpieces securely
4. (D) Phillips screwdriver
5. (B) Cutting wood into smaller pieces
6. (C) Adjustable wrench
7. (C) Carving decorative patterns
8. (C) Plane

9. (B) Sanding large surfaces quickly
10. (B) Slip-joint pliers
11. (B) Cutting metal
12. (C) Combination square
13. (B) Joining metal pieces together
14. (A) Drill press
15. (B) Sanding small areas

Mechanical Comprehension

1. (C) Wedge
2. (A) 4
3. (C) Third-class lever
4. (B) To change the direction of force
5. (C) 2π meters
6. (A) Frictional force
7. (A) Nutcracker
8. (A) 5

9. (C) To transfer rotational motion between shafts
10. (C) 15 centimeters
11. (C) Pulley
12. (A) It decreases the force needed to lift an object
13. (B) 2
14. (B) By decreasing the coefficient of friction
15. (D) 180 newtons

PRACTICE TEST 3 ANSWERS

General Science

1. (B) Absorbing water and nutrients
2. (B) DNA
3. (C) Circulatory system
4. (C) Cellular respiration
5. (B) Milky Way
6. (C) Crust
7. (D) Moon's gravity
8. (C) Full Moon
9. (C) Friction
10. (C) A compressed spring
11. (D) Ampere
12. (C) Law of Conservation of Energy
13. (B) C
14. (D) Helium
15. (C) Neutralization

Arithmetic Reasoning

1. (C) 43
2. (C) 35
3. (B) $\frac{3}{20}$
4. (B) 55
5. (B) 12
6. (D) $\frac{11}{12}$
7. (C) 49
8. (B) $2^2 * 3^2$
9. (B) 0.15
10. (B) 24
11. (D) 6
12. (C) 3
13. (A) $\frac{1}{2}$
14 (A) 4
15. (C) 56

Word Knowledge

1. (B) Consider
2. (C) Beautiful
3. (B) Lively
4. (B) Complex
5. (C) Evasive
6. (B) Sad
7. (C) Difficult
8. (A) Kind
9. (C) Brief
10. (B) Talkative
11. (C) Outdated
12. (C) Examine
13. (C) Boring
14. (B) Calm
15. (C) Adaptable

Paragraph Comprehension

1. (C) Egypt
2. (C) To assist humans in performing tasks more efficiently
3. (B) Finn
4. (A) The telephone
5. (B) Her advocacy for gender equality and racial justice
6. (C) The creation and operation of autonomous machines
7. (D) The components of the immune system
8. (D) Her achievements in aviation
9. (B) Chris
10 (A) James Watt

Mathematics Knowledge

1. (C) 3
2. (D) 24
3. (D) 4
4. (C) 31.4 cm
5. (B) 10x + 2
6. (C) 42
7. (A) -2
8. (C) 80
9. (B) 5
10. (D) 15 cm
11. (D) 4
12. (B) 216 cm²
13. (C) 4
14. (D) 26 cm
15. (D) 2

Electronics Information

1. (B) 1 Coulomb
2. (B) 9.11×10^{-31} kg
3. (C) Ampere
4. (B) Resistance
5. (D) V = I * R
6. (C) Rubber

7. (C) Rectifier
8. (C) Watt
9. (D) Kirchhoff's Current Law
10. (A) Parallel circuit
11. (A) V = IR

12. (B) Capacitor
13. (B) Current
14. (B) Alternating current (AC)
15. (C) Voltage

Auto and Shop

1. (C) Socket wrench
2. (B) To power pneumatic tools
3. (C) Feeler gauge
4. (A) Hydraulic floor jack
5. (B) To tighten bolts to a specific torque value
6. (D) Fuel injector
7. (B) To regulate fuel flow
8. (C) Engine oil

9. (A) Spark plug wrench
10. (C) To synchronize the opening and closing of engine valves
11. (A) Paper filter
12. (A) To regulate engine temperature
13. (A) Lug wrench
14. (A) To reduce harmful emissions
15. (C) Sway bar

Mechanical Comprehension

1. (B) Pulley
2. (B) 2
3. (B) Second-class lever
4. (D) To split objects apart
5. (B) $\pi/2$ meters
6. (A) Frictional force
7. (C) Fishing rod
8. (B) 4

9. (C) To transfer rotational motion between shafts
10. (B) 32 centimeters
11. (C) Pulley
12. (A) It decreases the force needed to lift an object
13. (D) 4
14. (B) By decreasing the coefficient of friction
15. (D) 320 newtons

PRACTICE TEST 4 ANSWERS

General Science

1. (C) Photosynthesis
2. (A) Hemoglobin
3. (B) Gene
4. (B) Fish
5. (C) Evaporation
6. (C) Igneous
7. (A) Venus
8. (A) Equator

9. (A) Newton
10. (A) Newton's First Law
11. (B) Joule
12. (B) Power
13. (B) Au
14. (C) Helium
15. (A) Ionic bond

Arithmetic Reasoning

1. (B) 17
2. (B) 420
3. (C) 72
4. (B) 32
5. (A) 33
6. (D) 66
7. (B) $\frac{3}{20}$
8. (B) 55

9. (B) 12
10. (D) $\frac{11}{12}$
11. (C) 49
12. (B) 0.15
13. (B) 6
14. (A) $\frac{1}{2}$
15 (B) 420

Word Knowledge

1. (B) Clever
2. (C) Nostalgic
3. (C) Enthusiastic
4. (B) Unusual
5. (B) Generous

6. (B) Convincing
7. (B) Central
8. (C) Temporary
9. (C) Thorough
10. (B) Soothe

11. (B) Persistence
12. (B) Chaotic
13. (B) Unyielding

14. (B) Generous
15. (B) Dull

Paragraph Comprehension

1. (B) It was haunted by a ghost
2. (C) Chris
3. (D) Isaac
4. (B) The importance of environmental conservation
5. (A) John Bardeen, Walter Brattain, and William Shockley

6 (C) The compass
7. (B) Her research on radioactivity
8. (B) The Wright brothers
9. (C) Her advocacy for girls' education
10. (C) The consequences of the Industrial Revolution

Mathematics Knowledge

1. (A) -7
2. (B) 36°
3. (A) x^2 - 4
4. (B) 21 cm
5. (C) -4
6. (B) 12
7. (A) 16x + 1
8. (C) 31.4 cm

9. (B) 2
10. (C) 113.04 cm³
11. (B) 18
12. (A) 2x + 11
13. (C) 37.68 cm
14. (A) 7
15. (C) 268.08 cm³

Electronics Information

1. (C) Electron
2. (B) Coulomb
3. (C) Copper
4. (C) Voltage
5. (B) Ohm
6. (B) Ohm's Law
7. (C) Circuit
8. (B) Capacitor

9. (B) Current
10. (A) Voltage
11. (A) Direct current (DC)
12. (C) Watt
13. (C) Diode
14. (B) Resistance
15. (A) Inductance

Auto and Shop

1. (B) Spark plug
2. (C) Tire gauge
3. (C) Engine oil
4. (D) Anti-Lock Braking System
5. (C) Torque converter
6. (B) Wrench
7. (B) LED
8. (B) Charges the battery

9. (A) Lug wrench
10. (A) Brake caliper
11. (B) To transfer power from the engine to various accessories
12. (A) Synthetic oil
13. (B) To reduce harmful emissions
14. (A) Shock absorber
15. (C) Heating, Ventilation, and Air Conditioning

Mechanical Comprehension

1. (B) Pulley
2. (C) 4
3. (B) Second-class lever
4. (D) To split objects apart
5. (B) 2π meters
6. (A) Frictional force
7. (C) Fishing rod
8. (C) 4

9 (C) To transfer rotational motion between shafts
10. (C) 40 centimeters
11. (B) Wedge
12. (A) It decreases the force needed to lift an object
13. (D) 5
14. (B) By decreasing the coefficient of friction
15. (D) 720 newtons

CONCLUSION

Congratulations!

You've made it this far, so you can do it! There's a couple of appendices to help you keep practicing, as well as the diagnostic test. There are also plenty of online resources to help you feel prepared prior to taking the test. Make sure to find out if what sections can help give you a boost toward any specialist role you want. While you should study all sections, especially the ones about ready and math, you don't need to spend a lot of time on sections where you don't have a particular interested. You can always retake the ASVAB once you are in the military if you want to switch roles or gain an interest in a different area.

Good luck on the test!

THE DIAGNOSTIC TEST

Have your times, pencil, and scratch paper ready. When you are ready, begin.

PART 1: GENERAL SCIENCE (GS)

25 Questions

Set the time for 11 minutes

This section tests you understanding of general science typically taught in high school. Circle the best answer.

1. What is the basic unit of life?

    (A) Atom
    (B) Cell
    (C) Molecule
    (D) Organ

2. Which organelle is responsible for producing energy in a cell?

    (A) Nucleus
    (B) Golgi apparatus
    (C) Mitochondria
    (D) Endoplasmic reticulum

3. What is the process by which plants produce nourishment for themselves?

    (A) Respiration
    (B) Photosynthesis
    (C) Fermentation
    (D) Digestion

4. Of the following, which kind of tissue is not present in the human body?

    (A) Epithelial tissue
    (B) Muscle tissue
    (C) Bone tissue
    (D) Genetic tissue

5. What is the term for the passing of traits from parents to offspring?

    (A) Adaptation
    (B) Mutation
    (C) Inheritance
    (D) Evolution

6. What is the main function of the respiratory system?

    (A) Pump blood around the body
    (B) Break down food into nutrients
    (C) Exchange gases with the environment
    (D) Control movement and coordination

7. Which of the following is a function of the nervous system?

    (A) Regulate body temperature
    (B) Digest food
    (C) Transmit signals between body parts
    (D) Pump blood

8. What is the role of white blood cells in the immune system?

    (A) Carry oxygen
    (B) Fight infection
    (C) Transport nutrients
    (D) Produce antibodies

9. What is the term for the liquid component of blood?

    (A) Plasma
    (B) Platelets
    (C) Hemoglobin
    (D) Serum

10. What is the function of the skeletal system?

    (A) Produce hormones
    (B) Protect internal organs
    (C) Filter blood
    (D) Digest food

11. Which of the following is a function of the kidneys?

    (A) Regulate blood sugar levels
    (B) Produce bile
    (C) Filter waste from the blood
    (D) Control body temperature

12. What is the main function of the digestive system?

    (A) Exchange gases with the environment
    (B) Pump blood around the body
    (C) Break down food into nutrients
    (D) Control movement and coordination

13. What is the purpose of the circulatory system?

    (A) Protect the body from infection
    (B) Transport oxygen and nutrients to cells
    (C) Maintain body temperature
    (D) Provide structural support

14. What is the name of the hormone that regulates blood sugar levels in the body?

    (A) Insulin
    (B) Adrenaline
    (C) Estrogen
    (D) Testosterone

15. What is the role of the liver in the body?

    (A) Produce bile
    (B) Filter waste from the blood
    (C) Pump blood around the body
    (D) Regulate body temperature

16. What is the function of the pancreas?

    (A) Produce insulin
    (B) Filter blood
    (C) Produce red blood cells
    (D) Control body temperature

17. What is the term for the movement of water across a semipermeable membrane?

    (A) Osmosis
    (B) Diffusion
    (C) Active transport
    (D) Filtration

18. What is the name of the process by which cells break down glucose to release energy?

    (A) Photosynthesis
    (B) Respiration
    (C) Fermentation
    (D) Transpiration

19. Which of the following is not a type of blood vessel?

    (A) Artery
    (B) Vein
    (C) Capillary
    (D) Cartilage

20. What is the term for the smallest unit of an element that retains the properties of that element?

    (A) Atom
    (B) Molecule
    (C) Compound
    (D) Proton

21. What is the name of the process by which plants lose water vapor through tiny pores in their leaves?

    (A) Osmosis
    (B) Transpiration
    (C) Respiration
    (D) Diffusion

22. What is the function of the ribosomes in a cell?

    (A) Produce energy
    (B) Store genetic information
    (C) Synthesize proteins
    (D) Transport materials

23. Which of the following is a function of the endocrine system?

    (A) Regulate body temperature
    (B) Control movement and coordination
    (C) Produce hormones
    (D) Exchange gases with the environment

24. What is the name of the structure that separates the chest cavity from the abdominal cavity?

    (A) Diaphragm
    (B) Pharynx
    (C) Trachea
    (D) Alveoli

25. What is the term for the outer layer of skin?

    (A) Epidermis
    (B) Dermis
    (C) Hypodermis
    (D) Cuticle

Go back and review your work until time runs out.

PART 2: ARITHMETIC REASONING (AR)

30 Questions

Set the time for 36 minutes

This section tests your arithmetic skills. Circle the best answer.

1. What is the value of 7 * 9 + 3?

    (A) 66
    (B) 60
    (C) 54
    (D) 46

2. If x = 5 and y = 3, what is $x^2 - y^2$?

    (A) 16
    (B) 12
    (C) 6
    (D) 2

3. What is the square root of 81?

    (A) 7
    (B) 8
    (C) 9
    (D) 10

4. If a = 4 and b = 7, what is $a^3 + b^3$?

    (A) 329
    (B) 347
    (C) 375
    (D) 411

5. What is the product of the first 5 prime numbers?

    (A) 80
    (B) 120
    (C) 210
    (D) 240

6. What is the absolute value of -25?

    (A) -25
    (B) 0
    (C) 25
    (D) 50

7. What is $\frac{2}{5}$ of 80?

    (A) 16
    (B) 24
    (C) 32
    (D) 40

8. If x = 8 and y = 3, what is $x^2 + 2xy + y^2$?

    (A) 157
    (B) 169
    (C) 178
    (D) 189

9 . What is the median of the following set of numbers: 5, 7, 3, 9, 2?

    (A) 3
    (B) 5
    (C) 7

    (D) 9

10. What is the value of 0.75 * 100?

    (A) 65
    (B) 75
    (C) 85
    (D) 95

11. What is $\frac{2}{3}$ of 120?

    (A) 60
    (B) 70
    (C) 80
    (D) 90

12. What is 3^4?

    (A) 9
    (B) 27
    (C) 64
    (D) 81

13. If x = 6 and y = 4, what is $x^2 - y^2$?

    (A) 20
    (B) 24
    (C) 28
    (D) 32

14, What is the cube root of 64?

    (A) 4
    (B) 6
    (C) 8
    (D) 10

15. What is $\frac{3}{4}$ of 96?

    (A) 57
    (B) 64
    (C) 72
    (D) 81

16. What is 6 * 4 + 8 ?

    (A) 32
    (B) 40
    (C) 48
    (D) 56

17. What is the sum of the first 15 positive odd integers?

    (A) 120
    (B) 135
    (C) 150
    (D) 165

18. What is $\frac{5}{8}$ of 80?

    (A) 40
    (B) 45
    (C) 50
    (D) 55

19. What is 2^5?

    (A) 16
    (B) 32
    (C) 64
    (D) 128

20. If x = 12 and y = 5, what is $x^2 - 2xy + y^2$?

    (A) 79
    (B) 89
    (C) 99
    (D) 109

21. Solve for 3x + 5 = 20

    (A) x = 5
    (B) x = 7
    (C) x = 8
    (D) x = 15

22. What is the value of | -10 |?

    (A) -10
    (B) 0
    (C) 10
    (D) 20

23. What is the result of $\frac{\frac{3}{5}*7}{8}$

    (A) $\frac{21}{40}$
    (B) $\frac{21}{56}$
    (C) $\frac{10}{21}$
    (D) $\frac{8}{15}$

24. What is the sum of the first 10 positive integers?

    (A) 45
    (B) 55
    (C) 65
    (D) 75

25. What is the square root of 144?

    (A) 10
    (B) 12
    (C) 14
    (D) 16

26. What is $\frac{7}{9} - \frac{2}{3}$?

    (A) $\frac{1}{9}$
    (B) $\frac{1}{6}$
    (C) $\frac{1}{3}$
    (D) $\frac{5}{9}$

27. If a = 3 and b = 5, what is the value of $a^2 + b^2$?

    (A) 20
    (B) 25
    (C) 34
    (D) 41

28. What is the prime factorization of 36?

    (A) $2^2 * 3^2$
    (B) $2^3 * 3$
    (C) $2 * 3^3$
    (D) $2^2 * 3$

29. What is 0.6 * 0.25?

    (A) 0.0125
    (B) 0.15
    (C) 0.25
    (D) 1.5

30. What is the value of 5!?

    (A) 120
    (B) 240
    (C) 360
    (D) 720

Go back and review your work until time runs out.

PART 3: WORD KNOWLEDGE (WK)

35 Questions

Set the time for 11 minutes

This section tests you vocabulary. Circle the answer that most closely means the same thing as the underlined word.

1. Accentuate most closely means

   (A) prominent or noticeable
   (B) visually stunning
   (C) a breath of fresh air
   (D) accurate shot

2. Boisterous most closely means

   (A) jittery and anxious
   (B) aloof and nonchalant
   (C) loud and unrestrained
   (D) manic and rambling

3. Tempo most closely means

   (A) a building tempest
   (B) a type of metronome
   (C) a type of note
   (D) rate of speed of a music piece

4. Labyrinth most closely means

   (A) a tunnel system underground
   (B) a myth
   (C) a type of sewer system
   (D) a complicated maze

5. Recede most closely means

   (A) retreat
   (B) decline
   (C) move back
   (D) slippery

6. Solace most closely means

   (A) a shoe tie
   (B) comfort in a time of grief
   (C) empathy for another
   (D) a mix of feelings

7. Formidable most closely means

   (A) intimidating strength
   (B) a great opponent
   (C) a force to be wary of
   (D) a type of form

8. Quandary most closely means

   (A) a moment of reflection
   (B) a quarrel with someone
   (C) a moment of realization
   (D) a state of confusion

9. Nuisance most closely means

   (A) an entity that's in the way

   (B) a dull object
   (C) someone/something being annoying
   (D) a dangerous some/something

10. Putrid most closely means

   (A) a unique scent
   (B) a horrid sight
   (C) a very unpleasant smell
   (D) a new sound

11. Vacillate most closely means

   (A) to facilitate
   (B) to be indecisive
   (C) too much gas
   (D) to be in control

12. Mutation most closely means

   (A) a change in DNA
   (B) a shift in spirit
   (C) an unexpected event
   (D) an exciting observation

12. Mutation most closely means

   (A) a change in DNA
   (B) a shift in spirit
   (C) an unexpected event
   (D) an exciting observation

13. Inference most closely means

   (A) to be clueless
   (B) having a gut feeling
   (C) jumping to conclusion
   (D) a conclusion drawn from evidence

14. Wretched most closely means

   (A) wicked or evil
   (B) miserable or unfortunate
   (C) tainted or befouled
   (D) risky or dangerous

15. Jaunty most closely means

   (A) to be shaking uncontrollably
   (B) having a jaggedly personality
   (C) having a cheerful and lively demeanor
   (D) a type of pirate song

16. Exult most closely means

   (A) jubilant
   (B) pessimistic
   (C) blase
   (D) energetic

17. Heritage most closely means

    (A) someone that is close to you
    (B) an item found in an unique location
    (C) something seen in a museum
    (D) something handed from one generation to the next

18. Correlate most closely means

    (A) to bring together
    (B) to establish a connection
    (C) to pick up and examine
    (D) to mess up royally

19. Omnipotent most closely means

    (A) able to do anything
    (B) to be hindered by
    (C) able to adapt
    (D) to give up on

20. Placate most closely means

    (A) Agitate
    (B) Soothe
    (C) Annoy
    (D) Provoke

21. When I walked into the extravagantly decorated mansion, I felt gauche because of my humble background.

    (A) awkward
    (B) excellent
    (C) proud
    (D) bored

22. Because he was so facetious in his approach, he caught more mistakes than anyone else.

    (A) misgiving
    (B) unempathetic
    (C) flippant
    (D) calm

23. The monotony of the meeting made me fall asleep.

    (A) dull activity
    (B) repeated daily routine
    (C) unimportant talks
    (D) hypnotic motion

24. I considered myself indispensable – my former employer did not.

    (A) high in quality
    (B) a dependable worker
    (C) another position filled
    (D) absolutely necessary

25. When my son becomes belligerent, I know it is past nap time.

    (A) barely staying awake
    (B) becoming unbearable
    (C) showing combativeness
    (D) showing signs of illness

26. With his first child, he approached childcare with assiduous attention.

    (A) giving a lot of attention and effort
    (B) seeing potential in
    (C) is suspicious of
    (D) trying to catch off guard

27. Since I saw the whole thing as perfunctory, I didn't care if we achieved our goal.

    (A) a priority
    (B) characterized by routine
    (C) something burdensome
    (D) a solo set of tasks

28. The travesty of the ruined wedding cake was just the first sign of trouble.

    (A) a grossly inferior imitation
    (B) a disastrous event
    (C) a nasty sight
    (D) a sad justification

29. No matter how he apologized, I felt that his actions were irredeemable.

    (A) not something to worry or be dismayed about
    (B) not able to be saved or improved
    (C) not easily forgivable or forgotten
    (D) not clear or understandable

30. The only alternative was to just finish the project on my own.

    (A) the next step
    (B) a different perspective
    (C) the only other option
    (D) altering the plan

31. My mother expedited the shipment so that I had my laptop before school started.

    (A) took an expedition
    (B) made the shipment go quicker
    (C) ordered online
    (D) bought the laptop

32. The coalition was ineffective because no one wanted to work together.

    (A) to coerce a different group
    (B) too many differing opinions
    (C) a group that's mismanaged
    (D) two or more groups working together

33. The bear had been dormant all winter and was ready to find food.

    (A) inactive then became active later
    (B) a slow time with things picking up again
    (C) played dead until spring
    (D) in hibernation

34. It was only after I was told that I was the valedictorian that I felt <u>vindicated</u> for focusing on school.

    (A) feeling victorious
    (B) to see things through
    (C) showing to be justified
    (D) being versatile

Go back and review your work until time runs out.

35. If you weren't so <u>impetuous</u>, you would have had all of the facts to make a better decision.

    (A) acting without thought
    (B) uptight and dismissive
    (C) being slothful
    (D) acting like an imp

PART 4: PARAGRAPH COMPREHENSION (PC)

15 Questions

Set the time for 13 minutes

This section tests your ability to logically complete incomplete statement and questions. Read the passage, then circle the answer that makes the most sense to complete the passage.

1. Emily had always been fascinated by astronomy. She spent countless nights gazing at the stars through her telescope, marveling at the vastness of the universe. Her favorite celestial body was Saturn, with its majestic rings and mysterious moons. One summer evening, while observing Saturn, Emily noticed a faint glow in the sky. As she focused her telescope on the source of the light, she realized that she was witnessing a rare meteor shower.

What is Emily's favorite celestial body?

    (A) Jupiter
    (B) Mars
    (C) Saturn
    (D) Venus

2. (As a marine biologist, David spent most of his days exploring the depths of the ocean. One day, while diving off the coast of Australia, he encountered a magnificent coral reef teeming with life. Mesmerized by the vibrant colors and diverse species of marine life, David knew that he had stumbled upon a hidden gem of the ocean.

Where was David diving when he encountered the coral reef?

    (A) Hawaii
    (B) Australia
    (C) Florida
    (D) California

3. Climate change refers to long-term shifts in temperature, precipitation, and other atmospheric conditions on Earth. Global warming is a result of human activity increasing the amount of greenhouse gases in the atmosphere, such as through the burning of fossil fuels and deforestation. Loss of biodiversity, harsh weather, and rising sea levels are some of the effects of climate change.

What is the main theme of the passage?

    (A) The causes of climate change
    (B) The effects of deforestation
    (C) The history of global warming
    (D) The impact of human activities on the environment

4. Maria was an aspiring chef who loved experimenting with new recipes in the kitchen. Her specialty was baking, and she spent hours perfecting her techniques to create mouthwatering desserts. One day, she decided to enter a baking competition, hoping to showcase her skills and passion for pastry.

What is Maria's specialty in the kitchen?

    (A) Grilling
    (B) Baking
    (C) Stir-frying
    (D) Boiling

5. Alex had a passion for photography and loved capturing the beauty of nature through his camera lens. One day, while hiking in the mountains, he stumbled upon a breathtaking vista overlooking a tranquil valley. Inspired by the scene before him, Alex snapped a series of photographs, each one more stunning than the last.

What is Alex passionate about?

    (A) Writing
    (B) Cooking
    (C) Photography
    (D) Gardening

6. Jason loved spending time outdoors, exploring the wilderness and observing wildlife. One day, while hiking in the mountains, he stumbled upon a hidden waterfall cascading down a rocky cliff. Mesmerized by its beauty, Jason decided to sit by the waterfall and sketch its serene surroundings in his notebook. As he sketched, he felt a sense of peace and tranquility wash over him, reminding him of the importance of connecting with nature.

What did Jason find while hiking in the mountains?

    (A) A hidden cave
    (B) A hidden waterfall
    (C) A hidden lake
    (D) A hidden meadow

7. Deoxyribonucleic acid, also known as DNA, is the molecule that houses the genetic instructions necessary for life. Adenine (A), Cytosine (C), Guanine (G), and Thymine (T) are its four chemical bases. The qualities and attributes of an organism are determined by the arrangement of these bases. A basic process known as DNA replication is the copying of DNA to create new cells during cell division.

What is the main theme of the passage?

    (A) The structure of DNA
    (B) The process of cell division
    (C) The functions of genetic instructions
    (D) The history of genetics

8. Mark loved the thrill of adventure and often sought out new challenges to conquer. One summer, he decided to embark on a solo backpacking trip through the rugged wilderness of Alaska. As he trekked through the towering mountains and dense forests, Mark felt a sense of freedom and exhilaration unlike anything he had ever experienced before.

Where did Mark go on his solo backpacking trip?

    (A) Amazon Rainforest
    (B) Sahara Desert
    (C) Alaskan Wilderness
    (D) Swiss Alps

9. Emma has a lifelong curiosity with the customs and civilizations of other people. She cherished visiting remote locations and embracing novel experiences. She went to Japan one summer to take part in a customary tea ceremony. Emma felt calmness go over her as she drank the aromatic green tea and was thankful for the chance to experience Japanese culture firsthand.

Where did Emma travel to participate in a traditional tea ceremony?

    (A) China
    (B) India
    (C) Japan
    (D) Thailand

10. The Amazon Rainforest is the largest tropical rainforest in the world, covering an area of over 6 million square kilometers across nine countries in South America. It is home to an estimated 10% of the world's known species and plays a crucial role in regulating the Earth's climate. Deforestation, primarily driven by agriculture and logging, poses a significant threat to the Amazon Rainforest and its biodiversity.

What is the main theme of the passage?

    (A) The importance of biodiversity
    (B) The geography of South America
    (C) The history of deforestation
    (D) The impact of climate change

11. The internet is a global network of computers that are connected to one another, facilitating communication and information sharing over great distances. The way individuals interact with each other, obtain information, and do business has all been completely transformed by it. But there are drawbacks to the internet as well, like cybersecurity risks, privacy issues when using it, and digital inequalities.

What is the main theme of the passage?

    (A) The history of computer networks

(B) The benefits of online communication
(C) The development of digital technology
(D) The challenges of the internet

12. Sarah, Emma, and Jake are siblings. Sarah is older than Emma, and Emma is older than Jake. If the statements are true, who is the youngest?

Who is the youngest sibling?

(A) Jake
(B) Emma
(C) Sarah
(D) Cannot be determined

13. From 1994 to 1999, Nelson Mandela, a political activist and revolutionary against apartheid from South Africa, led the country as president. Mandela won the Nobel Peace Prize in 1993 and gained international recognition for his work against racial segregation and his support of democracy and equality. What was Nelson Mandela's primary contribution to South Africa?

(A) His establishment of apartheid laws
(B) His promotion of racial segregation
(C) His support for colonial rule
(D) His advocacy for democracy and equality

14. Artificial intelligence (AI) is a branch of computer science that aims to create intelligent machines capable of performing tasks that typically require human intelligence. AI technologies, such as machine learning and natural language processing, are being integrated into various applications, including virtual assistants and autonomous vehicles.

What is the goal of artificial intelligence?

(A) To replace human intelligence
(B) To perform tasks beyond human capabilities
(C) To limit technological advancements
(D) To eliminate the need for human intervention

15. Four students—Oliver, Peter, Quinn, and Rachel—participated in a science fair. Oliver placed before Peter but after Quinn. Rachel placed before Peter. Who placed second?

Who placed second in the science fair?

(A) Quinn
(B) Peter
(C) Oliver
(D) Rachel

Go back and review your work until time runs out.

PART 5: MATHEMATICS KNOWLEDGE (MK)

24 Questions

Set the time for 25 minutes.

This section tests you ability to do basic math. Circle the best answer.

1. Solve for x: 2x+3=112x+3=11

    (A) 3
    (B) 4
    (C) 5
    (D) 6

2. What is the solution to the equation $x^2-4x+4=0$?

    (A) 0, 4
    (B) 2, -2
    (C) 4, -4
    (D) 2

3. Solve for y: y: $3y - 5 = 2y + 7$

    (A) 7
    (B) 9
    (C) 12
    (D) 5

4. Find the volume of a cube with side length 4 cm.

    (A) 16 cm³
    (B) 24 cm³
    (C) 64 cm³
    (D) 48 cm³

5. In a right triangle, if one angle is 30°, what is the measure of the other non-right angle?

    (A) 30°
    (B) 45°
    (C) 60°
    (D) 90°

6. Solve for x: $\frac{x-3}{4} = 5$

    (A) 17
    (B) 18
    (C) 19
    (D) 20

7. Find the length of the hypotenuse in a right triangle with legs of 6 cm and 8 cm.

    (A) 8 cm
    (B) 10 cm
    (C) 12 cm
    (D) 14 cm

8. What is the area of a circle with a diameter of 14 cm? (Use $\pi = \frac{22}{7}$)

    (A) 154 cm²
    (B) 77 cm²
    (C) 44 cm²
    (D) 308 cm²

9. In an equilateral triangle, what is the measure of each angle?

    (A) 45°
    (B) 60°
    (C) 90°
    (D) 120°

10. Solve for y: $y^2-16=0$

    (A) 0, 4
    (B) 4, -4
    (C) -4
    (D) 4

11. What is the measure of each interior angle of a regular pentagon?

    (A) 108°
    (B) 120°
    (C) 135°
    (D) 144°

12. Find the area of a triangle with a base of 10 cm and a height of 5 cm.

    (A) 25 cm²
    (B) 50 cm²
    (C) 75 cm²
    (D) 100 cm²

13. What is the circumference of a circle with a radius of 7 cm? ((Use $\pi = \frac{22}{7}$)

    (A) 22 cm
    (B) 44 cm
    (C) 14 cm
    (D) 28 cm

14. Simplify: 2a(3a - 4)

    (A) 6a - 8
    (B) $6a^2$ - 8a
    (C) $6a^2$ - 8
    (D) $6a^2$ - 4

15. Simplify: 5x - 2(3x - 4)

    (A) x - 8
    (B) x + 8
    (C) -x - 8
    (D) -x + 8

16. What is the perimeter of a rectangle with a length of 8 cm and a width of 3 cm?

    (A) 11 cm
    (B) 22 cm
    (C) 16 cm

(D) 24 cm

17. If f(x) = 3x + 2, what is f(4)?

    (A) 10
    (B) 12
    (C) 14
    (D) 16

18. What is the value of x in the equation $2x^2 - 8 = 0$?

    (A) 2
    (B) -2
    (C) 4
    (D) -4

19. What is the slope of the line passing through the points (2, 3) and (4, 7)?

    (A) 2
    (B) 3
    (C) 1
    (D) 4

20. What is the length of a side of a square with an area of 81 cm²?

    (A) 7 cm
    (B) 8 cm
    (C) 9 cm
    (D) 10 cm

Go back and review your work until time runs out.

21. Solve for x: 3x - 5 = 16

    (A)5
    (B)6
    (C)7
    (D)8

22. Simplify: 4(2x - 3)

    (A) 8x - 3
    (B)8x - 12
    (C)8x + 3
    (D)8x + 12

23 What is the area of a rectangle with a length of 10 cm and a width of 4 cm?

    (A)14 cm²
    (B)20 cm²
    (C)24 cm²
    (D)40 cm²

24. What is the measure of an interior angle of a regular hexagon?

    (A)100°
    (B)110°
    (C)120°
    (D)130°

PART 6: ELECTRONICS INFORMATION (EI)

20 Questions

Set the time for 9 minutes.

This section tests you understanding of electronic concepts. Circle the best answer.

1. What is the SI unit of electric charge?

(A) Ampere
(B) Coulomb
(C) Ohm
(D) Volt

2. What is the flow of electric charge through a conductor called?

(A) Voltage
(B) Current
(C) Resistance
(D) Capacitance

3. What component is used to oppose the flow of electric current in a circuit?

(A) Resistor
(B) Capacitor
(C) Diode
(D) Transistor

4. According to which law, the voltage across two points and the current flowing through a conductor between them are directly proportional?

(A) Faraday's Law
(B) Ohm's Law
(C) Kirchhoff's Law
(D) Coulomb's Law

5. Given a resistor with a resistance of 2 ohms and a voltage drop of 8 volts across it, what is the current that flows through it?

(A) 2 amperes
(B) 4 amperes
(C) 6 amperes
(D) 8 amperes

6. What is the resistance of a circuit with a voltage of 24 volts and a current of 6 amperes?

(A) 2 ohms
(B) 4 ohms
(C) 6 ohms
(D) 8 ohms

7. What is the voltage across a circuit with a 5 ohm resistance and a 2 ampere current flowing through it?

(A) 2 volts
(B) 5 volts
(C) 7 volts
(D) 10 volts

8. What is the formula to calculate voltage (V) in a circuit?

(A) V = I * R
(B) V = I / R
(C) V = R / I
(D) V = R + I

9. What is the term for the opposition to the flow of electric current in a circuit?

(A) Voltage
(B) Current
(C) Resistance
(D) Conductance

10. Which material has very low resistance and is commonly used in electrical wiring?

(A) Aluminum
(B) Copper
(C) Silver
(D) Gold

11. What is the unit of electrical resistance?

(A) Ampere
(B) Ohm
(C) Volt
(D) Watt

12. What is the formula to calculate current (I) in a circuit?

(A) I = V * R
(B) I = V / R
(C) I = R / V
(D) I = R + V

13. What is the function of a capacitor in an electrical circuit?

(A) To store electrical charge
(B) To oppose changes in voltage
(C) To regulate current flow
(D) To amplify signals

14. Which type of current flows in one direction only?

(A) Direct current (DC)
(B) Alternating current (AC)
(C) Bipolar current (BC)
(D) Variable current (VC)

15. What is the measure of the potential difference between two points in a circuit?

(A) Current
(B) Resistance
(C) Voltage
(D) Power

16. What is the term for the rate at which electric energy is transferred by an electric circuit?

   (A) Voltage
   (B) Current
   (C) Resistance
   (D) Power

17. Which law states that the algebraic sum of currents entering and leaving a junction in a circuit must be zero?

   (A) Ohm's Law
   (B) Faraday's Law
   (C) Kirchhoff's Current Law
   (D) Gauss's Law

18. Given a resistor with a resistance of 2 ohms and a voltage drop of 6 volts across it, what is the current that passes through it?

   (A) 2 amperes
   (B) 3 amperes
   (C) 4 amperes
   (D) 6 amperes

19. What is the voltage drop across a resistor with a resistance of 8 ohms and a current of 4 amperes flowing through it?

   (A) 2 volts
   (B) 4 volts
   (C) 8 volts
   (D) 16 volts

20. If a circuit has a voltage drop of 12 volts across it and a resistance of 6 ohms, what is the current flowing through it?

   (A) 2 amperes
   (B) 4 amperes
   (C) 6 amperes
   (D) 12 amperes

Go back and review your work until time runs out.

PART 7: AUTO AND SHOP INFORMATION (AS)

25 Questions

Set the time for 11 minutes.

This section tests your knowledge of basic shop auto and shop concepts. Circle the best answer.

1. What is the purpose of a torque wrench in automotive work?

   (A) To measure tire pressure
   (B) To tighten bolts to a specific torque value
   (C) To adjust engine timing
   (D) To check battery voltage

3. Which tool is used to remove and install spark plugs in a car engine?

   (A) Socket wrench
   (B) Pliers
   (C) Screwdriver
   (D) Spark plug wrench

4. What does ABS stand for in relation to car brakes?

   (A) Automatic Braking System
   (B) Advanced Braking System
   (C) Anti-Lock Braking System
   (D) Anti-Brake System

5. What is the primary purpose of a welding machine in a workshop?

   (A) Cutting metal
   (B) Joining metal pieces together
   (C) Smoothing metal surfaces
   (D) Bending metal

6. What is the function of a lathe in a workshop?

   (A) Cutting metal
   (B) Shaping cylindrical objects
   (C) Sanding wood
   (D) Measuring angles

7. Which device is commonly used to remove excess material from a workpiece?

   (A) Lathe
   (B) Planer
   (C) Bandsaw
   (D) Drill press

8. Which tool is commonly used for gripping and turning objects?

   (A) Needle-nose pliers
   (B) Slip-joint pliers
   (C) Locking pliers
   (D) Diagonal cutting pliers

9. What is the purpose of a miter saw in woodworking?

   (A) Sanding wood
   (B) Cutting metal
   (C) Cutting precise angles in wood
   (D) Drilling holes

10. What is the primary function of a table saw in woodworking?

   (A) Sanding wood
   (B) Cutting wood into smaller pieces
   (C) Drilling holes in wood
   (D) Painting wooden surfaces

11. What tool is used to measure tire pressure in a car?

   (A) Lug wrench
   (B) Jack
   (C) Tire gauge
   (D) Socket wrench

12. What is the purpose of an air compressor in an automotive workshop?

   (A) To inflate tires
   (B) To power pneumatic tools
   (C) To cool the engine
   (D) To charge the battery

13. Which tool is used on a car to remove and replace the wheels?

   (A) Pliers
   (B) Screwdriver
   (C) Lug wrench
   (D) Hammer

14. What is the function of a carburetor in an internal combustion engine?

   (A) To filter air
   (B) To regulate fuel flow
   (C) To ignite the air-fuel mixture
   (D) To control exhaust emissions

15. What tool is used to remove rust and old paint from metal surfaces?

   (A) Drill press
   (B) Bench grinder
   (C) Wire brush
   (D) Sander

16. In woodworking, what is the main function of a jointer?

   (A) Sanding wood
   (B) Planing wood surfaces
   (C) Cutting precise angles
   (D) Joining wood edges

17. What is the purpose of a rotary tool (such as a Dremel) in a workshop?

   (A) Painting surfaces
   (B) Sanding small areas

(C) Cutting large pieces of wood
(D) Measuring angles

18. Which tool is used to hold workpieces securely in place while working on them?

(A) Pliers
(B) Screwdriver
(C) Vise
(D) Clamp

19. What is the primary function of a router table in a workshop?

(A) Cutting metal
(B) Sanding wood
(C) Shaping edges of wood
(D) Welding pieces together

20. What tool is used to fasten pieces of wood together with nails or staples?

(A) Screwdriver
(B) Hammer
(C) Drill press
(D) Pliers

21. What is the purpose of a bandsaw in a workshop?

(A) Cutting metal
(B) Measuring angles
(C) Welding pieces together

Go back and review your work until time runs out.

(D) Painting surfaces

22. What is the function of a bench grinder in a workshop?

(A) Cutting metal
(B) Sharpening tools
(C) Smoothing rough surfaces
(D) Measuring angles

23. What does HVAC stand for in automotive terminology?

(A) High-Voltage Alternating Current
(B) Hot Vapors Air Conditioning
(C) Heating, Ventilation, and Air Conditioning
(D) Hybrid Vehicle Access Control

24. Which device is used to create holes in metal or wood with precision?

(A) Hammer drill
(B) Circular saw
(C) Jigsaw
(D) Drill press

25. What tool is used to accurately measure lengths and angles in woodworking?

(A) Tape measure
(B) Protractor
(C) Combination square
(D) Compass

PART 8: MECHANICAL COMPREHENSION (MC)

25 Questions

Set the time for 19 minutes.

This section tests your understanding of basic mechanics and physics. Circle the best answer.

1. Which simple machine is commonly used to lift heavy objects using a rope or cable?

    (A) Lever
    (B) Pulley
    (C) Inclined plane
    (D) Wedge

2. What is the mechanical advantage of a lever if the effort arm is 8 meters long and the load arm is 2 meters long?

    (A) 2
    (B) 3
    (C) 4
    (D) 5

3. Which type of lever has the fulcrum placed between the effort and the load?

    (A) First-class lever
    (B) Second-class lever
    (C) Third-class lever
    (D) It depends on the specific setup

4. What is the primary function of a wedge in a mechanical system?

    (A) To amplify force
    (B) To change the direction of force
    (C) To increase the distance over which force is applied
    (D) To split objects apart

5. If a wheel has a diameter of 1 meter and is rotated through an angle of 90 degrees, what is the distance traveled by a point on its circumference?

    (A) $\pi/2$ meters
    (B) π meters
    (C) 2π meters
    (D) 4π meters

6. If two objects are sliding against one other, what type of force opposes it?

    (A) Frictional force
    (B) Tension force
    (C) Gravitational force
    (D) Magnetic force

7. Which of the following is an example of a third-class lever?

    (A) Nutcracker
    (B) Wheelbarrow
    (C) Fishing rod
    (D) Bottle opener

8. What is the mechanical advantage of an inclined plane if its length is 6 meters and its height is 2 meters?

    (A) 2
    (B) 3
    (C) 4
    (D) 5

9. What is the purpose of a gear in a mechanical system?

    (A) To change the direction of force
    (B) To amplify force
    (C) To transfer rotational motion between shafts
    (D) To reduce friction between moving parts

10. If a screw has 8 threads per centimeter and is turned through 4 complete rotations, how far does it advance?

    (A) 4 centimeters
    (B) 16 centimeters
    (C) 32 centimeters
    (D) 64 centimeters

11. Which simple machine is best suited for splitting wood?

    (A) Lever
    (B) Wedge
    (C) Pulley
    (D) Inclined plane

12. What is the advantage of using a lever?

    (A) It decreases the force needed to lift an object
    (B) It increases the distance over which force is applied
    (C) It changes the direction of force
    (D) It allows for precise control of force

13. In a wheel and axle system, what mechanical benefit does a system with a radius of five meters for the wheel and one meter for the axle offer?

    (A) 1
    (B) 2
    (C) 3
    (D) 5

14. How does lubricating a machine reduce friction?

    (A) By increasing the coefficient of friction
    (B) By decreasing the coefficient of friction
    (C) By increasing the weight of the moving parts
    (D) By increasing the force of friction

15. If a lever has a mechanical advantage of 6 and the effort force applied is 120 newtons, what is the load force that can be lifted?

    (A) 20 newtons

(B) 40 newtons
(C) 60 newtons
(D) 720 newtons

16. What is the mechanical advantage of a lever if the effort arm is 12 meters long and the load arm is 3 meters long?

(A) 2
(B) 3
(C) 4
(D) 5

17. Which type of lever has the load situated between the fulcrum and the effort?

(A) First-class lever
(B) Second-class lever
(C) Third-class lever
(D) It depends on the specific setup

19. If a wheel has a diameter of 0.5 meters and is rotated through an angle of 180 degrees, what is the distance traveled by a point on its circumference?

(A) $\pi/4$ meters
(B) π meters
(C) 2π meters
(D) 4π meters

20. When two items slide against one another, what type of force prevents this motion?

(A) Frictional force
(B) Tension force
(C) Gravitational force
(D) Magnetic force

Go back and review your work until time runs out.

21. Which of the following is an example of a second-class lever?

(A) Nutcracker
(B) Wheelbarrow
(C) Fishing rod
(D) Bottle opener

22. What is the advantage of using an inclined plane?

(A) It decreases the force needed to lift an object
(B) It increases the distance over which force is applied
(C) It changes the direction of force
(D) It allows for precise control of force

23. What type of lever has the load positioned between the fulcrum and the effort?

(A) First-class lever
(B) Second-class lever
(C) Third-class lever
(D) It depends on the specific setup

24. If a screw has 12 threads per centimeter and is turned through 3 complete rotations, how far does it advance?

(A) 36 centimeters
(B) 72 centimeters
(C) 108 centimeters
(D) 144 centimeters

25. Which simple machine is best suited for splitting objects apart?

(A) Lever
(B) Inclined plane
(C) Wedge
(D) Pulley

PART 9: ASSEMBLING OBJECTS (AO)

26 Questions

Set the time for 15 minutes.

This section tests your skills at recognizing patterns and predicting how something will look when it is assembled. Circle the best answer.

1. Which figure best represents how the objects in the first box appear if put together?

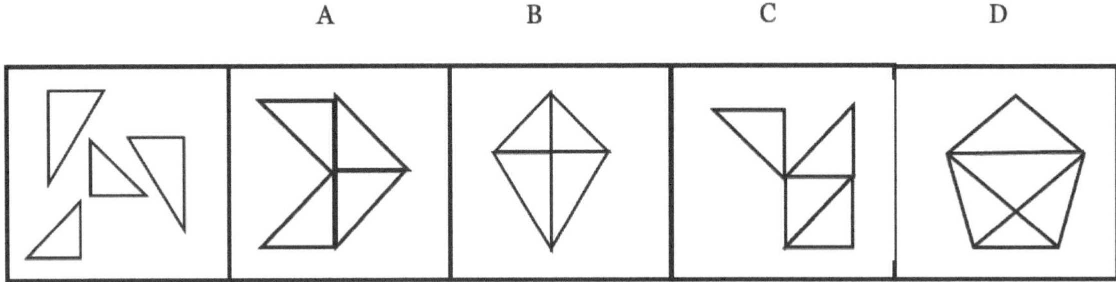

2. Which figure best represents how the objects in the first box appear if put together?

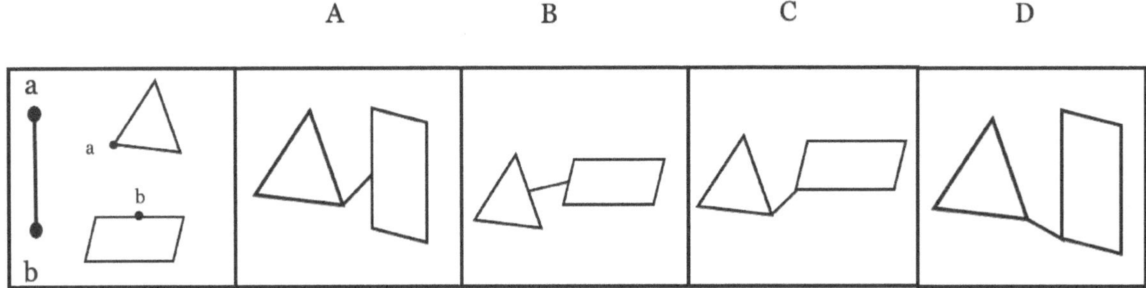

3. Which figure best represents how the objects in the first box appear if put together?

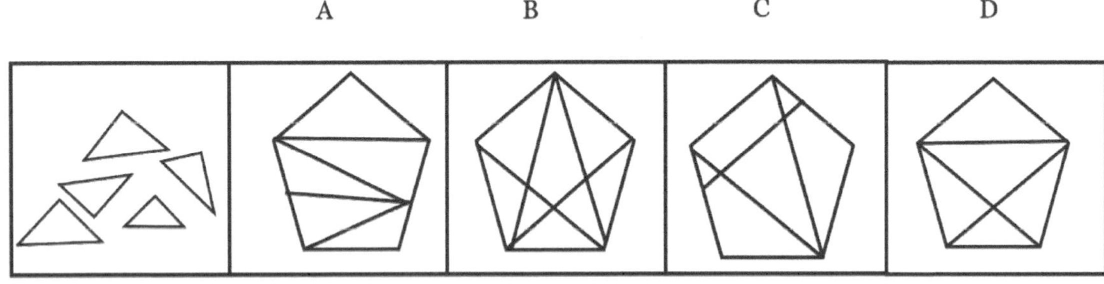

4. Which figure best represents how the objects in the first box appear if put together?

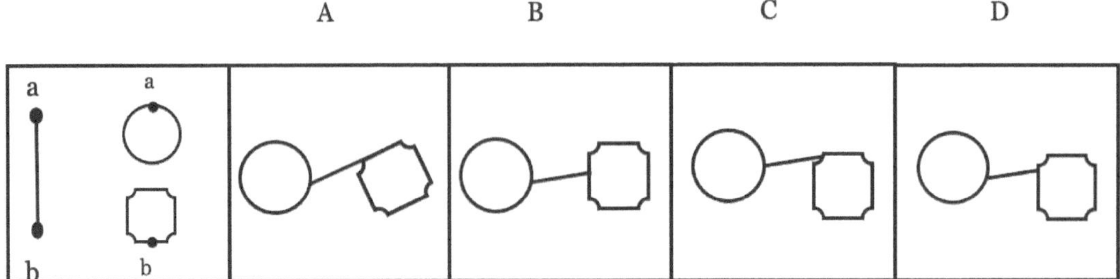

5. Which figure best represents how the objects in the first box appear if put together?

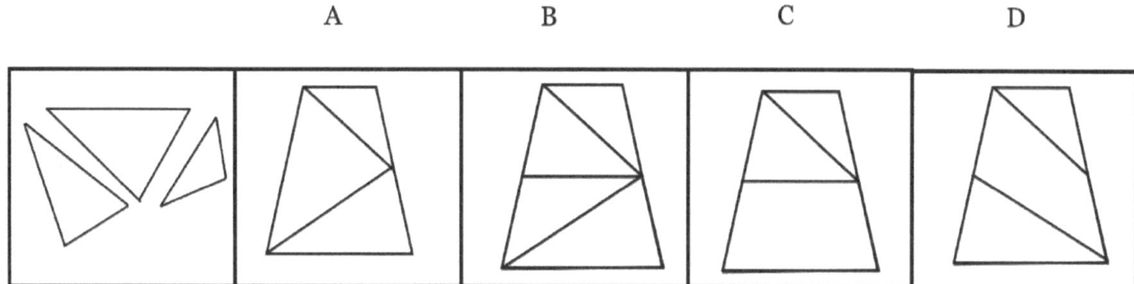

6. Which figure best represents how the objects in the first box appear if put together?

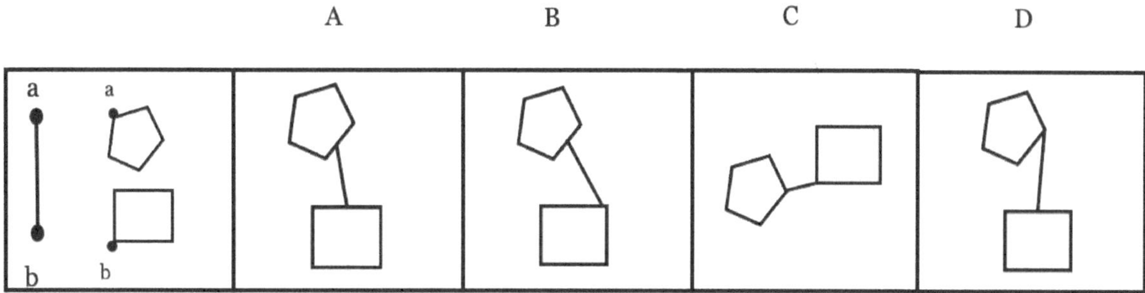

7. Which figure best represents how the objects in the first box appear if put together?

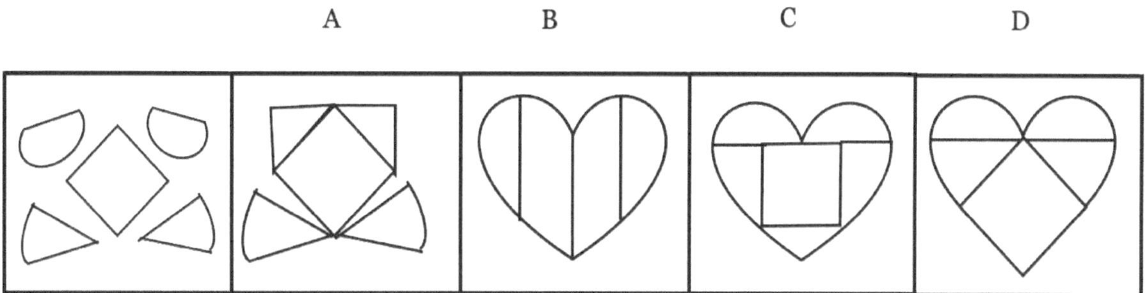

8. Which figure best represents how the objects in the first box appear if put together?

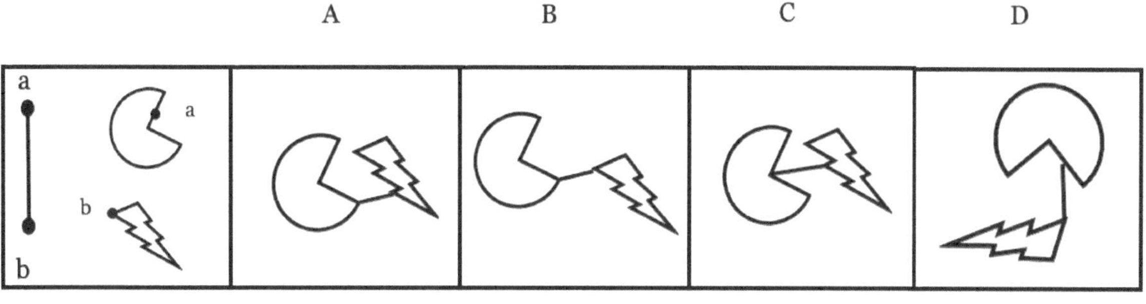

9. Which figure best represents how the objects in the first box appear if put together?

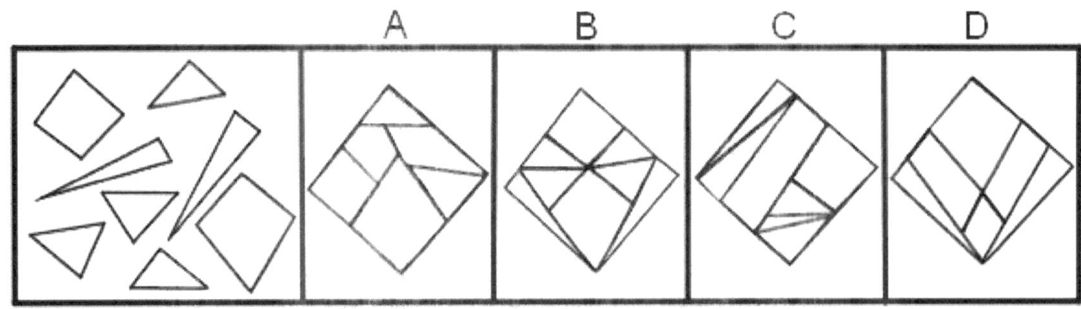

10. Which figure best represents how the objects in the first box appear if put together?

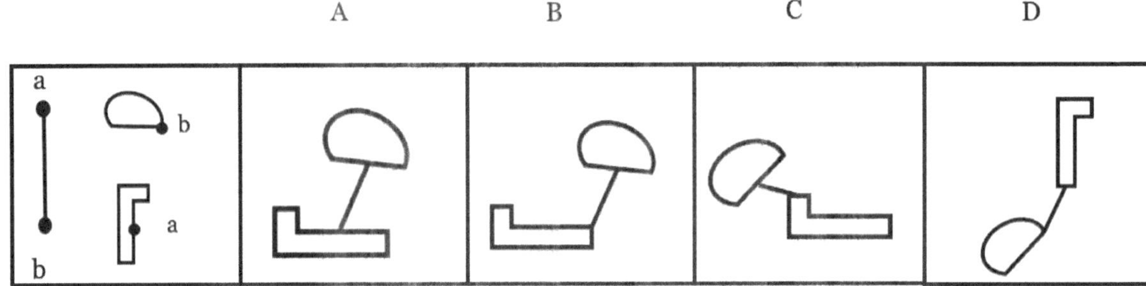

11. Which figure best represents how the objects in the first box appear if put together?

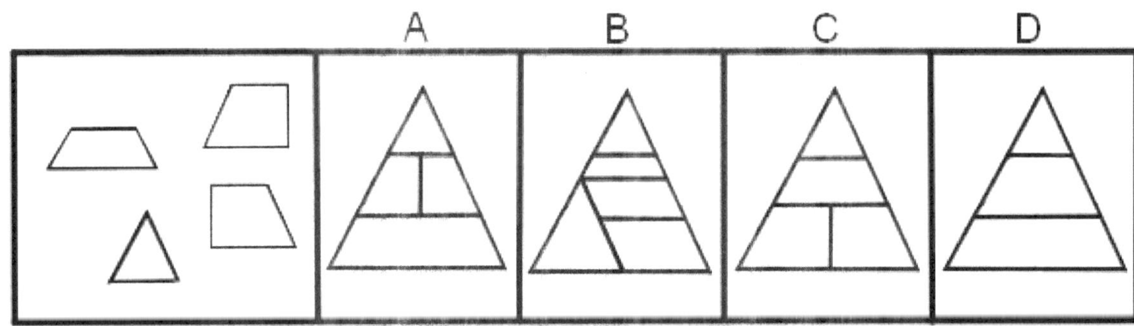

12. Which figure best represents how the objects in the first box appear if put together?

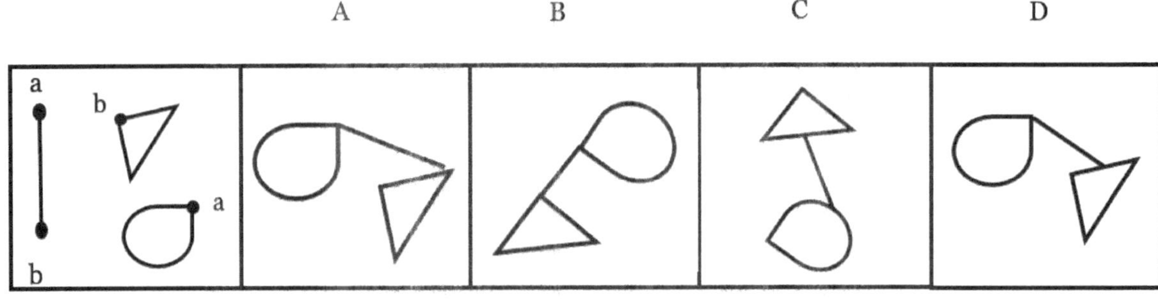

13. Which figure best represents how the objects in the first box appear if put together?

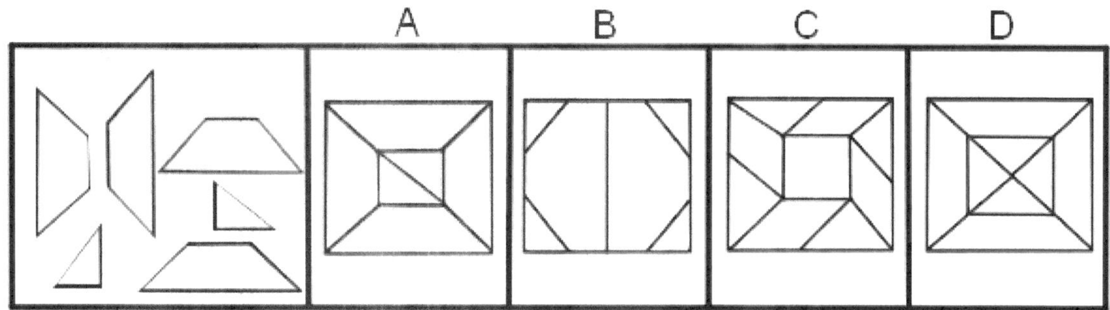

14. Which figure best represents how the objects in the first box appear if put together?

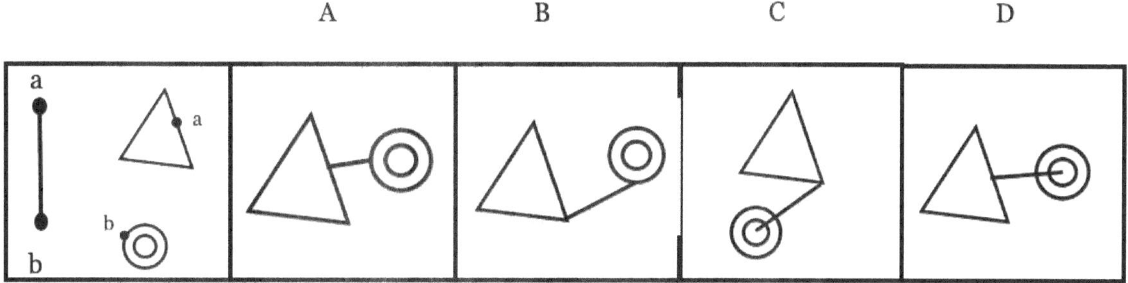

15. Which figure best represents how the objects in the first box appear if put together?

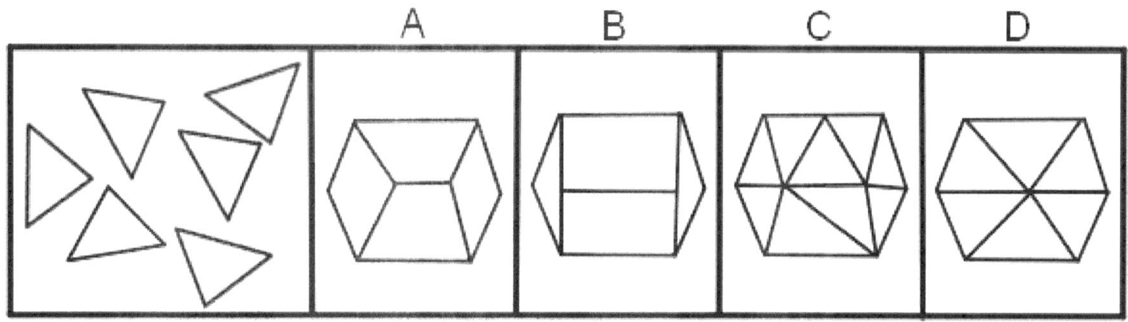

16. Which figure best represents how the objects in the first box appear if put together?

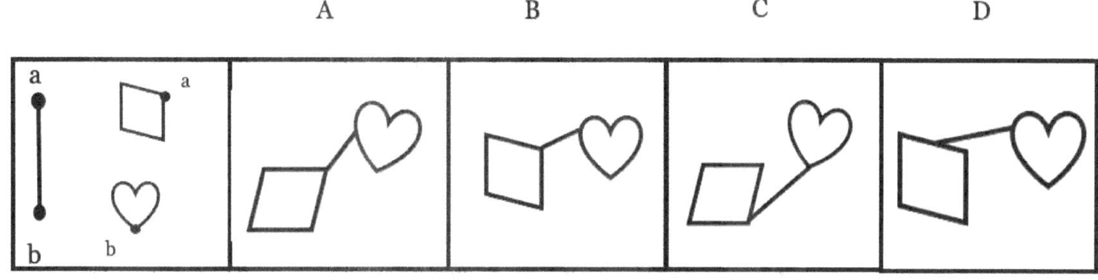

17. Which figure best represents how the objects in the first box appear if put together?

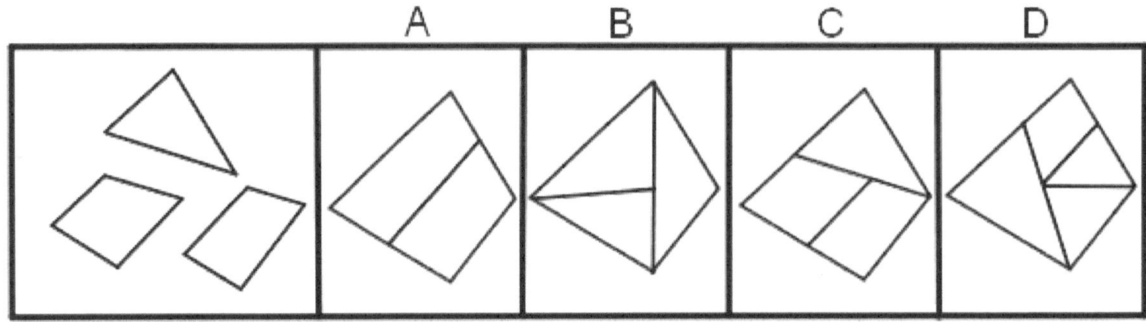

18. Which figure best represents how the objects in the first box appear if put together?

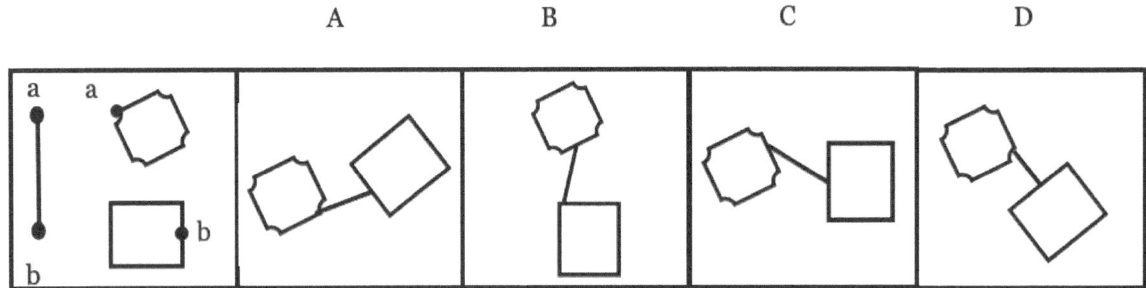

19. Which figure best represents how the objects in the first box appear if put together?

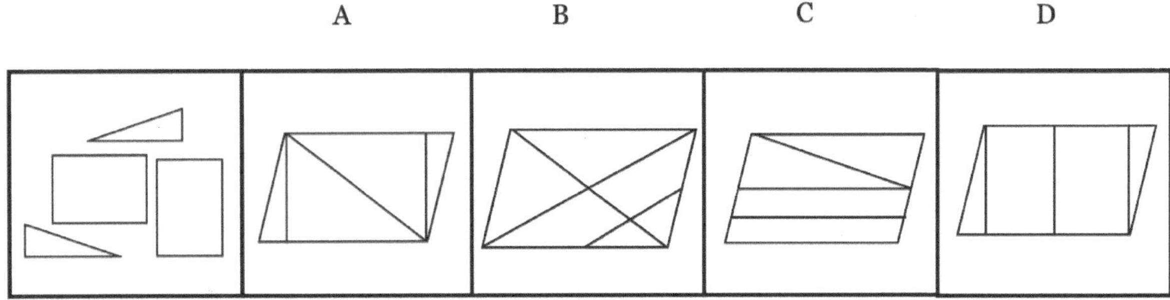

20. Which figure best represents how the objects in the first box appear if put together?

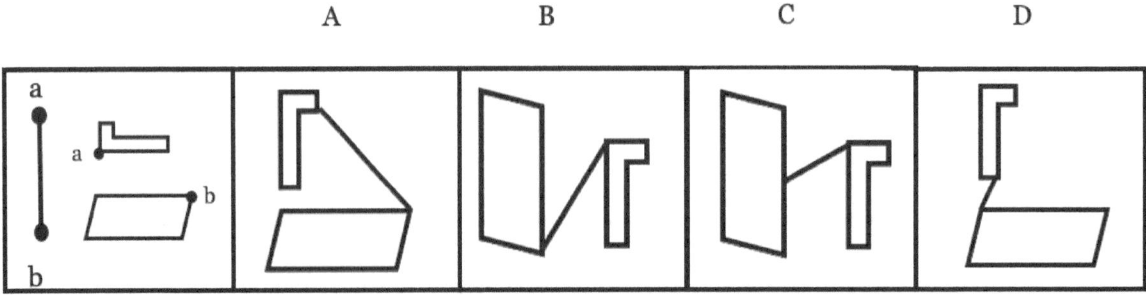

21. Which figure best represents how the objects in the first box appear if put together?

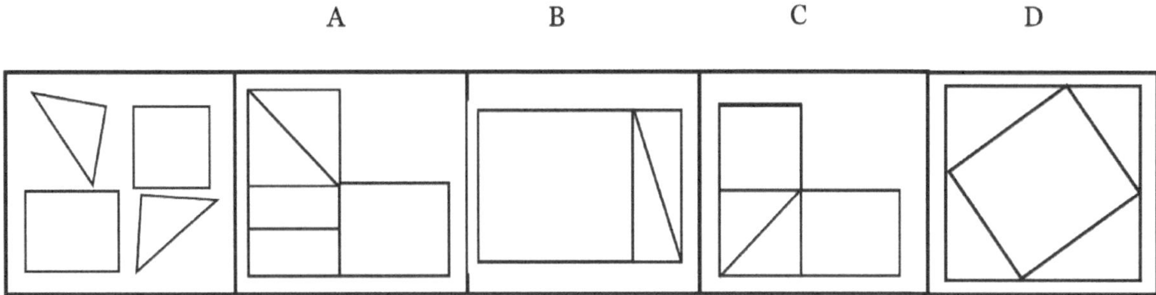

22. Which figure best represents how the objects in the first box appear if put together?

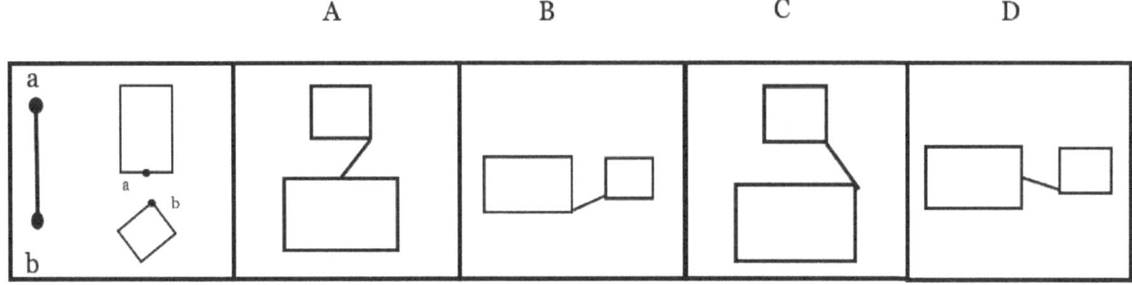

23. Which figure best represents how the objects in the first box appear if put together?

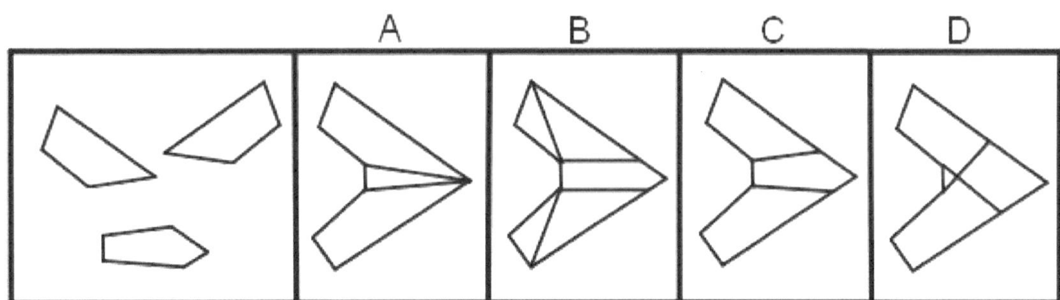

24. Which figure best represents how the objects in the first box appear if put together?

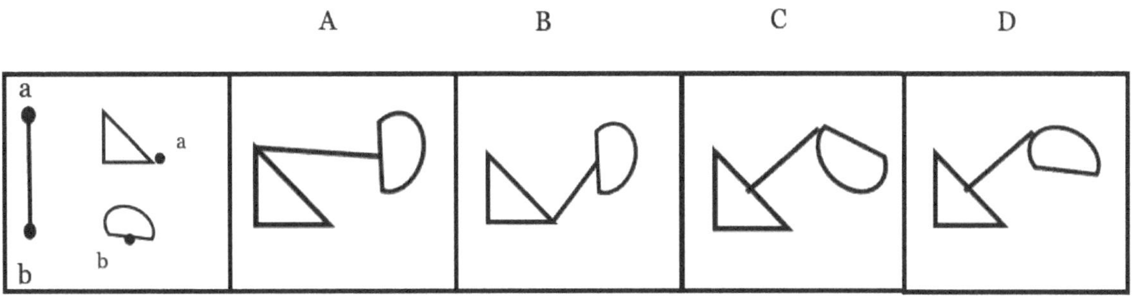

25. Which figure best represents how the objects in the first box appear if put together?

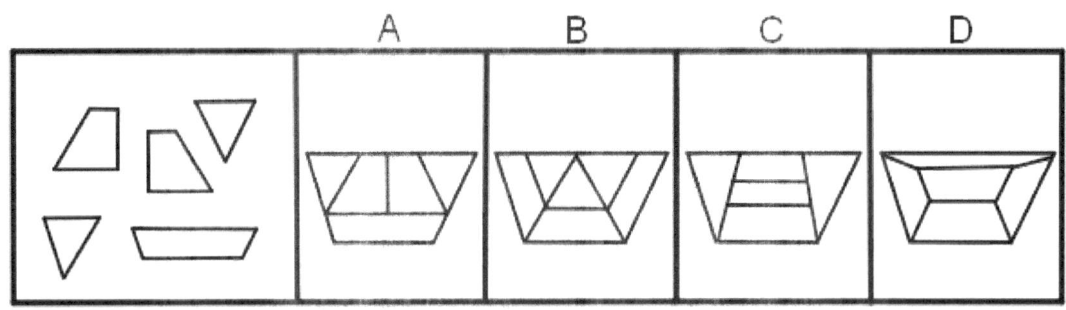

26. Which figure best represents how the objects in the first box appear if put together?

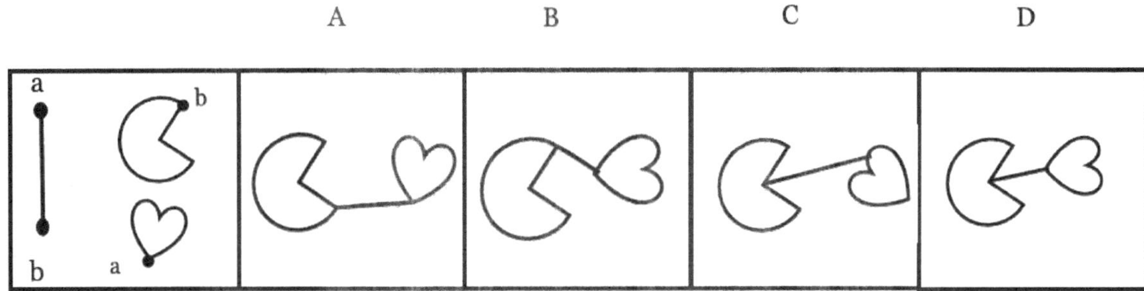

Go back and review your work until time runs out.

Congratulations! You've finished your first test – now you have an idea of what the experience is like.

Now it's time to score it.

ANSWER SHEET

Now that you've finished the test, use the following answering sheet to score it.

PART 1: GENERAL SCIENCE (GS)

1. (B) Cell
2. (C) Mitochondria
3. (B) Photosynthesis
4. (D) Genetic tissue
5. (C) Inheritance
6. (C) Exchange gases with the environment
7. (C) Transmit signals between body parts
8. (B) Fight infection
9. (A) Plasma
10. (B) Protect internal organs
11. (C) Filter waste from the blood
12. (C) Break down food into nutrients
13. (B) Transport oxygen and nutrients to cells
14. (A) Insulin
15. (A) Produce bile
16. (A) Produce insulin
17. (A) Osmosis
18. (B) Respiration
19. (D) Cartilage
20. (A) Atom
21. (B) Transpiration
22. (C) Synthesize proteins
23. (C) Produce hormones
24. (A) Diaphragm
25. (A) Epidermis

PART 2: ARITHMETIC REASONING (AR)

1. (A) 66
2. (A) 16
3. (C) 9
4. (B) 347
5. (C) 210
6. (C) 25
7. (C) 32
8. (B) 169
9. (B) 5
10. (B) 75
11. (D) 80
12. (D) 81
13. (C) 28
14. (A) 4
15. (C) 72
16. (C) 48
17. (D) 165
18. (A) 40
19. (B) 32
20. (B) 89
21. (B) x = 5
22. (C) 10
23. (A) $\frac{21}{40}$
24. (B) 55
25. (B) 12
26. (A) $\frac{1}{9}$
27. (D) 34
28. (A) $2^2 * 3^2$
29. (B) 0.15
30. (A) 120

PART 3: WORD KNOWLEDGE (WK)

1. (A) prominent or noticeable
2. (C) loud and unrestrained
3. (D) rate of speed of a music piece
4. (D) a complicated maze
5. (C) move back
6. (B) comfort in a time of grief
7. (A) intimidating strength
8. (D) a state of confusion
9. (C) someone/something being annoying
10. (C) a very unpleasant smell
11. (B) to be indecisive
12. (A) a change in DNA
13. (D) a conclusion drawn from evidence
14. (B) miserable or unfortunate
15. (C) having a cheerful and lively demeanor
16. (A) jubilant
17. (D) something handed from one generation to the next
18. (B) to establish a connection
19. (A) able to do anything
20. (B) Soothe

21. (A) awkward
22. (C) flippant
23. (B) repeated daily routine
24. (D) absolutely necessary
25. (C) showing combativeness
26. (A) giving a lot of attention and effort
27. (B) characterized by routine
28. (A) a grossly inferior imitation
29. (B) not able to be saved or improved
30. (C) the only other option
31. (B) made the shipment go quicker
32. (D) two or more groups working together
33. (A) inactive then became active later
34. (C) showing to be justified
35. (A) acting without thought

PART 4: PARAGRAPH COMPREHENSION (PC)

1. (C) Saturn
2. (B) Australia
3. (C) The impact of human activities on the environment
4. (B) Baking
5. (C) Photography
6. (B) A hidden waterfall
7. (C) The functions of genetic instructions
8. (C) Alaskan Wilderness
9. (C Japan
10. (A) The importance of biodiversity
11. (D) The challenges of the internet
12. (A) Jake
13. (D) His advocacy for democracy and equality
14. (B) To perform tasks beyond human capabilities
15. (A) Quinn

PART 5: MATHEMATICS KNOWLEDGE (MK)

1. (B) 4
2. (D) 2
3. (A) 12
4. (C) 64 cm³
5. (C) 60°
6. (B) 18
7. (B) 10 cm
8. (A) 154 cm²
9. (B) 60°
10. (B) 4, -4
11. (A) 108°
12. (A) 25 cm²
13. (B) 44 cm
14. (B) $6a^2 - 8a$
15. (D) -x + 8
16. (B) 22 cm
17. (C) 14
18. (A) 2
19. (A) 2
20. (C) 9 cm
21. (C) 7
22. (B) 8x - 12
23 (D) 40 cm²
24. (C) 120°

PART 6: ELECTRONICS INFORMATION (EI)

1. (B) Coulomb
2. (B) Current
3. (A) Resistor
4. (B) Ohm's Law
5. (B) 4 amperes
6. (B) 4 ohms
7. (D) 10 volts
8. (A) V = I * R
9. (C) Resistance
10. (B) Copper
11. (B) Ohm
12. (B) I = V / R
13. (A) To store electrical charge
14. (A) Direct current (DC)
15. (C) Voltage
16. (D) Power
17. (C) Kirchhoff's Current Law
18. (C) 4 amperes
19. (D) 16 volts
20 (B) 4 amperes

PART 7: AUTO AND SHOP INFORMATION (AS)

1. (B) To tighten bolts to a specific torque value
2. (A) Holding workpieces securely
3. (D) Spark plug wrench
4. (C) Anti-Lock Braking System
5. (B) Joining metal pieces together
6. (B) Shaping cylindrical objects
7. (B) Planer
8. (B) Slip-joint pliers
9. (C) Cutting precise angles in wood
10. (B) Cutting wood into smaller pieces

11. (C) Tire gauge
12. (B) To power pneumatic tools
13. (C) Lug wrench
14. (B) To regulate fuel flow
15. (C) Wire brush
16. (D) Joining wood edges
17. (B) Sanding small areas
18. (C) Vise
19. (C) Shaping edges of wood
20. (B) Hammer
21. (A) Cutting metal
22. (B) Sharpening tools
23. (C) Heating, Ventilation, and Air Conditioning
24. (D) Drill press
25. (C) Combination square

PART 8: MECHANICAL COMPREHENSION (MC)

1. (B) Pulley
2. (C) 4
3. (A) First-class lever
4. (D) To split objects apart
5. (B) π meters
6. (A) Frictional force
7. (C) Fishing rod
8. (C) 4
9. (C) To transfer rotational motion between shafts
10. (C) 32 centimeters
11. (B) Wedge
12. (A) It decreases the force needed to lift an object
13. (D) 5
14. (B) By decreasing the coefficient of friction
15. (D) 720 newtons
16. (C) 4
17. (C) Third-class lever
18. (B) To change the direction of force
19. (C) 2π meters
20. (A) Frictional force
21. (B) Wheelbarrow
22. (A) It decreases the force needed to lift
23. (C) Third-class lever
24. (B) 72 centimeters
25. (C) Wedge

PART 9: ASSEMBLING OBJECTS (AO)

1. B
2. A
3. D
4. B
5. A
6. C
7. D
8. D
9. B
10. A
11. A
12. B
13. A
14. A
15. D
16. C
17. C
18. A
19. D
20. B
21. C
22. D
23. C
24. B
25. A
26. B

APPENDIX B

ADDITIONAL PRACTICE TESTS

This chapter is meant to help you continue practicing the four critical subjects:

- WK
- PC
- AR
- MK

WK ADDITIONAL PRACTICE TESTS

Here are two practice sets with 15 words each.

Practice 1

1. Prodigious most closely means

    (A) Small
    (B) Mediocre
    (C) Extraordinary
    (D) Ordinary

2. She showed great <u>aplomb</u> during the chaotic situation.

    (A) Panic
    (B) Clumsiness
    (C) Confidence
    (D) Indifference

3. Ubiquitous most closely means

    (A) Rare
    (B) Omnipresent
    (C) Obsolete
    (D) Unique

4. Loquacious most closely means

    (A) Talkative
    (B) Silent
    (C) Boring
    (D) Friendly

5. Taciturn most closely means

    (A) Talkative
    (B) Shy
    (C) Reserved
    (D) Energetic

6. She had a <u>penchant</u> for collecting rare books.

    (A) Dislike
    (B) Indifference
    (C) Weakness
    (D) Preference

7. Ephemeral most closely means

(A) Long-lasting
(B) Endless
(C) Short-lived
(D) Ancient

8. The weather forecast was <u>auspicious</u> for the outdoor wedding.

(A) Unfavorable
(B) Promising
(C) Uncertain
(D) Clear

9. Candor most closely means

(A) Deception
(B) Openness
(C) Secretiveness
(D) Hesitation

10. Tenacious most closely means

(A) Weak
(B) Determined
(C) Indifferent
(D) Fragile

11. Benevolent most closely means

(A) Kind
(B) Cruel

Practice 2

1. Arduous most closely means

(A) Difficult
(B) Delightful
(C) Easy
(D) Relaxing

2. The professor's lecture was so <u>erudite</u> that most students found it difficult to follow.

(A) Simple
(B) Scholarly
(C) Boring
(D) Irrelevant

3. Facilitate most closely means

(A) Ease
(B) Complicate
(C) Hinder
(D) Destroy

4. The artist's <u>idiosyncratic</u> style made her work instantly recognizable.

(A) Common
(B) Unique
(C) Outdated
(D) Modern

(C) Generous
(D) Compassionate

12. Her <u>taciturn</u> nature made it difficult to get to know her.

(A) Talkative
(B) Shy
(C) Quiet
(D) Rude

13. Cursory most closely means

(A) Thorough
(B) Superficial
(C) Detailed
(D) Comprehensive

14. His <u>mellifluous</u> voice was perfect for the role of the radio announcer.

(A) Harsh
(B) Sweet-sounding
(C) Loud
(D) Monotonous

15. Gregarious most closely means

(A) Shy
(B) Outgoing
(C) Hostile
(D) Rude

5. Lucid most closely means

(A) Clear
(B) Obscure
(C) Coherent
(D) Bright

6. Ambivalent most closely means

(A) Certain
(B) Decisive
(C) Unchanging
(D) Uncertain

7. Nefarious most closely means

(A) Heroic
(B) Wicked
(C) Innocent
(D) Noble

8. The criminal's <u>nefarious</u> activities were eventually exposed by the detective.

(A) Heroic
(B) Lawful
(C) Wicked
(D) Foolish

9. The new policy was met with <u>vociferous</u> opposition from the public.

    (A) Silent
    (B) Loud
    (C) Mild
    (D) Agreeable

10. Epitome most closely means

    (A) Anomaly
    (B) Exception
    (C) Embodiment
    (D) Deviation

11. Palpable most closely means

    (A) Intangible
    (B) Evident
    (C) Hidden
    (D) Inconspicuous

12. His <u>magnanimous</u> gesture towards his opponent earned him widespread praise.

    (A) Petty
    (B) Generous
    (C) Selfish
    (D) Aggressive

13. Pragmatic most closely means

    (A) Theoretical
    (B) Realistic
    (C) Impractical
    (D) Idealistic

14. His <u>laconic</u> replies indicated that he was not interested in the conversation.

    (A) Detailed
    (B) Brief
    (C) Enthusiastic
    (D) Rude

15. Succinct most closely means

    (A) Verbose
    (B) Brief
    (C) Lengthy
    (D) Elaborate

PC ADDITIONAL PRACTICE TESTS

Here are two practice sets with 15 words each.

Practice 1

1. Evolution is the process by which species change over time through genetic variation, natural selection, and adaptation. It is the central concept in biology and explains how life on Earth has diversified and evolved over billions of years. The theory of evolution, first proposed by Charles Darwin in the 19th century, has since been supported by a vast body of scientific evidence.

What is the main theme of the passage?

    (A) The history of biology
    (B) The theory of natural selection
    (C) The process of evolution
    (D) The contributions of Charles Darwin

2. Rosa Parks was an African American civil rights activist known for her pivotal role in the Montgomery bus boycott of 1955. Born in Alabama in 1913, Parks' refusal to give up her bus seat to a white passenger sparked a nationwide movement against racial segregation in the United States. What event propelled Rosa Parks into the forefront of the civil rights movement?

What event propelled Rosa Parks into the forefront of the civil rights movement?

    (A) Her participation in the March on Washington
    (B) Her leadership of the Montgomery bus boycott
    (C) Her arrest for protesting racial segregation
    (D) Her advocacy for voting rights legislation

3. Five people—Sam, Tom, Ursula, Victor, and Wendy—each have a different favorite fruit: apple, banana, cherry, orange, and strawberry. Sam's favorite fruit is cherry. Tom's favorite fruit is not an apple or a banana. Ursula's favorite fruit is not a cherry or a strawberry. Whose favorite fruit is an orange?

Whose favorite fruit is an orange?

    (A) Sam
    (B) Tom
    (C) Ursula
    (D) Victor

4. The printing press, invented by Johannes Gutenberg in the 15th century, transformed the way information was disseminated and preserved. Gutenberg's movable type printing technology made books more accessible, leading to the spread of literacy and the democratization of knowledge. Who invented the printing press?

Who invented the printing press?

    (A) Alexander Graham Bell
    (B) Johannes Gutenberg
    (C) Isaac Newton
    (D) Galileo Galilei

5. Plants and other living things employ a process called photosynthesis to transform light energy into chemical energy. Sugars and other carbohydrate molecules, which are created from carbon dioxide and water and release oxygen as a byproduct, are stores of this energy.

What is the byproduct of photosynthesis that is released?

    (A) Carbon dioxide
    (B) Water
    (C) Oxygen
    (D) Carbohydrates

6. Albert Einstein, one of the most influential physicists of the 20th century, is best known for developing the theory of relativity. His work fundamentally changed our understanding of space, time, and energy.

What is Albert Einstein best known for?

    (A) Developing the theory of relativity
    (B) Discovering gravity

(C) Developing quantum mechanics
(D) Inventing the telephone

7. The human respiratory system consists of the airways, the lungs, and the muscles that enable breathing. Its main function is to supply the blood with oxygen and remove carbon dioxide.

What is the main function of the human respiratory system?

(A) To circulate blood
(B) To digest food
(C) To protect against infections
(D) To supply oxygen and remove carbon dioxide

8. The Pythagorean Theorem is a fundamental principle in geometry, stating that in a right-angled triangle, the square of the length of the hypotenuse is equal to the sum of the squares of the lengths of the other two sides.

What does the Pythagorean Theorem state?

(A) The angles in a triangle add up to 180 degrees
(B) The square of the hypotenuse equals the sum of the squares of the other two sides
(C) The area of a triangle is half the base times the height
(D) The perimeter of a triangle is the sum of its sides

9. The mitochondrion is commonly referred to as the cell's powerhouse since it produces the majority of the cell's adenosine triphosphate (ATP), which is used as a source of chemical energy.

Why is the mitochondrion referred to as the powerhouse of the cell?

(A) It produces oxygen
(B) It generates ATP, the cell's energy source
(C) It stores nutrients
(D) It protects the cell from damage

10. Shakespeare's "Romeo and Juliet" is a tragic love story about two young lovers whose deaths ultimately reconcile their feuding families. The play explores themes of love, fate, and the impact of family conflicts.

What is one of the main themes explored in "Romeo and Juliet"?

(A) Wealth and poverty
(B) Love and family conflict
(C) War and peace
(D) Adventure and discovery

11. The American Civil War, fought from 1861 to 1865, was a significant event in United States history. The war was primarily about the issues of slavery and states' rights, leading to the abolition of slavery and the preservation of the Union.

What were the primary issues that led to the American Civil War?

(A) Economic policies and tariffs
(B) Slavery and states' rights
(C) Foreign invasion and colonization
(D) Territorial expansion and exploration

12.The circulatory system, commonly known as the cardiovascular system, distributes nutrition, oxygen, and hormones to cells throughout the body while removing metabolic wastes. It is composed of up of the heart, blood, and blood vessels.

What does the circulatory system transport to cells?

(A) Oxygen and hormones
(B) Carbon dioxide and waste products
(C) Digested food and enzymes
(D) Nerve signals and impulses

13. The Hubble Space Telescope, launched in 1990, has provided invaluable data about our universe. It has captured detailed images of distant galaxies, nebulae, and other astronomical phenomena, significantly advancing our understanding of the cosmos.

What has the Hubble Space Telescope significantly advanced our understanding of?

(A) The ocean depths

(B) The human brain
(C) The Earth's core
(D) The cosmos

14. In George Orwell's dystopian novel "1984," the protagonist, Winston Smith, lives in a society under constant surveillance by the government, known as Big Brother. The novel explores themes of totalitarianism, censorship, and individuality.

What is one of the main themes explored in George Orwell's "1984"?

(A) Democracy and freedom
(B) Totalitarianism and surveillance
(C) Love and romance
(D) Adventure and exploration

15. The Magna Carta, signed in 1215, was a significant document in the history of democracy. It limited the powers of the king and established certain legal rights for the people, laying the foundation for modern constitutional governance.

What did the Magna Carta establish?

(A) Unlimited powers for the king
(B) Legal rights for the people and limits on the king's powers
(C) A new religion
(D) The foundation of a new empire

Practice 2

1. The water cycle, also known as the hydrological cycle, is the ongoing movement of water on, above, and beneath the Earth's surface. Evaporation, condensation, precipitation, and infiltration are some of the processes involved.

What does the water cycle describe?

(A) The formation of mountains
(B) The movement of water on, above, and below the Earth's surface
(C) The growth of plants
(D) The aging of rocks

2. In Harper Lee's novel "To Kill a Mockingbird," the main character, Scout Finch, learns valuable life lessons about empathy, justice, and human dignity as her father, Atticus Finch, defends a black man falsely accused of raping a white woman in the racially charged South.

What valuable life lessons does Scout Finch learn in "To Kill a Mockingbird"?

(A) Empathy, justice, and human dignity
(B) Wealth and power
(C) Adventure and exploration
(D) Art and literature

3. The Grand Canyon, located in Arizona, USA, is a natural wonder formed over millions of years by the erosion caused by the Colorado River. It is known for its immense size, stunning vistas, and geological significance.

How was the Grand Canyon formed?

(A) By volcanic activity
(B) By glacial activity
(C) By tectonic plate movements
(D) By erosion caused by the Colorado River

4. The Declaration of Independence, written primarily by Thomas Jefferson in 1776, announced the American colonies' separation from British rule. It emphasizes the principles of individual liberty and government by consent of the governed.

What does the Declaration of Independence emphasize?

(A) The power of the monarchy
(B) Individual liberty and government by consent of the governed

    (C) The importance of trade agreements
    (D) The need for military alliances

5. The solar system consists of the Sun and the objects that orbit it, including eight planets, their moons, and numerous smaller objects such as asteroids and comets. The Sun, a star at the center, provides the gravitational force that keeps the system together.

What provides the gravitational force that keeps the solar system together?

    (A) The planets
    (B) The moons
    (C) The Sun
    (D) The asteroids

6. The United States Constitution, adopted in 1787, is the supreme law of the United States. It establishes the framework of the federal government and outlines the rights and freedoms of American citizens.

What does the United States Constitution establish?

    (A) The laws of individual states
    (B) The framework of the federal government
    (C) International treaties
    (D) The British monarchy's powers

7. A cultural movement known as the Harlem Renaissance took place in the 1920s and 1930s to honor African American contributions to theater, music, art, and writing. Its focal point was New York City's Harlem district.

Where was the Harlem Renaissance centered?

    (A) Chicago
    (B) Los Angeles
    (C) New York City
    (D) Atlanta

8. Plate tectonics is a theory that describes how Earth's lithospheric plates move. These movements result in earthquakes, volcanic activity, and the formation of mountain ranges. Evidence supporting the theory includes the spread of earthquakes and fossils.

What does plate tectonics theory explain?

    (A) The formation of weather patterns
    (B) The movement of Earth's lithospheric plates
    (C) The origin of species
    (D) The cycle of seasons

9. Ecosystems consist of all the living organisms in a particular area, along with the non-living components of their environment, interacting as a system. These interactions sustain the flow of energy and the cycling of nutrients necessary for life.

What do ecosystems consist of?

    (A) Only plants and animals
    (B) Non-living components only
    (C) Living organisms and non-living components
    (D) Human-made structures

10. The French Revolution, which began in 1789, was a period of radical social and political change in France. It led to the downfall of the monarchy, the rise of Napoleon Bonaparte, and the establishment of a republic based on the principles of liberty, equality, and fraternity.

What did the French Revolution lead to?

    (A) The rise of the British Empire
    (B) The downfall of the French monarchy
    (C) The discovery of America
    (D) The beginning of World War I

11. The Internet has revolutionized communication and information sharing since its inception in the late 20th century. It allows for instant access to vast amounts of information and connects people around the globe through social media, email, and other platforms.

What has the Internet revolutionized?

    (A) Transportation
    (B) Communication and information sharing
    (C) Agriculture
    (D) Space exploration

12. The Great Wall of China is an ancient series of walls and fortifications built to protect against invasions from nomadic tribes. It stretches over 13,000 miles and is one of the most famous landmarks in the world.

What was the primary purpose of the Great Wall of China?

    (A) To serve as a trade route
    (B) To facilitate transportation
    (C) To mark territorial boundaries
    (D) To protect against invasions

13. The periodic table, developed by Dmitri Mendeleev, organizes chemical elements based on their atomic number, electron configurations, and recurring chemical properties. This organization helps scientists predict the properties of elements and their compounds.

What does the periodic table organize elements based on?

    (A) Their atomic number and electron configurations
    (B) Their mass and volume
    (C) Their color and density
    (D) Their temperature and pressure

14. Mount Everest, the highest peak in the world, stands at 29,032 feet above sea level. Located in the Himalayas, it attracts climbers from around the world who attempt to reach its summit, despite the harsh and dangerous conditions.

Where is Mount Everest located?

    (A) The Alps
    (B) The Andes
    (C) The Himalayas
    (D) The Rockies

15. The Roman Empire was one of the largest and most powerful civilizations in history, lasting from 27 BC to AD 476. Its legacy includes significant contributions to law, government, architecture, and language that influence the modern world.

What is one of the significant contributions of the Roman Empire?

    (A) The discovery of electricity
    (B) Advances in medical technology
    (C) Contributions to law and government
    (D) The invention of the printing press

AR ADDITIONAL PRACTICE TESTS

Here are two practice sets with 15 words each.

Practice 1

1. What is 15% of 200?

   (A) 20
   (B) 25
   (C) 30
   (D) 35

2. Solve for x: 4x - 7 = 17.

   (A) 4
   (B) 5
   (C) 6
   (D) 7

3. If $\frac{2}{3} + \frac{1}{6}$ = x, what is x?

   (A) $\frac{1}{2}$
   (B) $\frac{2}{6}$
   (C) $\frac{5}{6}$
   (D) $\frac{1}{3}$

4. What is the least common multiple (LCM) of 4 and 5?

   (A) 10
   (B) 15
   (C) 20
   (D) 25

5. Calculate the mean of the following set of numbers: 5, 10, 15, 20, 25.

   (A) 12
   (B) 15
   (C) 18
   (D) 20

6. What is the median of the following set of numbers: 7, 13, 5, 10, 8?

   (A) 7
   (B) 8
   (C) 10
   (D) 13

7. What is the mode of the following set of numbers: 3, 5, 3, 8, 10?

   (A) 3
   (B) 5
   (C) 8
   (D) 10

8. What is the prime factorization of 60?

   (A) $2^2 * 3 * 5$
   (B) $2 * 3 * 10$
   (C) $2^3 * 5$
   (D) $2 * 3 * 5^2$

9. Simplify: 7 - (3 - 2) + 4.

   (A) 6
   (B) 8
   (C) 10
   (D) 12

10. Solve for x: 3(x - 4) = 18.

   (A) 6
   (B) 8
   (C) 10
   (D) 12

11. Convert 0.75 to a fraction.

   (A) $\frac{1}{2}$
   (B) $\frac{3}{4}$
   (C) $\frac{4}{5}$
   (D) $\frac{3}{5}$

12. If a rectangle has a length of 12 units and a width of 5 units, what is its area?

   (A) 17 square units
   (B) 30 square units
   (C) 60 square units
   (D) 120 square units

13. What is the greatest common divisor (GCD) of 24 and 36?

   (A) 6
   (B) 8
   (C) 12
   (D) 18

14. Calculate $\frac{\frac{5}{8} * 2}{3}$.

   (A) $\frac{5}{24}$
   (B) $\frac{5}{12}$
   (C) $\frac{10}{24}$
   (D) $\frac{10}{18}$

15. If 5y - 3 = 2y + 12, what is the value of y?

   (A) 3
   (B) 4
   (C) 5
   (D) 6

Practice 2

1.What is 25% of 80?

    (A) 15
    (B) 20
    (C) 25
    (D) 30

2. Solve for x: 2x + 5 = 17.

    (A) 4
    (B) 5
    (C) 6
    (D) 7

3. If $\frac{3}{4} + \frac{2}{5}$ = x, what is x?

    (A) $\frac{11}{20}$
    (B) $\frac{19}{20}$
    (C) $\frac{23}{20}$
    (D) $\frac{31}{20}$

4. What is the least common multiple (LCM) of 6 and 8?

    (A) 12
    (B) 18
    (C) 24
    (D) 30

5. Calculate the mean of the following set of numbers: 8, 12, 16, 20, 24.

    (A) 12
    (B) 16
    (C) 18
    (D) 20

6. What is the median of the following set of numbers: 14, 9, 13, 16, 10?

    (A) 10
    (B) 13
    (C) 14
    (D) 16

7. What is the mode of the following set of numbers: 6, 9, 6, 8, 10?

    (A) 6
    (B) 8
    (C) 9
    (D) 10

8. What is the prime factorization of 72?

    (A) $2^3 * 3^2$
    (B) $2^2 * 3^3$
    (C) $2^4 * 3$
    (D) $3^2 * 2^2$

9. Simplify: 9 - (4 - 3) + 6.

    (A) 10
    (B) 12
    (C) 14
    (D) 16

10. Solve for x: 5(x - 3) = 25.

    (A) 3
    (B) 5
    (C) 6
    (D) 8

11. Convert 0.125 to a fraction.

    (A) $\frac{1}{2}$
    (B) $\frac{1}{4}$
    (C) $\frac{1}{8}$
    (D) $\frac{1}{10}$

12. If a rectangle has a length of 14 units and a width of 6 units, what is its area?

    (A) 60 square units
    (B) 70 square units
    (C) 84 square units
    (D) 100 square units

13. What is the greatest common divisor (GCD) of 45 and 60?

    (A) 5
    (B) 10
    (C) 15
    (D) 20

14. Calculate $\frac{\frac{7}{12}*4}{5}$

    (A) $\frac{7}{30}$
    (B) $\frac{14}{30}$
    (C) $\frac{7}{15}$
    (D) $\frac{28}{60}$

15. If 4y - 8 = 12, what is the value of y?

    (A) 4
    (B) 5
    (C) 6
    (D) 7

MK ADDITIONAL PRACTICE TESTS

Here are two practice sets with 15 words each.

Practice 1

1. A rectangle has a length of 8 units and a width of 3 units. What is the perimeter of the rectangle?

    (A) 11 units
    (B) 16 units
    (C) 22 units
    (D) 24 units

2. The area of a triangle is given by the formula $A = \frac{1}{2} * base * height$. If the base of a triangle is 10 units and the height is 6 units, what is the area of the triangle?

    (A) 20 square units
    (B) 30 square units
    (C) 40 square units
    (D) 60 square units
    3. Solve for x if 3x + 5 = 20.
    (A) 3
    (B) 5
    (C) 7
    (D) 15

4.The circumference of a circle is given by the formula C = 2πr. If the radius of the circle is 4 units, what is the circumference?

    (A) 8π units
    (B) 12π units
    (C) 16π units
    (D) 20π units

5.A right triangle has legs of lengths 6 units and 8 units. What is the length of the hypotenuse?

    (A) 10 units
    (B) 12 units
    (C) 14 units
    (D) 16 units

6. If y = 2x + 3 and y = 11, what is the value of x?

    (A) 3
    (B) 4
    (C) 5
    (D) 6

7. The volume of a rectangular prism is given by the formula V = l * w * h. If the length is 5 units, the width is 3 units, and the height is 2 units, what is the volume of the prism?

    (A) 10 cubic units
    (B) 15 cubic units
    (C) 20 cubic units
    (D) 30 cubic units

8. he quadratic equation x^2 - 4x - 5 = 0 can be solved using the quadratic formula. What are the solutions to the equation?

    (A) x = -5 and x = 1
    (B) x = 5 and x = -1
    (C) x = -5 and x = -1
    (D) x = 5 and x = 1

9. The slope of a line is calculated using the formula m = $\frac{y_2 - y_1}{x_2 - x_1}$. What is the slope of the line passing through the points (2, 3) and (4, 7)?

    (A) 1
    (B) 2
    (C) 3
    (D) 4

10. A circle has an area given by the formula A = πr^2. If the radius of the circle is 5 units, what is the area of the circle?

    (A) 10π square units
    (B) 15π square units
    (C) 20π square units
    (D) 25π square units

11. Solve for y if 5y - 7 = 2y + 8.

    (A) 2
    (B) 3
    (C) 4
    (D) 5

12. The Pythagorean Theorem states that $a^2 + b^2 = c^2$. If a right triangle has legs of 9 units and 12 units, what is the length of the hypotenuse?

    (A) 15 units
    (B) 18 units
    (C) 21 units
    (D) 24 units

13. If f(x) = 3x + 4 and f(x) = 10, what is the value of x?

    (A) 1
    (B) 2
    (C) 3
    (D) 4

14. The surface area of a cylinder is given by the formula SA = 2πr(h + r). If the radius is 3 units and the height is 5 units, what is the surface area of the cylinder?

    (A) 24π square units
    (B) 36π square units
    (C) 48π square units
    (D) 60π square units

15. Solve for x if 4(x + 2) = 3x + 10.

    (A) 2
    (B) 4
    (C) 6
    (D) 8

Practice 2

1. The coordinates of the midpoint of a segment are given by the formula $M = (\frac{x_1+x_2}{2}, \frac{y_1+y_2}{2})$. What is the midpoint of the line segment with endpoints (2, 3) and (6, 7)?

    A. (3, 4)
    B. (4, 5)
    C. (5, 6)
    D. (4, 6)

2. Solve for x in the equation 3(x + 2) = 18.

    A. 2
    B. 4
    C. 6
    D. 8

3. The area of a circle is given by $A = \pi r^2$. If the radius of the circle is 7 units, what is the area of the circle?

    A. 14π square units
    B. 21π square units
    C. 49π square units
    D. 77π square units

4. If y = 3x - 5 and y = 10, what is the value of x?

    A. 3
    B. 4
    C. 5
    D. 6

5. The legs of a right triangle are nine and twelve units long. How long is the hypotenuse?

    A. 10 units
    B. 15 units
    C. 18 units
    D. 21 units

6. The slope of a line passing through the points 1, 2 and 3, 6 is given by $m = \frac{y_2+y_1}{x_2+x_1}$. What is the slope of the line?

    A. 1
    B. 2
    C. 3
    D. 4

7. The volume of a cylinder is given by $V = \pi r^2 h$. If the radius is 3 units and the height is 5 units, what is the volume of the cylinder?

    A. 15π cubic units
    B. 30π cubic units
    C. 45π cubic units
    D. 60π cubic units

8. Solve for x if 2x + 7 = 19.

    A. 4
    B. 5
    C. 6
    D. 7

9. The surface area of a sphere is given by $SA = 4\pi r^2$. If the radius is 4 units, what is the surface area of the sphere?

    A. 16π square units
    B. 32π square units
    C. 48π square units
    D. 64π square units

10. The equation of a line is y = 2x + 3. What is the y-intercept of the line?

    A. 1
    B. 2
    C. 3
    D. 4

11. The length of a rectangle is three times its width. If the perimeter of the rectangle is 48 units, what is the width of the rectangle?

    A. 4 units
    B. 6 units
    C. 8 units
    D. 10 units

12. The sum of the interior angles of a quadrilateral is always 360 degrees. If three of the angles are 90 degrees, 85 degrees, and 95 degrees, what is the measure of the fourth angle?

    A. 80 degrees
    B. 90 degrees
    C. 95 degrees
    D. 100 degrees

13. Solve for x in the equation $x^2 - 9 = 0$.

    A. -3
    B. 3
    C. ±3
    D. 9

14. A parallelogram's height is six units, while its base is ten units. What is its area?

    A. 30 square units
    B. 40 square units
    C. 50 square units
    D. 60 square units

15. If 5x - 2 = 3x + 6, what is the value of x?

    A. 1
    B. 2
    C. 3
    D. 4

CHECKING YOUR ANSWERS

Now it's time to see how well you did.

Practice 1

1. (C) Extraordinary
2. (C) Confidence
3. (B) Omnipresent
4. (A) Talkative
5. (C) Reserved
6. (D) Preference
7. (C) Short-lived
8. (B) Promising
9. (B) Openness
10. (B) Determined
11. (A) Kind
12. (C) Quiet
13. (B) Superficial
14. (B) Sweet-sounding
15. (B) Outgoing

Practice 2

1. (A) Difficult
2. (B) Scholarly
3. (A) Ease
4. (B) Unique
5. (A) Clear
6. (D) Uncertain
7. (B) Wicked
8. (C) Wicked
9. (B) Loud
10. (C) Embodiment
11. (B) Evident
12. (B) Generous
13. (B) Realistic
14. (B) Brief
15. (B) Brief

PC ADDITIONAL PRACTICE TESTS

Practice 1

1. (C) The process of evolution
2. (C) Her arrest for protesting racial segregation
3. (D) Victor
4. (B) Johannes Gutenberg
5. (C) Oxygen
6. (A) Developing the theory of relativity
7. (D) To supply oxygen and remove carbon dioxide
8. (B) The square of the hypotenuse equals the sum of the squares of the other two sides
9. (B) It generates ATP, the cell's energy source
10. (B) Love and family conflict
11. (B) Slavery and states' rights
12. (A) Oxygen and hormones
13. (D) The cosmos
14. (B) Totalitarianism and surveillance
15. (B) Legal rights for the people and limits on the king's powers

Practice 2

1. (B) The movement of water on, above, and below the Earth's surface
2. (A) Empathy, justice, and human dignity
3. (D) By erosion caused by the Colorado River
4. (B) Individual liberty and government by consent of the governed
5. (C) The Sun
6. (B) The framework of the federal government
7. (C) New York City
8. (B) The movement of Earth's lithospheric plates
9. (C) Living organisms and non-living components
10. (B) The downfall of the French monarchy
11. (B) Communication and information sharing
12. (D) To protect against invasions
13. (B) Their atomic number and electron configurations
14. (C) The Himalayas
15. (C) Contributions to law and government

AR ADDITIONAL PRACTICE TESTS

Practice 1

1. (C) 30
2. (D) 6
3. (C) $\frac{5}{6}$
4. (C) 20
5. (B) 15
6. (B) 8
7. (A) 3
8. (A) $2^2 * 3 * 5$

9. (C) 10
10. (B) 10
11. (B) $\frac{3}{4}$
12. (C) 60 square units
13. (C) 12
14. (B) $\frac{5}{12}$
15. (D) 5

Practice 2

1. (B) 20
2. (A) 6
3. (D) $\frac{31}{20}$
4. (C) 24
5. (B) 16
6. (B) 13
7. (A) 6
8. (A) $2^3 * 3^2$

9. (D) 14
10. (C) 8
11. (C) $\frac{1}{8}$
12. (C) 84 square units
13. (C) 15
14. (C) $\frac{7}{15}$
15 (A) 5

MK ADDITIONAL PRACTICE TESTS

Practice 1

1. (C) 22 units
2. (B) 30 square units
3. (A) 5
4. (C) 16π units
5. (A) 10 units
6. (B) 4
7. (D) 30 cubic units
8. (B) x = 5 and x = -1

9. (B) 2
10. (D) 25π square units
11. (D) 5
12. (A) 15 units
13. (B) 2
14. (D) 60π square units
15. (A) 2

Practice 2

1. B. (4, 5)
2. C. 4
3. C. 49π square units
4. C. 5
5. B. 15 units
6. B. 2
7. C. 45π cubic units
8. B. 6

9. D. 64π square units
10. C. 3
11. B. 6 units
12. A. 90 degrees
13. C. ±3
14. D. 60 square units
15. D. 4

APPENDIX C

PREFIXES, SUFFIXES, AND ROOTS

This is the first appendix that will help you start memorizing things you need to know to do well on the ASVAB.

PREFIXES, SUFFIXES, AND ROOTS

This is the first appendix that will help you start memorizing things you need to know to do well on the ASVAB.

PREFIXES

The following table covers most of the prefixes in the English language.

Prefix	Meaning
a(n)-	not , without
ab-	away
ac-	to, toward
acr(o)-	high, up
aden(o)-	of a gland
aer(o)-	air
agr(o)-	relating to farming
an(a)-	out of
andr-	man
Anglo-	English or British
ante-	before
anthrop(o)-	relating to human beings

Prefix	Meaning
ant(i)-	against
aut(o)-	self
bar(o)-	atmosphere
bathy-	deep
be-	completely, thoroughly; excessively; on; around; about; used to form transitive verbs
bi-	two
bi(o)-	life
bibli(o)-	relating to books
brady-	slow
bronch(o)-	relating to breathing
cardi(o)-	heart
cent-, centi -	hundred or hundredth
chron(o)-	time
circum-	around
co-	together
com-, con-, col-, cor-	with, together
contr(a)-, contr(o)	against, opposite
counter-	against, opposite
cry(o)-	frost , icy cold
crypt(o)-	hidden
cyt(o)-	hollow, receptacle, cell
de-	taking something away, the opposite
dec(a)-, dek(a)-	ten
deci -	one tenth
dem(o)-	people, nation
derm(o)-, derm(a)-, dermat-	skin
di-	two , double
didact(o)-	apt at teaching
dynam(o)-	power , force
dis-	reverse , opposite
dox(o)-	Doxology
dys-	bad, deformed, abnormal
eco-	house
ecto(s)-	outside
electr(o)-	electricity

Prefix	Meaning
embry(o)-	full
encephal(o)-	brain
enne(a)-	nine
eo-	dawn, early
ep(i)-, eph-	above
eu-	good
ex-	former
ex(o)-	outside
extra-	very, more than usual, outside, beyond
flor(i)-	relating to flowers
fore-	before, in advance
gyn-	female
hemi-	half
hex(a)-	six
hyper-	excessive, (least to greatest in order: hypo, sub, super, hyper)
hyp(o)-	below, (least to greatest in order: hypo, sub, super, hyper)
in-	extremely
in-, il-, im-, ir-	not
inter-	between, from one to another
intra-	within, interior
kilo -	thousand
mal-	bad, badly
maxi-	most, very, large
mega -, megal-	million , very large
meta-	after, beyond
micro -	one millionth, very small
mid-	in the middle of
milli -	thousandth
mini-	small
mis-	bad , wrong
mon(o)-	one , single
multi-	many
non-	not
non-	nine
oct(o)-, oct(a)-	eight

Prefix	Meaning
oo- (" oo -oh")	egg, ovum
out-	more, to a greater degree
over-	more than normal, too much
pent(a)-	five
post-	after
pre-	before
pro-	for, in favour of
quadr-, quart-	four
quin(que)-	five , into five parts
quint(i)-	fifth , five
re-	again, repeatedly
rect(i)-	proper, straight
scler(o)-	hard
semi-	half
sept(a)-	seven
sex(i)-	six
Sino-	Chinese
sphygm(o)-	pulso
schiz(o)-, schist(o)-	split
styl(o)-	pillar
sub-	below, less than, under,(least to greatest in order: hypo, sub, super, hyper)
super-	extremely, more than, (least to greatest in order: *hypo, sub, super, hyper*)
syn-	along with, together, at the same time
tach(y)-, tach(o)-	fast, speed
tel(e)-, tel(o)-	far, over a long distance
tel(o)-, tel(e)-	end, complete, mature
trans-	across, beyond
tri-	three
ultra-	extremely, beyond a certain limit
un-	not, opposite, take something away
uni-	one , single
ur-	first, original
zoo- (" zoo -oh")	relating to animals

Suffixes

The following table covers many of the suffixes in the English language.

Suffix	Meaning	Part of Speech
-able	possible to do	adjective/adverb
-acy	state or quality	noun
-age	condition of being	noun
-al	like, suitable for	adjective/adverb
-an	person who identifies with a place or practice	adjective/adverb
-ance	state, quality	adjective/adverb
-ant	state of being	adjective/adverb
-ant	Person	noun
-ar	relating to	adjective/adverb
-ar	person who does	noun
-ary	place where	noun
-ate	to make	verb
-ation	condition	noun
-dom	place or state of being	noun
-ee	person	noun
-en	made from	adjective/adverb
-en	Become	verb
-ence	state, quality	adjective/adverb
-ent	person who does	noun
-er	More	adjective/adverb
-er	activity, characterized by	verb
-er	person who does or lives somewhere	noun
-fic	characterized by	adjective/adverb
-fy	make or become	verb
-ible	possible to do	adjective/adverb

-ice	condition, quality	noun
-id	state, condition	adjective/adverb
-ify	make or become	verb
-ile	relating to	adjective/adverb
-ion	state of being	noun
-ious	characterized by	adjective/adverb
-ism	act, state, or doctrine	noun
-ist	person who	noun
-ity	quality, degree, or state	noun
-ive	relating to	adjective/adverb
-ize/-ise	to make or become	verb
-less	without	adjective/adverb
-ment	state of being	noun
-or (person)	person who does	noun
-or (thing)	thing that does	noun
-ory	place for	noun
-ous	full of	adjective/adverb
-some	tending to	adjective/adverb
-tude	condition	noun
-ty	state, condition	noun
-ward	direction, course	adjective/adverb

Roots

There are two primary word roots in the English language, Greek and Latin.

Root	Meaning	Source
ab	to move away	Latin
acer, acri	bitter	Latin
ambi	both	Latin
anthropo	man; human; humanity	Greek
anti	against	Greek
aqu	water	Latin
aqua	water	Latin
ast(er)	star	Greek
aud	to hear	Latin
audi	hear	Latin
auto	self	Greek
auto	self	Greek
bene	good	Latin
bene	good	Latin
biblio	book	Greek
bio	life	Greek
bio	life	Greek
brev	short	Latin
cent	one hundred	Latin
chrome	color	Greek
chron	time	Greek
chrono	time	Greek
circ	round	Latin
circum	around	Latin
contra/counter	against	Latin
dict	say	Latin
dict	to say	Latin
doc	teach	Latin
duc	lead, make	Latin
duc/duct	to lead	Latin
dyna	power	Greek
dyna	power	Greek
dys	bad; hard; unlucky	Greek
fac	to do; to make	Latin
form	shape	Latin
fort	strength	Latin
fract	to break	Latin

fund	bottom	Latin
gen	to birth	Latin
geo	earth	Greek
gno	to know	Greek
gram	thing written	Greek
graph	write	Greek
graph	writing	Greek
hab	to have	Latin
hetero	different	Greek
homo	same	Greek
hydr	water	Greek
hydr	water	Greek
hypo	below; beneath	Greek
ject	throw	Latin
jud	judge	Latin
jur	law	Latin
kinesis	movement	Greek
lev	to lift	Latin
log	thought	Greek
logos	word, study	Greek
logy	study of	Greek
luc, lum	light	Latin
mal	bad	Latin
manu	hand	Latin
mater	mother	Latin
meter/metr	measure	Greek
micro	small	Greek
mis, mit	send	Latin
mis/miso	hate	Greek
mit	to send	Latin
mono	one	Greek
morph	form; shape	Greek
mort	death	Latin
multi	many	Latin
narc	sleep	Greek
nym	name	Greek
omni	all	Latin
pac	peace	Latin
pater	father	Latin
path	feel	Greek
phil	love	Greek

phil	love	Greek
phobia	fear	Greek
phon	sound	Greek
phon	sound	Greek
photo	light	Greek
photo/phos	light	Greek
port	carry	Latin
port	to carry	Latin
pseudo	FALSE	Greek
psycho	soul; spirit	Greek
quit	silent, restive	Latin
rupt	to break	Latin
schem	plan	Greek
scope	viewing instrument	Greek
scrib, script	to write	Latin
scrib/scribe	to write	Latin
sect/sec	to cut	Latin
sens	to feel	Latin
sent	to feel; to send	Latin
spect	to look	Latin
struct	to build	Latin
syn	together, with	Greek
techno	art; science; skill	Greek
tele	far	Greek
tele	far off	Greek
terr	earth	Latin
therm	heat	Greek
tim	to fear	Latin
tropos	turning	Greek
vac	empty	Latin
vid, vis	to see	Latin
vid/vis	to see	Latin
voc	voice; to call	Latin

Made in United States
Troutdale, OR
09/12/2024

22756656R00170